SCIENCE
and
LIBERATION

Editors

Rita Arditti
Pat Brennan
Steve Cavrak

SOUTH END PRESS   **Boston MA**

The following articles have appeared in prior publications:

"The Myth of the Neutrality of Science" appeared in *Impact of Science on Society*, #2, 1971.

"Sociobiology" is a longer version of an earlier article from *SftP*, May 1977.

"Genetics as a Social Weapon" appeared in *SftP*, March 1974, with the title "A History of Eugenics and the Class Struggle."

"How Poverty Breeds Overpopulation" appeared in *Ramparts* #31, 1975.

"Sterilization Abuse" is adapted from a talk given by Helen Rodriguez-Trias at Barnard College in November 1976 when she was a Women's Center Reid Lecturer and is reprinted from the current Women's Center Publication.

"Asbestos, Science for Sale" is an updated version of an earlier article that appeared in *SftP*, September 1975.

"Recombinant DNA" appeared in *Mother Jones*, February/March 1977.

"Dealing with the Experts" appeared in *SftP*, September/October 1977.

"Scholars for Dollars" is an adaptation by the Science for the People editorial committee of the pamphlet "Academics in Government and Industry" by Charles Schwartz, and it appeared in *SftP*, January 1976.

"The Yale Professor" by Mary Mackey appeared in the book *One Night Stand*, Effie's Press, 1976, Emeryville, CA.

"The Science Establishment" appeared in *Center* magazine, May/June 1976.

"Ladies in the Lab" appeared in *Women: A Journal of Liberation*, Vol. 2, #3, 1971.

"Women in Chemistry" appeared in *SftP*, July 1972.

"The Scientist as Worker" appeared in *Liberation*, May/June 1974.

"Equality for Women in Science" appeared in *SftP*, August 1970.

"Science and Black People" appeared as an editorial in *Black Scholar*, March 1974.

"Science Teaching" appeared in *SftP*, September 1972.

"People's Science" appeared in *Liberation*, March 1972.

"Science, Technology and Black Liberation" appeared in the *Radicalisation of Science* edited by Hilary and Steven Rose, MacMillan Press, London, 1976.

"History of Science for the People: A Ten Year Perspective" appeared in *SftP*, January/February 1979.

"Computers in South Africa: A Survey of U.S. Companies" is a pamphlet produced by The Africa Fund (198 Broadway, NY, NY 10038) in cooperation with the Interfaith Center on Corporate Responsibility.

Cover design by Bonnie Acker

Typesetting, design, and paste-up done by South End Press

Printed at Maple Vail, U.S.A.

ISBN 0-89608-022-6 paper
ISBN 0-89608-023-4 cloth
Library of Congress Card #79-64087

NOTES ON CONTRIBUTORS

Gar Allen is an associate professor of biology at Washington University in St. Louis, MO where he teaches History of Science. He has been involved for a number of years with CAR (Committee Against Racism), has written an introductory biology text *The Study of Biology* and just published a biography of Thomas Morgan with Princeton University Press.

S. E. Anderson is a member of the Black New York Action Committee, Black Liberation Press and a contributing and advisory editor of the *Black Scholar* magazine. He is also a mathematician, poet and essayist residing in New York.

Rita Arditti is a long-standing member of Science for the People with a special interest in women and science. She is one of the co-owners of new Words, a women's bookstore in Cambridge, MA, and is on the faculty of Union Graduate School.

Maurice Bazin is a French physicist active in developing a pedagogy of the oppressed in the area of Science and Technology. He is currently residing and struggling in Brazil.

Pat Brennan is a member of Science for the People and is currently completing a doctorate on women in science at Boston University.

Bob Broedel works as an electronics technician in Tallahassee, FL, and is a long-time member of Science for the People. He is now working on a directory of social change groups for the southeastern region of the U.S.

Steve Cavrak works as an applied mathematician and computer consultant at the University of Vermont, Burlington in the Academic Computer Center. He is a long-standing member of Science for the People.

Barry Commoner is director of the Center for the Biology of Natural Systems at Washington University, St. Louis, MO, and chairman of the Board of Directors of SIPI (Scientists Institute for Public Information). He is the author of several books, among them *The Closing Circle* and *The Poverty of Power*.

Barbara Chasin is an associate professor of sociology at Monclair State College, NJ. She has a book coming out in the fall of 1979 with co-author Richard W. Franke on the West African famine, *The Political Economy od Ecological Destruction—Drought, Famine and Development in the West African Sahel* (published by Allanhdld, Osmun and Co., Montclair, NJ).

Ana Berta Chepelinsky does research on molecular biology and is interested in women's issues.

Andre Gorz is a major French political theorist, interested in issues of work and the environment. He has published several articles and books, among which are *Strategy for Labor* (Beacon Press) and *Ecology As Politics*, forthcoming from South End Press.

Kathy Greeley is a member of Science for the People and has been active with the alternative technology group of that organization. She has worked as coordinator of the Boston chapter.

David Kotelchuck has been a staff member for the past five years of the Health Policy Advisory Center in New York City (Health-Pac) and has a special interest in occupational health. He is presently studying Occupational Health at the Harvard School of Public Health.

Richard Leonard is a former executive with the Africa Fund—American Committee on Africa. He is now a freelance writer and researcher for various organizations. He is currently involved in a project for the International University Exchange Fund in Geneva, Switzerland.

Marian Lowe is an associate professor of chemistry at Boston University and has recently co-edited *Genes and Gender II—Pitfalls in Research on Sex and Gender* with Ruth Hubbard (Gordian Press, 1979).

Bart Meyers is an associate professor in psychology at Brooklyn College in New York. He has worked with the Mobilization for Survival in New York for the last year and a half.

Angela C. Murphy has worked as a laboratory technician in Memorial Hospital in New York City, the Rockefeller Institute and Yale Medical School.

David Noble teaches at M.I.T. in the Program on Science, Technology and Society and has recently published *America by Design—Science, Technology and the Rise of Corporate Capitalism* (Knopf, NY, 1977).

Bob Park is a member of Science for the People and has been involved in the genetics and social policy group of Science for the People.

Len Radinsky is professor of anatomy and evolutionary biology and chairman of the anatomy department at the University of Chicago. He teaches Biology and Social Issues for undergraduates.

Jeremy Rifkin is co-director of the People Business Commission in Washington, D.C. He has written several books, most recently, *Who Should Play God?* with co-author Ted Howard.

Helen Rodriguez-Trias is a New York based pediatrician, currently director of a children's health program and a founding member of the Committee to End Sterilization Abuse (CESA).

Hilary Rose is professor of Applied Social Studies at the University of Bradford, England. She has written *The Housing Problem* and *Doctors, Patients and Pathology* with Brian Abel Smith and co-edited *The Political Economy of Science and the Radicalisation of Science* (MacMillan Press Ltd., London, 1976).

Steven Rose is a professor of biology at the Open University, England. He has written *The Chemistry of Life* and *The Conscious Brain*, co-edited *Chemical and Biological Warfare* and *Biochemistry and Mental Disorder* with Leslie Iversen, and co-edited *The Political Economy of Science and the Radicalisation of Science* with Hilary Rose.

Mel Rothenberg is a faculty member in the mathematics department at the University of Chicago and a long-time movement activist.

iv

Charles Schwartz is professor of physics at the University of California, Berkeley and is active in the anti-nuclear movement and the Berkeley chapter of Science for the People.

Sue Tafler has long been active in the science teaching group of the Boston chapter of Science for the People and is now active in the food and nutrition group.

Scott Thatcher is a member of Science for the People and has been involved in the genetics and social policy study group of SftP.

Thimann Laboratory is a group of biologists concerned with developing new ways of working together in research laboratories.

UCNWLCP (University of California Nuclear Weapons Lab Conversion Project) is a statewide coalition of various groups and concerned citizens formed in October 1976 that focused on the renewal of U/C's five-year contracts with the Energy Research and Development Administration (ERDA) to operate two nuclear weapons laboratories.

Martha Verbrugge teaches courses on the history of science and medicine at Bucknell University in Lewisburg, PA, and is doing research on women's health in the 19th century.

Robert Yaes was an assistant professor of physics at Memorial University of Newfoundland in Canada. Presently he is a student in the School of Medicine.

Bill Zimmerman was campaign manager for Tom Hayden for U.S. Senate in 1976 in California. He is the author of *Airlift to Wounded Knee* and is currently director of *LOUDSPEAKER* in Los Angeles, a political management and media group that works with left and progressive organizations.

TABLE OF CONTENTS

INTRODUCTION   1

I THE MYTH OF THE NEUTRALITY OF SCIENCE   15

The Myth of the Neutrality of Science
—*Hilary Rose and Steven Rose*   17
Sociobiology, a Pseudo-Scientific Synthesis
—*Barbara Chasin*   33
Genetics as a Social Weapon—*Gar Allen*   48
Corporate Roots of American Science—*David Noble*   63
How Poverty Breeds Overpopulation—*Barry Commoner*   76

II SCIENCE AND SOCIAL CONTROL   91

The University of California Operation of the Lawrence
Livermore and Los Alamos Scientific Laboratories
—*U/California Nuclear Weapons Labs Conversion Project*   93
Sterilization Abuse—*Helen Rodriguez-Trias*   113
Asbestos, Science for Sale—*David Kotelchuck*   128
Recombinant DNA—*Jeremy Rifkin*   145
Dealing With the Experts—*Scott Thatcher and Bob Park*   157
Scholars for Dollars—*Charles Schwartz*   171
Computers in South Africa: A Survey of U.S. Companies
—*Richard Leonard*   191

III WORKING IN SCIENCE   215

The Science Establishment—*Robert Yaes*   217
What is a People's E-Tech?—*Bob Broedel*   239
"Ladies" in the Lab—*Angela Corigliano Murphy*   247
Women in Chemistry—*Ana Berta Chepelinsky et al.*   257
The Scientist as Worker—*Andre Gorz*   267

IV TOWARDS A LIBERATING SCIENCE   281

Declaration: Equality for Women in Science
—*Women's Group from Science for the People*   283
Science and Black People—*Editorial from Black Scholar*   287
Science Teaching: Towards an Alternative
—*Science Teaching Group from Science for the People*   290
People's Science—*Bill Zimmerman et al.*   299
Towards a Liberated Research Environment
—*Thimann Laboratory Group*   320
Science, Technology and Black Liberation
—*S. E. Anderson and Maurice Bazin*   330
Feminism and Science—*Rita Arditti*   350
History of Science for the People: A Ten Year Perspective
—*Kathy Greeley and Sue Tafler*   369

BIBLIOGRAPHY   383
INDEX   389

Introduction

To most people of the United States it is heresy to suggest that science is not absolute truth, or that it is undertaken in such a way as to reinforce a certain set of values. Science is touted as the main ingredient in the mixing bowl of this country's greatness. It is viewed by most as *the* essential ingredient for what is called "progress." A recent Harris survey places scientific research at the very top of the list of major factors that Americans consider will be more important in the next twenty-five years than in the last, and identifies "technological genius" as the key to future national success.[1] Even young school children when asked how specific problems might be solved, suggest more scientific research as the answer. Certainly a country that put a man on the moon can do anything, solve any problem, with just some more science.

Science in this society has become deified. Characterized as "pure" truth, as fact and as the law of nature, it offers simplistic solutions to all problems. Our society is characterized as technological, pragmatic, and scientific. Yet in the popular mind, science is remote and inaccessible.

1

Since WW II, however, and even more so since the Vietnam war, a growing number of people, scientists and citizens alike, have begun to question the role of science in this society. Blasphemous as it may be, these people have suggested that science is not always "pure," truthful, objective, and absolute. They suggest that it frequently, if not often, has a negative impact on the everyday lives of most people, and that it is value laden and hence political. These people, growing in number, have come to question the notion of "science done for its own sake" and have begun to point out that more often science is done to fulfill the needs of a society that is warlike, competitive, hierarchical and unequal.

This book is about these questions. It is about science and its relation to societal problems and social and economic structures. The ideas put forward in this work originate in the social unrest of the late 1960s and early 1970s in the U.S.—the result of the civil rights, anti-war, and women's movements—and are based on the conviction that the practice of science here today serves the interests of small groups in power rather than the interests of the majority of North Americans.

The natural sciences have long been considered politically neutral, value free, and objective. Indeed for some the very word scientific has become synonymous with objective. For many of us, the war in Vietnam brought our first questioning of this vision. This war dramatically altered our level of awareness about the involvement of the scientific establishment in supporting government policy. It became clear during that time that there was a contradiction between the possibilities for human welfare offered by science and what was actually being produced with scientific knowledge. This contradiction was not just a matter of the misuse of scientific knowledge by a few evil minded scientists, but was actually a part of the basic relationship between science and the powerful groups (government and industry) that funded it. Examination of this relationship makes the connections quite clear.

More than half of all scientific research and development (R & D) in the U.S. is funded by the government (53%), and the rest, for the most part, is supported by industry.[2] Despite the government being the prime funding source, industry performs the vast majority of scientific research and development (68%) and focuses on translating scientific ideas into marketable products.[3] What is the estimated $40.8 billion dollars that is budgeted for scientific R & D by both government and industry spent on? What kind of science does it support?

More than half of the federal government's 1977 allocation of $23.5 billion was spent for scientific R & D in defense and space programs.[4] When broken down to more detailed expenditures, $12 billion went for defense-related R & D, while approximately $3 billion went to space R & D, which is usually defense-related too. Commenting on the relationship between defense and the space program, one arms conversion expert stated that the space program was "the largest single scientific endeavor ever undertaken in U.S. history for purposes of developing military space systems and political prestige."[5]

The remainder of the government's science funding, $8.5 billion, is divided among many competing areas such as health, energy, "science and technology base," environment, transportation, natural resources, food fiber, agricultural products and education, with each receiving a decreasing portion of the budget pie.[6] Industry's spending on scientific R & D also seems to focus on science for defense. In 1977 industry spent approximately $17.5 billion on defense-related areas such as electrical and communications equipment, aircraft, missiles, machinery and motor vehicles.

These data clearly indicate that the majority of dollars expended for science in the U.S. each year is spent on science for defense and not on science to help people build better or healthier lives. The scientific norm is a science done to find new and better ways to kill, destroy and defend. It is science for the military and it does not represent a mere aberration from the mainstream of total scientific endeavor. On the contrary, science for the military *is* the mainstream. It is the science establishment.

The relationship between the funding source, the scientist, science and its products becomes abundantly clear in the realm of weaponry. For example, during the Vietnam war, the work of Jason, an elite group of academic physicists, led to the development of the "electronic battlefield," a system of sensors, communication links, antipersonnel weapons and bombs invented to stop transport of soldiers and supplies from North to South Vietnam.[7] The Jason study group, 47 of the United States' most distinguished scientists, met under the auspices of the Institute of Defense Analysis, a powerful independent consulting organization which worked for the Pentagon on research problems of particular interest to the military. One of the supposedly successful projects that resulted from Jason's work was Project Igloo White, which ran for about 3 years and cost approximately a billion dollars per year.

With this project, acoustic and seismic sensors were strewn from high-speed aircraft over the Ho Chi Minh Trail. The seismic sensor with its antenna camouflaged to look like a tropical plant embedded itself in the ground and picked up the earth's vibrations, while the acoustic sensor picked up noises via a minute radio-microphone. Both sensors transmitted signals that were picked up by patrol planes which relayed them to a computer center in Thailand where summaries of the data were prepared for planning strikes by bombers. This project was regarded as a resounding success because it was supposed to have been able to find and destroy 80% of all traffic coming down the trail (approximately 12,000 trucks and a large number of troops) without putting a single U.S. soldier on the ground.[8] The development of this sophisticated weaponry with such a capacity to kill can hardly be rationalized as value free and politically neutral.

Every conceivable branch and subdiscipline of science is brought to bear on the solution of military problems. The research directorates of all the military services fund and purposefully choose the direction of thousands of research projects even in the most basic and abstract realms of science.[9] An example of this so called basic and "pure" research was that done by the University of Wisconsin's Army Mathematics Research Center (AMRC). In August 1970 attention was called to the work at the center by a bombing in which the center's building was destroyed and a research assistant accidentally killed. Center apologists argued that AMRC did only "pure" mathematics research, but the contract between the University of Wisconsin and the Army suggested a different story. The clearly stated objectives of the center were as follows:

1. to provide a group of highly qualified mathematicians who would conduct mathematical research. The emphasis of this research is to be on long-range investigations with the intention of discovering mathematical techniques that may have application to the scientific and technical needs of the Army. The research is to supplement, not replace that of existing Army facilities.

2. to provide for the Army a source of advice and assistance on mathematical techniques, mathematical programs and mathematical problems.

3. to provide a center for stimulating scientific contact and cooperation between Army scientific personnel and other scientists.

4. to increase the reservoir of mathematicians that may be called upon by the government for assistance in the event of national emergency, by acquainting mathematicians with problem areas relevant to Army needs.[10]

Not only was the work of AMRC not "pure and neutral" but it had a decided set of values explicitly related to war. The Center advised the Department of Defense (DOD) on guerilla warfare by preparing equations which simulated as accurately as possible what actually happens in combat. It also assisted various Army bases in producing strategies for counterinsurgency and chemical and biological warfare, and as the war ebbed, it moved beyond the development of new weapons. By war's end the Center, still funded by DOD, had expanded its involvement to include areas such as economic demography, analyses of social systems and simulation of new methods for social control.[11]

How is the image of neutrality for scientific research maintained?

One way has been to use a title for the project that shields the source of the funding as well as the purpose of the research. For example in 1972 a group of students at Stanford University in California, the Stanford Workshop on Political and Social Issues (SWOPSI), examined the role of DOD at their university and found that it funded about 25% of all contracts and grants at the university. However, DOD actively obscured this fact by giving projects two titles, one for the public and one for the military.[12] It was not unusual to find that a publicly-titled project called "High Power Broadly Tunable Laser Action in Ultra Violet Spectrum" at the university, was known as "Weaponry-Lasers for Increased Damage Effectiveness" in the private storage center of DOD in Washington, the Defense Documentation Center.

It is also possible that scientists working on particular DOD funded research have not always been aware of its military relevance and application. Although in the event that they are aware—which is not too infrequent—DOD actively discourages them from acknowledging the military uses for their research, and thereby reinforces the presentation of science as politically neutral.

Another means of promoting this image of scientific neutrality has been the use of some of this name change and subterfuge in order to cover, co-opt, appease and cajole the public and its leaders into acceptance of certain products and ideas. For example, the Department of Defense uses rather positive sounding phrases like

"Enhanced Radiation Warheads" (ERW) to refer to the devastatingly harmful neutron bombs and a term like "immediate transient incapacitation" to refer to the gradual, painful, slow death via physical decay caused by the radiation from such bombs.[14] But despite deceptive names and secretive DOD funding, scientists continue to point to this work as benefiting humankind (e.g. for pollution control, improving traffic conditions, etc.), or they rationalize the work by claiming it is basic, fundamental research, far removed from immediate application for military use. In doing so, both scientists and their universities effectively eschew responsibility for this science which is done against people.

Many scientists who actually do the basic research upon which such weapons are built claim innocence about the uses of their work. Even scientists who work for the military view their science as "pure," "objective," unsullied "truth" which is not surprising when one considers that they, like most scientists, have been trained to accept the myth of the neutrality of science. The myth claims that scientific truth is "objective" truth regardless of its uses or applications. It is not surprising that after years of breaking down problems into their components, analyzing them and replicating experiments to validate results, most scientists view science and its products as "truth," separate from and not connected to or deriving from the values and problems of society. Instead, science is seen as an international endeavor, above national politics, and its practitioners are seen as disinterested seekers of the "truth."

Studying science within a social and political context is not part of current scientific training in the U.S. Instead, scientists are imbued with an ideology of "knowledge for knowledge's sake." They work within a social structure of competition and reward that encourages pursuit of peer recognition and prestige at the expense of humanness.[15]

Scientists undergo long years of training. PhD's, for example, do approximately 3-5 years of post-undergraduate work and frequently an additional 2-3 years of post-doctoral study before starting a job at the lowest level of the so-called professional hierarchy. From the beginning the scientist is trained in the scientific method, which teaches one to isolate an object for study, to analyze its internal workings, to formulate hypotheses about it and to develop laws or truths. Both the isolation of the object and the method used to examine it, i.e., looking at parts rather than the whole, reinforces piecemeal rather than holistic thinking. It would not be possible for science to appear "neutral" were it viewed as part of the cultural and social institutions that support, create and encourage it. Science as it is currently practiced is concerned with validity and not with values. It is pursuit of knowledge regardless of ethical consideration.

Long years of training in reductionist thinking produces scientists who are experts, special people who are thought to be more knowledgeable, rational, precise and brighter than others— high priests of technological truth. The scientist becomes a "professional" isolated from common people and often unable to converse with others unlike him/herself. This "professionalism" allows the scientist to disdain any questioning of his/her work that does not have the aura of scientific research and helps to stifle questioning from both inside and outside the scientific community. It allows scientists to keep a professional distance between themselves and non-scientists as well as between selves and subordinates. And it supports the social hierarchy within science itself. It is not unusual that scientists at the top can organize and direct work of subordinates and not even be questioned about it.

Professionalism allows for a distance between people and ensures an attitude of moral superiority that in turn justifies elitism. Clearly the training and development of "professional" scientists produce persons for the most part who do not and perhaps cannot examine the social values of their work, and who not only shut out criticism, but silence it. Given such long years of training and socialization in the ideology and behaviors of science, it is not surprising that scientists who do military research perceive their work as value free.

The military application of so much scientific research and development, though overwhelming, ubiquitous and so clearly value laden, is the mainstay of the physical sciences and math, but it does not account for all scientific endeavor. The biological sciences are also value laden and buttress predominant values in society. Some biological research has been and continues to be used as a means of providing supposedly "natural"and therefore unquestionable explanations of social inequities in race, sex, health and wealth. This research attempts to define these differences as "human nature," something we just have to adapt to since it is programmed into our "genes." In other words, biology is destiny.

Scientists who espouse this biological determinism bring their prestige and an aura of scientific objectivity to their theories. One such theory which has been widely publicized is that on IQ and race.* It suggests that intelligence is inherited through the genes and is somehow related to skin color. One example of the negative effects of such a theory occurred with the publication of a paper on this subject by one of the leading proponents of the genetic basis for inequalities, Arthur Jensen. Subsequent to its publication, William Shockley, a Nobel prize winner and co-inventor of the transistor, drew up a horrifying legislative proposal calling for the sterilization of all persons with IQs below 100.[16] Defining a social construct (IQ) as biological clearly could have dire consequences in society.

In a cogent paper on the political uses of biological science to reinforce social prejudice, Inez Smith Reid, a black lawyer, describes how innocent people become victims of decisions made by scientists and policy makers, and how research by persons like Jensen, Schockley and Hernnstein is actually more political than scientific.[17] She describes how scientists, physicians and educators quickly jumped on the sickle cell anemia and hypertension bandwagons before they carefully examined the possible consequences of their decisions to educate the public about the frequency of these diseases in the United States black population. Unfortunately, those with these diseases became victims of hasty decisions based on unsubstantiated data. These people were labeled "learning disabled," were refused job opportunities, were exposed to demands for genetic counseling prior to issuance of a marriage license, and were recipients of demands for mandatory pregnancy screening. Ultimately, they faced increased costs in insurance premiums.[18]

Casting hypotheses into truth via unsubstantiated data is the hallmark of biological determinists. One such example of this was a study at Boston Hospital for Women undertaken to identify an extra Y chromosome in newborn boys.[19] The study was based on the hypothesis that an extra Y chromosome was related to increased antisocial and even criminal behavior in persons who possessed it. Although the data upon which this study was based had been widely criticized and its validity questioned, the study proceeded to identify newborns who might be "at risk" of developing some rather ill-defined "behavior problems." Fortunately, this study was eventually halted through the valiant efforts of concerned activists who recognized that labeling a child as "at risk" of being a behavior problem or a criminal might ensure that this in fact happened—a self-fulfilling prophecy.

But for every study like this one that is stopped, a few more spring up in its place. For example, a particularly popular and

---

* For a full discussion of this topic, see *Biology As A Social Weapon* (Ann Arbor Science for the People Collective, Burgess Publishing, 1977).

flourishing area of research nowadays is that which is attempting to relate criminal or antisocial classroom behavior to brain dysfunction or so-called neurological problems. A recent study by a professor at the University of Rhode Island attempts to identify specific malfunctioning parts of the brain of "delinquent" children in an attempt to identify a biological basis for this behavior, and another study conducted by a team from Children's Hospital Medical Center in Boston reported findings that youth in detention centers showed "symptoms of minor brain dysfunctions."[19] These attempts to find biological rather than social and cultural determinants for "abnormal" behavior are especially insidious in a society like ours which refuses to examine itself. These studies provide a convenient and "scientific" way to blame the victim of social inequity for their so-called abnormalities and also provide a rationale for continuing to study the problem rather than trying to find solutions.

Unfortunately, theoretical hypotheses for biological determinism have recently made a big, flashy, Ivy League comeback with the introduction of a supposed new area of study—sociobiology. This new "ology" attempts to combine biology and sociology and postulates a genetic basis for many aspects of our society, such as the nuclear family, the capitalist market economy, male aggression, competitiveness, and dominance, sex roles, ethnic and racial prejudice, cheating, spite and altruism.[20] Sociobiology describes as universal and genetic those human behaviors which are surprisingly similar to those of persons of affluence in society. In the text by sociobiology's foremost proponent, Harvard professor E.O. Wilson, *Sociobiology*, warfare, genocide and male dominance are depicted as practically unavoidable, given our human nature.[21]

Other areas of biological science also suffer from a lack of socio-political scrutiny. Sometimes excitement over a new biological discovery clouds possibilities of examining it in a broader context. One such recent discovery is the technology of "DNA Recombination," the insertion or combining of genetic material (DNA) from one organism with the genetic material of another organism, producing a new organism with never-before-used genetic information. Proponents of this new technology claim that it will revolutionize health care by introducing new means of curing genetic diseases resulting from an incomplete or missing piece of DNA (a gene). Other claims are that it will help solve world food problems by producing bumper crops of corn and wheat capable of producing their own nitrogen fertilizer, or that new organisms will be invented to do all kinds of specific work, such as clean up oil spills, produce hormones like insulin or possibly cure cancer.[22] In other words, claims for the value of genetic recombination range from curing disease to solving world food problems, with the additional suggestion that knowledge accrued from this research will lead to a clearer understanding of the function of genes in high organisms.

Opponents of DNA research argue that this technology contains unknown and perhaps devastating possibilities for human destruction and that the costs far outweigh any positive benefits. One of the foremost criticisms of the technology is that research is being done with an organism (E. coli) that naturally inhabits the intestines of all humans. Critics say that if one of these newly created bacteria, actually a variation of E. coli, were to accidentally get into the human digestive tract, there is no way of foretelling the possible consequences.[23] It could mean the extinction of the human race. Proponents counter that this could not possibly happen since the E. coli bacteria being used in the research has been so weakened that it couldn't even infect a person, and secondly, that such a bacteria would never escape from the special high security labs where such research is done.

Such arguments hold no water with critics, however, who point to the recent (September 1978) escape of a virulent virus from a similar high security federal animal disease center on an island close to Long Island, New York.[24] Despite the fact that disease center officials describe their facility as P4, "the highest rating developed by the National Institutes of Health for gene splicing experiments," the virus escaped the lab and infected cattle on the island. Not surprisingly, this event aroused fear and consternation among residents of neighboring Long Island, who now worry about the Center's next project with Rift Valley fever, a disease carried by mosquitos that can infect humans as well as animals.[25]

Another chink in the armor of Recombinant proponents is that although the "enfeebled" new bacteria are themselves unable to infect, they can transfer their genetic material (DNA) to other healthy bacteria which could in turn infect and possibly cause a cancer epidemic, a subtle unknown infection throughout the human population, an environmental disaster, or any number of other possible Orwellian catastrophes. What could be done if a bacteria, newly invented to eat oil spills, came in contact with the oil that lubricates airplanes or heats our home? Unfortunately, there is no encouraging answer, for such a "new" organism cannot be recalled like a new car. Once it exists and penetrates the environment, there is no reverse. In view of this, a noted scientist asks, "Do humans have the right to counteract irreversibly the evolutionary wisdom of millions of years in order to satisfy the curiosity of a few scientists?"[26]

Unfortunately, "right" or no right, this new technology is being used and is perhaps the hottest area of research in the U.S. today. Although scientists and citizens have spoken out in fear and outrage against it (see articles in this book), the powerful voices of pragmatism, efficiency and "progress" have prevailed. Already the technology has moved out of the university and into big business.[27] Eli Lilly & Co., the major supplier of animal insulin has already contracted for the right to produce and market the new human insulin produced by "new" bacteria (September 1978).[28] Unfortunately, it is difficult to imagine that this first marketable product of recombination technology will be offered at a low price or will benefit anyone other than those with access to top quality medical care. Indeed, one is left only with the hope that the people who stand to be detrimentally affected by this new, awesome and perhaps fatal technological instrument—all of us—will organize against this new technology that stands to benefit the few.

A fundamental tenet of U.S. ideology today is that new, better and more scientific technology will solve societal problems. Science is seen as the hallmark of American knowhow, the means by which new technologies will be developed to fix any problem. This perspective, often referred to as the "technological fix," has gained widespread acceptance in most parts of affluent Western society, and it spans a great diversity of areas from health care to space travel. With profit as the primary incentive, present day technologists bring us fabricated food, plastic money, high technology medicine, and a myriad of other products of technological wizardry. In each case, a large social problem has been redefined in such a way that a technological solution supposedly can be found. The fact that there are many spinoffs from this solution that have never been taken into account means that in most cases, new problems— and more serious ones—are created.

Such is the case with the Green Revolution. In this case, scientists set out to tackle the problem of hunger in certain heavily-populated countries. What was in fact a multi-faceted problem of food distribution, social and economic inequities, primitive technology and so on was viewed simply as a question of developing new hybrids of high yielding seeds.[29] The developers of new seeds, scientist and politician alike, had not adequately considered the socio-political context into which this new scientific development was to be placed. The revolution never happened. The root of the hunger problem, unequal distribution of wealth, could not be solved via introduction of new seeds.[30] Even when some bumper crops were produced, they often did not reach the masses of people since storage, transportation and marketing mechanisms were not geared for such distribution. Moreover, only a few of the most prosperous land owning farmers could afford to pay for the expensive fertilizers, pesticides, herbicides and irrigation necessary to raise the high yielding crops.[31] In retrospect, it is clear that scientists, politicians and vested interests sold these countries a bright, shiny, technological package that would only work with the trappings of U.S. high technology farming.

Given the situation just described, what do scientists do who are interested in social change? How can scientists overcome the isolation and privileges of their role? What is the role of women in this effort? One of the important tasks ahead is to address the issue of "professionalism," the way it is instilled and the way it is played out in the workplace and in social relations. Progressive scientists can make cracks in the barrier of professionalism, and start sharing their ideas and their work with the community at large. Clearly, to be a scientist committed to social change will require that we live with many contradictions, for one must live and work in society as it now is, and changes are not made overnight. It will be a long, hard process to rearrange priorities in our society, but a new society will require a new science and in turn a new science will contribute to a new society. Only we can do it!

The purpose of this book is to discuss the role of science and scientists in maintaining social oppression and to present ideas and concrete examples of the movement—such as it is—toward a new science: a science of liberation. The book is divided into four sections. The first deals with the myth of the neutrality of science, the idea that science exists and is practiced independently of social and political concerns. The articles in this section challenge the idea of the "neutrality" and "purity" of science and point to connections between scientific theories and the maintenance of oppression in society. Section two points to the ways science is used to ensure power for those already powerful and how its mystification can be a means of controlling people. It also describes how science affects people, their health, everyday lives and possible future existences.

The third section presents specific examples of what it is like to work in science as technicians, assistants and as women—the people who actually do most of the everyday work of science. These are the ordinary people who work in science, not the Nobel prize winners, but people whose stories belie claims that a profession in science is the most exciting and most noble activity one can perform. The last section of the book shows how science teachers, researchers, black people, women and Science for the People members are attempting to initiate change. It presents alternatives toward a new science, one in which theory and practice are not split, and it introduces ways to think about sharing knowledge and creating meaningful alliances to demand that science serve the people.

FOOTNOTES

1. Daddario, Emilio Q., "Science and Its Place in Society," *Science*, Vol. 200, April 21, 1978, p. 263.

2. National Science Foundation, *National Patterns of R&D Resources: Funds and Manpower in the U.S. 1953-1977*, U.S. Govt. Printing Office, April 1977, p. vi.

3. *Ibid*, p. vi.

4. National Science Foundation, *An Analysis of Federal R&D Funding by Function, Fiscal Years 1969-1977*, U.S.-Govt Printing Office, September 1976, p. vii and viii.

5. Leitenber, Milton, "The Conversion Potential of Military Research and Development Expenditures," *The Bulletin of Peace Proposals*, Offprint Vol. 5, 1974, Universitetsforlaset, Tromso, Sweden, p. 83.

6. NSF, *Analysis of R&D*, Sept/76, *op. cit.*, p. vi.

7. Berkeley Scientists and Engineers for Social and Political Action, *Science Against the People: The Story of Jason*, Berkeley Science for the People, December 1972, p. 3.

8. *Ibid.*, p. 8.

9. Leitenber, Milton, "The Role of Military Technology Today," *International Social Science Journal*, (UNESCO), Vol. 25, no. 3, September 1973, p. 336.

10. Madison Science for the People Collective, "The AMRC Papers: Excerpts," *Science for the People*, Vol. vi, no. 1, January 1974, p. 30.

11. *Ibid.*, p. 35.

12. Shapley, Deborah, "Defense Research: The Names Are Changed to Protect the Innocent," *Science*, Vol. 175, 25 February 1972, pp. 866-868.

13. *Ibid.*, p. 866.

14. Kistiakowsky, George B., "Weaponry: The Folly of the Neutron Bomb," *The Atlantic*, Vol. 24, no. 6, June 1978, pp. 4 and 9.

15. Barber, Bernard, "Ethics of Experimentation with Human Subjects," *Scientific American*, February 1976, Vol. 234, no. 2, p. 31.

16. Woodward, Val, "IQ and Scientific Racism," *Biology as a Social Weapon*, Ann Arbor Science for the People Collective, Burgess Publishing Co., Minneapolis MN, 1977, p. 40.

17. Reid, Inez Smith, "Science, Politics and Race," *SIGNS*, Vol. 1, no. 2, 1975, University of Chicago Press, p. 412.

18. *Ibid.*, p. 400.

19. The Genetic Engineering Group, "The XYY Controversy Continued," *Science for the People*, Vol. vii, no. 4, July 1975, pp. 28-32.

20. Chasin, Barbara, "Sociobiology, A Sexist Synthesis," *Science for the People*, Vol. 9, no. 3, May 1977, pp. 27-31.

21. Wilson, Edward O., *Sociobiology: The New Synthesis*, Harvard University Press, Cambridge, 1975, p. 553.

22. King, Jonathan, "Summary Remarks," *Tech Talk*, MIT, June 1976.

23. Loechler, Edward, "Social and Political Issues in Genetic Engineering," *Genetic Engineering*, ed. A. M. Chakrabarty, Chemical Rubber Co., August 1977.

24. Wade, Nicholas, "Accident and Hostile Citizens Beset Animal Disease Laboratory," *Science*, Vol. 202, no. 17, November 1978, p. 723.

25. *Ibid.*, p. 723.

26. Chargaff, Edwin, "On Dangers of Genetic Meddling," *Science*, Vol. 192, April 1976, p. 40.

27. Rifkin, Jeremy, "DNA," *Mother Jones*, February/March 1977, pp. 23-26.

28. Editorial Comment, "Human Insulin: Seizing the Golden Plasmid," *Science News*, Vol. 114, no. 12, 16 September 1978, p. 195.

29. Science for the People Collective, "The Green Revolution: A Critique, *Science for the People*, Vol. V, no. 6, November 1973, p. 41.

30. Foreign Commentary, "The Green Revolution Yields Bitter Fruit," *Business Week*, 21 November 1970, p. 35.

31. Singh, Narenda, "Is Our Food Problem Due to Overpopulation?" *BAWI*, Vol. 5, no. 4, April 1975.

I
THE MYTH OF THE
NEUTRALITY OF SCIENCE

Introduction

> *Good afternoon, ladies and gentlemen. This is your pilot speaking. We are flying at an altitude of 35,000 feet and a speed of 700 miles an hour. I have two pieces of news to report, one good and one bad. The bad news is that we are lost. The good news is that we are making excellent time.*
>
> *—Author unknown*

Over and over science is presented to us as a "value-free" activity, independent of the rest of society, a source of rationality in a chaotic and difficult world. Science, they tell us, will solve our environmental problems, will give us the correct information about what human beings are like and will also contribute to our welfare by finding new knowledge useful to us. When a difficult situation arises, we are told, science will take care of the problem, will provide a scientific solution, a perfect answer.

How is it that a particular human activity, rooted in society, has come to be regarded as separate, endowed with characteristics that put it above the other activities of the society? How is it that science is regarded as politically neutral, impartial and unbiased, pure and objective?

We contend that scientific knowledge is now an instrument of power, that science in our society sides with the government, with the military and the industrial world, that the separation of mind from emotion characteristic of Western patriarchal thought has resulted in an absurd science that follows its own narrow logic, regardless of social consequences. We see this science as a closed system that creates and perpetuates its rules and maintains certain beliefs—most importantly the belief in its own neutrality and objectivity. The thrust of this section is the examination of this myth—the myth of the neutrality of science. The common thread that ties the five articles in this section is the belief that science is not absolute truth, that it is not an objective or detached activity. Quite to the contrary, science is made by human beings (more specifically white males) and the scientific theories produced by these people play a critical role in shaping the ideology of society. Popular belief in the neutrality of science protects the scientific establishment from critical examination. It lies at the core of science's "value-free" image. And it is the subject of analysis in this section of our volume.

The Myth of the Neutrality of Science
—Steven and Hilary Rose

This article chronicles more than three centuries of philosophical debate on the issue of social and political neutrality of science and asks provoking questions about what kind of research gets done and why.

One of the key items of intellectual baggage which most practicing scientists carry over from their training—there are not many such, for few scientific training programs have more than a passing reference to theory—is one labelled "the neutrality of science." Briefly stated the phrase may be explained thus: "The activities of science are morally and socially value-free. Science is the pursuit of natural laws, laws which are valid irrespective of the nation, race, politics, religion or class position of their discoverer. Although science proceeds by a series of approximations to a never-attained objective truth, the laws and facts of science have an immutable quality. The velocity of light is the same whoever makes the experiment which measures it. Because this is the case, although the uses to which society may put science may be good or evil, the scientist carries no special responsibility for those uses, save as a normal citizen. The two-edged sword of science is fashioned for whomsoever will pick it up and wield it."

Thus, a recent analysis of the social responsibility of the scientist by a Nobel laureate, E. B. Chain, italicized the statement that "science, as long as it limits itself to the descriptive study of the laws of nature, has no moral or ethical quality and this applies to the physical as well as the biological sciences."[1]

It is this set of beliefs and ideas concerning the neutrality of science which has begun to wear the aspect of a myth, which while at present ubiquitous in socialist and capitalist, industrialized and non-industrialized societies, none the less is of relatively recent origin. Our purpose here is to challenge this myth, to reveal it for what it is, for it is our feeling that the unthinking, unquestioning acceptance of it as gospel has been to a large degree responsible for the anti-social applications—the non-human or inhuman uses—of science, which have strongly contributed to many of the world's major problems.

If we treat the neutrality of science as a myth, we are then committed to attempting to discuss a series of questions: How has the myth emerged? What role has it played? How has the myth been challenged? How far does the future of science and of human survival itself depend on our ability to transcend and refashion such myths of science? These are dauntingly large questions, yet it is inescapably clear that they form the agenda of increasing numbers of socially aware scientists, and that they reflect the intellectual and moral crisis of contemporary science.

The Emergence of the Myth of Pure Science

It is widely accepted that the modern activity of science emerged and began to find its social and philosophical articulation in post-Renaissance Europe, and rather specifically in seventeenth-century Britain. That science was seen as a progressive force, closely linked to the enhancement of human welfare, is clear from the writings of Francis Bacon and the founders of the Royal Society. For Bacon, knowledge for knowledge's sake, or for power's sake, was subordinate to knowledge for charity's sake, for "of charity there can be no excess." And for the chemist, Sir Robert Boyle, science was for "the greater glory of God and for the good of mankind."

But this clarity of purpose did not survive into the eighteenth century in Britain, where science became an activity for a cultured, moneyed and leisured class. It is no accident that the term "scientist" did not exist in Britain until the mid-nineteenth century; the phrase actually used was "a cultivator of science." In France, too, rigid social stratification made the Academie des Sciences a cultured gentleman inventor's club until its degeneration and transformation following the 1789 Revolution. And the discrepancy between "aristocratic" and "citizen's" science even then contributed to the death of Lavoisier.

The Industrial Revolution in the nineteenth century saw the continuation and hardening of this gulf. The rigid amateurism of the Royal Society precipitated the formation of alternative institutions more clearly integrated with the needs of burgeoning capitalism, such as the Midland industrialists' Lunar Society, which embraced Wedgwood, Boulton, Watt and Dalton, and later Babbage's British Association for the Advancement of Science. Yet such was the forward thrust of the scientific temper that Lunar Society members were attacked for their radical political views and sympathy to the French Revolution, while the debates at the early British Association raised such fundamental questions as to convince a magistrate that "science and learning, if universally diffused, would speedily overturn the best-constituted government on earth." The movement in Britain towards working-class scientific education in the mid-nineteenth century reflected a constant tension between Establishment fears of the revolutionary potential of scientific thought and recognition of the needs of capitalist society for a more skilled worker.

The conflicts were equally in evidence in the rest of Europe. France saw the emergence of the strange scientific religion of St. Simonism, whilst in Germany the *naturphilosophen*, with their speculative attempts to synthesize science and natural knowledge into an all-embracing scheme of the world, conflicted with the more arid traditions of the university scholar. Eventually, the universities digested this development; "pure" science emerged as a university discipline uncontaminated with relevance to anything, and the *Technische Hochschulen* concentrated on the practical side of science and industry's needs, contributing to the success of the German science-based industries, which served the nation so well right into the First World War.

The development of socialist materialist ideas in the later nineteenth century, and particularly those of Marxism, set these tensions into a new frame; Marxism claimed to be a scientific socialism. It applied the techniques and dialectical methods of the natural sciences to the disordered world of human affairs. Socialism was seen to be on the side of history, and science was on the side of socialism. Major theoretical scientific advances contributed to the interpretation of society; Marx, like the conservative Herbert Spencer, saw the value of a biological metaphor for human activities. It is well known that Marx wished to dedicate one of the volumes of *Das Kapital* to Darwin, who declined the honor. At the same time Darwinism found itself recruited to service the needs of racists and eugenicists.

Nor was it just biology which was seen as both deriving impetus from, and contributing to, social analysis. The inmost heart of science itself—fundamental physics—could be interpreted in terms of dialectical materialism and in the *Dialectics of Nature* Engels attempts precisely this major task: a theoretical re-interpretation of the whole of science, as then perceived, as exemplifying the workings of the dialectic—a degree of grand theorizing which has daunted most later Marxists.

Engels' attempt to interpret the activities of science as conforming to the workings of the dialectic was extended by later Marxists, but could only expect to achieve recognition in practical policies subsequent to the Soviet Revolution of 1917. But even in the Soviet Union, it was not until ten years after the revolution, when the at times uneasy truce of experts and professionals with the Party which typified the period of the New Economic Policy was coming to an end, that a specific attempt was made to break with the ideology of pure science which still dominated the Academy of Sciences. The period of Five-Year Plans and the march towards socialism was typified, as Joravsky has documented, by a fresh attempt to define a "socialist physics," "socialist biology," and so forth.[2]

The form of such a socialist science was uncertain and open to debate. What was clear was that it ought to be different from "capitalist" science, and that in this scheme of things, there was no such thing as "pure" science.

"Pure" science was simultaneously under ideological attack not only from the left, but from the right, for the emergent Nazism of Germany was beginning to talk of "Aryan" and "non-Aryan" (i.e. Jewish) science. Some branches of physics, notably Einsteinian relativity and quantum theory, were under attack for their non-Aryan quality, and there were cases when the same type of physics was attacked both for its non-Aryanism and for its non-socialist, "idealist" nature.

The reaction to these attacks upon its self-perceived integrity by the majority of the scientific community was predictable. In Germany and the Soviet Union, the response was one which Haberer has characterized as the politics of "prudential acquiescence."[3] This acquiescence led in Germany to the dismissal of

Jewish university scientists and the official acceptance of an openly racialist biology, justifying the organized and monstrous slaughter of the concentration camps. In the Soviet Union, it took the milder form of prudential acquiescence in the systematic destruction of a particular field of science, genetics, and the exile or silencing of its protagonists.

In pre-war Britain, the debate took a different turn. Traditional empiricism reduced a discussion of whether the internal logic of science itself was ideologically determined to one concerning the harnessing of science to human welfare. Radical and Marxist scientists generated a wave of optimism about the prospects and significance of science as a factor in the liberation of humankind. This opinion is typified by Bernal, writing just before the Second World War.

> Science puts into our hands the means of satisfying our material needs. It gives us also the ideas which will enable us to understand, to co-ordinate and to satisfy our needs in the social sphere. Beyond this, science has something important to offer: a reasonable hope in the unexplored possibilities of the future, an inspiration which is slowly but surely becoming the dominant driving force of modern thought and action.[4]

To achieve this effect science must be rationally planned and organized. In an appropriate social structure, science would inevitably proceed so as to enhance human welfare. Even this proposition, however, resulted in much of the academic community closing ranks against the onslaught, proclaiming its self-interested purity with virginal fastidiousness. Polanyi provided the philosophical rationale and Baker the organizational steam for a Society for Freedom in Science, whose exponents demanded independence for the Republic of Science. This claim found voice in the complaint that the Bernalians proposed to direct science to such an extent that a microscopist would not even be allowed to choose what color to stain his preparations.

War, and the mobilization of the scientific community in the prosecution of war, made the debate meaningless. Even the most prudential of the German scientists found themselves involved in the war effort—some Jewish scientists even managed to survive the war in their laboratories. In Britain, the purest of academics were registered and drafted into war research. And in the United States, the biggest scientific mobilization the world had ever seen, the Manhattan Project, took place. The Republic of Science became a dream of peace which science, even when the peace at length came, was never really to recover.

By the end of the war, the era of Big Science had arrived, ushered in by the explosions at Hiroshima and Nagasaki, and the debates of the 1930s seemed strangely irrelevant. Science was paid for by the government; the largest part of this payment was for war (defense) science, but governmental and industrial research contracts permeated the universities, too. At first opposed, at least in Britain, by the end of the 1950s such a permeation was being actively welcomed. In the United States, of course, it always has been, for there the concept of academic freedom was one of freedom for the scientific entrepreneur, typified by the archetypal Dr. Grant Swinger, the mythical money-hunting scientist created by Dan Greenberg, of *Science* magazine. The key to American economic and scientific success was seen in precisely this articulation by university, industry and government. Its devotees urged the creation, in Britain and elsewhere in Europe, of local versions of the impressive "Route 128," the highway which is lined with science-based industries which have sprung out of research at the mighty Massachusetts Institute of Technology.

The acceptance and rationalization of this new conventional wisdom may be plotted as a function of the emergence in many nations, over the late 1950s and through the 1960s, of science policy-making committees and Ministers of Science—an inconceivable prospect in the 1930s except amongst the Bernalians, though advocated by some prescient individuals as far back as the 1850s. By 1970, indeed, something like an academic discipline (metadiscipline) of "science policy" has emerged. For such policy-makers, the suggestion that science should *not* be locked into the national economy, making its appropriate contribution to national goals, is risible. National goals may even be specifically technological, such as the commitment of Kennedy to put an American on the moon by 1970. And in a technologically based society, the preservation of the scientocracy and technocracy's neutrality and independence from governmental process has become impossible.

Meanwhile science and technology are judged by their economic and political pay-off for the nation. British participation in the proposed 300 GeV particle accelerator at CERN was ruled out by the Labor Government in 1968 precisely on the grounds that no economic pay-off in the short or middle term could be perceived from it; a cheaper version was ruled in again by the Conservative

Government in 1970 on the grounds of the national advantage to be gained politically by close scientific links with Europe.

The Present Status of the Myth

What then of the present status of the myth of the neutrality of science? Surprisingly, it has not withered *pari passu* with the emergence of Big Science. The reasons for this seem to relate back in part to the debate of the 1930s and 1940s on neutrality. For many scientists the debate effectively ended during a confrontation in Moscow between the agronomist Lysenko and his critics. Lysenko, defending his own "Michurinist" (or modified Lamarckian) biology, against the "idealist" and "quasi-fascist" genetics attributed to Mendel, Morgan and Weissman, and against the rump of Soviet geneticists, finally clinched his argument by revealing that his views had the approval of Stalin himself. The revelation of the distortions introduced by the imposition of Lysenkoism in Soviet biology threw Marxist Western scientists into a state of intellectual disarray. Some left the Party, like Haldane. Most retreated from an outspoken defense of the prospects of a socialist science into a more neutralist position.

A second, and perhaps still more important factor in strengthening the neutrality of science idea arose from the dilemma faced by many physicists who had taken part, many with the noblest of intentions, in the Manhattan Project. Believing that Hitler was developing the Bomb and that Hitler must be stopped and Fascism destroyed, what choice was there but to attempt to provide a Bomb for the Allies?

The fact that Hitler was not on the way to getting an atomic bomb, and that the ones that the physicists had made were used on Hiroshima and Nagasaki, provoked a crisis of conscience. If their profession was to be saved, and their consciences salved, they needed to discriminate between the effects of this use of their physics, and the physics itself: the physics had to be neutral; only the use to which it is put need be condemned. And as more and more it became apparent that, in the West, science was being applied to evil ends, the need to maintain the distinction between the subject and its use became sharper. For precisely those who in the past had argued that the link between science and human progress was inevitable, the retreat back to the laboratory and its neutrality became a necessity *if they were not to stop doing science altogether.*

An interesting parallel evolution seems to have taken place in the Soviet Union. Again, it may be related both to the failure of Lysenkoism and to the success of the Bomb. It is exemplified by the changing Soviet view on nuclear weapons. In the early 1950s the official view still reflected a belief that technology could be made to serve man's ends. The Bomb was seen as an adjunct to, but not a transformation of, the class war, which by definition must be victorious. The language of nuclear holocaust, the Doomsday machine, and the jargon of the United States arms control experts was repugnant. But by 1970, this view was no longer held in the Soviet Union. The Bomb is now seen as transforming war.[5]

The jargon of the two super powers is a mutual one, the language of the strategic arms limitation talks (SALT). Technology is seen to be just as sweetly inevitable to the Soviets as it was to Oppenheimer, when asked his views on the American H-bomb. And the corollary of technological inevitability is neutrality. "Science," a distinguished Soviet physicist assured one of us recently, "is neutral; it is how one uses it which determines its good or evil potential."

The Reaction Against Science

It is only in the last few years that this ubiquitous acceptance of a science simultaneously closely articulated into the bureaucratic, military and industrial machine of contemporary society, and yet freed from responsibility, of the ethos of the god-given right of the scientist to ask for the money he wants from the government, and to do research on what the government and he like when he gets it, has been effectively challenged. Partly, the challenge has taken the form of a wave of anti-science and the deliberate advocacy of non-rational forms of truth, of a refusal by many young people to do science—a phenomenon anxiously analyzed as the "swing from science" by governmental committees in several nations. Partly, it has been a rediscovery of the ideological issues by science and engineering students, who have become increasingly politicized by the general ferment in the universities which is occurring as practically a world-wide phenomenon.

And partly, it has undoubtedly been catalyzed by the repercussions of events following the Cultural Revolution in China in the late 1960s. For the Maoists in China resuscitated old arguments about the relationship of science and scientists to the people. With their insistence on egalitarianism, the down-grading of the expert, and the demand that his/her work should "serve the people," they made a demand whose echoes were those of the populist scientist artisans of the French Revolution, who closed the Academie des Sciences and helped send Lavoisier to the guillotine.

For Western Maoists, science is valid—and correct—only in so far as it is ideologically sound; its internal logic is totally subordinated to the ideological demands made upon it. It is scarcely surprising that virtually the only overt defenders of Lysenko to be found today, in or out of the Soviet Union, are the Maoist students.

For the less ideologically coherent, the rejection of neutral science has been enlarged to embrace a rejection of all science as irrevocably linked with the instruments of State oppression. The relationship perceived by Marcuse between the products of contemporary physics and the needs of IBM and the U.S. Atomic Energy Commission has, we have argued elsewhere, resulted in a specific rejection, not only of the sciences of the Establishment, but of the whole methodological apparatus and rationality of science itself.[6]

It is against this background, the re-opened questions of the continuance of science itself and of its ideological inputs, that the 1960s and 1970s have seen the emergence in several countries of Societies for Social Responsibility in Science. These groups have disparate views, but they are all linked in a common attempt to restate the values and the methods of science and the possibilities of harnessing it for constructive purposes, to recognize the ideological inputs inherent in science which the consensus politics of West and East of the last fifteen years have obscured, and to recreate a critical science. The objectives of such groups must be to use the critical methods of science to analyze and where necessary combat the social implications of particular kinds of science. This means taking issue with certain bits of science itself.

The Philosophical Issues

This compressed discussion of more than three centuries of philosophical, methodological and ideological debate has of necessity had to make a number of sweeping generalizations. Having both chronicled and asserted the validity of a particular point of view in this debate, we now attempt to state more clearly what we mean when we question the validity of the claim that science is neutral.

One way of doing this is to document the nature of the

constraints that operate on the activity of science within the present system. But, if we recognize that Big Science is State financed and that there is always more possible science than actual science, as well as more ideas about what to do than people or money to do them, the debate is, in a sense, short-circuited. Science policy means making choices about what science to do. Whoever makes these choices, by definition they cannot be ideology- or value-free, they imply an acceptance of certain directions for science and not others. Putting a man on the moon means not doing other sorts of things. Such choices are inherent in any system. And as they are clearly not neutral choices, the science they generate cannot be neutral.

This is easiest to see, of course, in such obviously applied fields as space technology. But it can be seen in other fields as well. In biology, for example, when an American woman scientist was awarded the United States Army's highest civilian award for the development of a more effective form of rice-blast fungus, specifically suited to Southeast Asian conditions, this award was clearly not given for neutral science.

But what about the basic science that led up to this work, which may not have been done under military contract or in a defense establishment? Take the development of the tear-gas CS, now extensively used in Vietnam, for example. This gas was developed in the mid-1950s by Britain's chemical defense establishment at Porton Down, as a result of a recognition of the inadequacies of the then-used tear gas, CN, on a number of technical criteria. Researchers at Porton began their search for a new agent as a result of a specific directive from the British Ministry of Defense. In the course of screening a number of possible agents they came across CS. Bulk production followed. Thus the work was done for a specific objective; it was clearly not value-free, by definition not neutral. Porton work was plainly mission-oriented. As a mission cannot be neutral, the science done in achieving it cannot either.

But what about the work from which the Porton studies were derived? Are we to indict Corson and Stoughton, back in the 1920s, for the initial observation that ortho-chlorobenzylidene-malonitrile was a lachrymator, just because thirty years later Porton picked it up and used it for a new purpose?

If this is the case, we should indict not only Corson and Stoughton but also all the other hundreds or thousands of academic researchers doing their "pure" research in the laboratories of the 1920s, churning out their three papers a year on the properties of odd chemicals and the behaviour of model systems, simply because they were working in the ambience of a society whose structure imposes a consequence not in harmony with human welfare. We would have to indict not merely Rutherford and Einstein for the atomic bomb, but practically all the chemists, physicists, mathematicians and biologists who have published research in the present century. Plainly, this is a *reductio ad absurdum*.

It might then be argued that there is a cut-off point in the neutrality debate: non-mission-oriented basic research, whose immediate application is not apparent, might seem excluded from it. At a pragmatic level, such a common sense idea might seem acceptable, provided the research was not sponsored by a fund-giving agency or organization with a mission other than the support of basic science for its own sake, such as a Department of Defense or industry. Particle physics and molecular biology might seem to come into this category.

While such a view might appear sound enough as a rule of thumb for everyday practical purposes, it avoids recognition of the possible interconnections between science as a cognitive system and the social system. That is, it assumes that, whatever the goal choices made by funding agencies, within those financial constraints the actual content of the science which is done depends only on the objective accretion of data, facts and theories. It thus implies a purely "internalist" view of the nature of scientific knowledge, as a set of ever-advancing and self-consistent absolute approaches to a statement of "truth" about the universe.

This view of science, though, is one that has come under serious challenge recently from the philosophers of science. Thus, the activity of scientists can be divided, according to the illuminating insight of Kuhn, into "normal" science and "revolutionary" science.[7] Normal science is what most scientists do all the time, and all scientists do most of the time; it is solving a set of puzzles about the natural world. The puzzles are designed and solved in terms of a paradigm, a gestalt view of the world, which provides a framework for normal science. At certain periods in science, and for a variety of only partially understood reasons, Kuhn argues, there occurs a paradigm switch, a change in the gestalt view of the world among working scientists which alters the puzzles they set themselves. This paradigm-switching is what Kuhn calls revolutionary science: problem-solving instead of puzzle-solving.

Kuhn's concept of revolutionary science may be suspect; what is sure is that his view of normal puzzle-solving science strikes an answering chord from many working scientists who recognize it as an accurate description of their activity, so different from the pompous proposals of verification, falsification and hypothesis-making which they were always supposed to be doing by philosophers in the past.

Corson and Stoughton's work, like that of the great majority of scientists, is essentially of the puzzle-solving kind. What we wish to propose here is that, while it is not possible to ascribe a "value," a measurement of non-neutrality, to all single pieces of puzzle-solving basic science of this type, it is possible to ascribe a value to the paradigm within which they are conducted. Puzzle-solving basic science of itself, unlike mission-oriented science or development, cannot have either neutrality or non-neutrality ascribed to it; the concepts are irrelevant. They are relevant only to the paradigm within which the puzzle solving activity is being conducted.

But a paradigm is never value-free. A paradigm is never neutral. Hence while we do not have to search for non-neutrality, or its moral congener, responsibility, in the work of a particular puzzle-solving scientist, we find it without difficulty in the paradigm within which s/he is working. To put it in other terms, and to answer the question which we raised earlier, while we cannot indict Corson and Stoughton and other puzzle-solving basic scientists for their "pure" research because of the uses to which their results are put independently of them, we assert that their work is non-neutral because the paradigm within which it is done is non-neutral.

Let us take some perhaps obvious examples of the non-neutrality of some biological paradigms. The framework of evolutionary biology set by the Darwinian revolution of the 1850s was one in which the central metaphors, drawn from society, and in their turn interacting with society, were of the competition of species, the struggle for existence, the ecological niche, and the survival of the fittest—a set of metaphors which closely reflected the norms of the society in which they arose, and contributed to the subsequent development of its ideology by giving it a seemingly "inevitable" biological base. These metaphors replaced earlier ones which located biology firmly into an explanation of a god-designed universe. Using these metaphors, a vast amount of puzzle-solving research has been conducted.

Today's central biological metaphors are framed in a different language system; they refer to control, communication and community, feedback and interaction, metaphors certainly more appropriate to the managed society in which we move today—and indeed, by virtue of a group of determined ethological extrapolators such as Morris and Lorenz, going some way to provide it with a biological rationale as well. And once again, today's puzzle-solving science is conducted within the framework provided by the metaphors.

The paradigm—the metaphor—sets the questions we ask of our subject, and hence the answers we seek from our materials. If a behaviour geneticist asks the question, "How much does heredity determine intelligence?" the person has limited the answers to the question before beginning the empirical research—that is, s/he has located the answer within a particular paradigm. It is the question and its framework, not so much the answer, which are non-neutral, their historical antecedents belonging in the line of eugenics stretching back to Galton and beyond. And we should not be surprised to find this type of research providing certain types of answers, which are then clearly related to certain social and political purposes.

It is not that the question should not be asked, but that we need to be very clear about the nature of the paradigm which sponsors it and which, within the limits of the data, specifies the answer, as it must for all puzzle-solving research. For much of science, the analysis of possible non-neutral components in specific paradigms is very difficult. It may well be that it is only in periods of social and intellectual crisis that we can glimpse the interconnections between science and the social system.

To suggest the non-neutrality of science is, we realize, not merely to recognize the passing of a myth—a myth which has nurtured academic science, and its considerable achievements, for more than a hundred years. To reject the myth also comes close to taking away another leg from the already wobbling basic tenet of scientific philosophy—that of the objectivity of the natural sciences. Admittedly this is a tenet that the philosopher-physicists of the Copenhagen School have in their time taken a healthy crack at and been attacked as idealists in consequence. It is a tenet, too, which has been under attack in contemporary biology, for the seeming success of the reductionist, objective techniques in such fields as molecular biology has begun to generate a counter-attack, an anti-reductionism which again poses the central problems of the relationship, in biology, between observer and observed.[8]

None the less, this is not to deny the validity of science's inner logic, that the advance of science, despite the shifting perspective of its paradigms, presents successively more accurate approximations to comprehensive statements about the nature of the universe. The way in which people see the world may vary with their viewpoint, but the variation will be confined within the crucial limitation that, for all of us, our viewpoint is human; there is a finite specification to the way in which our brains function, and to the relationship of this functioning to the environment.

Hence, in so far as objectivity specifies a public viewpoint that is more or less common to all people, there is an objective internal logic to science, which new techniques, experimentations and paradigms help refine. The universe may be (can be?) solved in this sense, granted time—and a paradigm which does not precipitate disaster. But the language in which the solution will be framed will depend upon the paradigm; of the infinity of questions we can ask about the universe, we will choose some of them and not others from within the ideological framework set by the paradigm.

When, as they now are, issues of human survival are on the agenda, it means that the time has come when it is right to assert, as a necessary counter-myth, that it is possible to use the techniques and methods of science for people-centered and not people-destroying purposes; that we must build human relevance into the paradigms of science itself.

From Abstraction to Action

How do we even begin to translate those abstractions into a concrete program for action? There are four distinct areas of activity possible.

First, scientists themselves have to become aware of the social, political and economic pressures affecting the development of science. They must learn the significance of the old saying, "he who pays the piper calls the tune," for all science in general and for their own science in particular. They have to find out how decisions are made, in and for science, and, where these decisions are concealed and undemocratic, expose them, working with the community at large to open the decision-making processes and make them democratically accountable.

At the same time, scientists will be wise to recognize that their criticisms will not necessarily endear them to those who pay for their research. They cannot afford to be as naive as those mathematicians in the United States who publicly deplored the massive abuse of science and technology in Vietnam and were then surprised when the Department of Defense threatened to cut off their research funds.

Second, scientists must learn to communicate not only with the community at large but also with their colleagues. They must be willing—and able—to explain what work they are doing, and why, and to speak out plainly if they feel that the social implications of their work are ambiguous or dangerous. They cannot afford to leave this communication task to the professional communicators alone; communication is not merely a matter of professional expertise but of moral responsibility. It is no longer adequate for the scientists to retire to their ivory-towered laboratories, emerging only to share the mutually congratulatory blandnesses of the scientific establishment. They have to learn to speak out.

But speaking out does not mean cheap sensationalism or wild prophesy of doom. To preach a sort of technological fatalism, of the inevitability of scientific and technological advance in this or that direction, is not merely irresponsible, it is to misunderstand the nature of science and its links with the social order. It is to subscribe to the myth of the scientific and technological imperative.

Nor does speaking out imply the carrying of apparently bulky but in reality lightweight burdens of specious moral responsibility about potential future abuses, the sort of attitudinizing morality of several biologists who have become concerned at prospects of "genetic engineering" for example. It is not the scientists' responsibility to carry this burden—if indeed it is a real one—and it distracts attention from the real issues of survival that are on today's agenda.

Third, the problem of science education and course content must be tackled. We cannot continue to educate as scientists individuals who are not taught to be aware of the pressures on, and real value of the activity by which they will subsequently earn their living. Science must cease to be taught, as it is in almost all countries of the world, in a sort of moral and social vacuum, as if there were no relations between the activities of the laboratory and the "real world" outside. Present experiments in which ways of introducing social relevance into science education are being attempted, need to be evaluated, improved and extended.

Finally, and most important, even having done all this, the scientists cannot merely sigh with relief, return to their laboratories with a "business as usual" sign up and get back to their favored research topics. They must instead ask themselves the question, "How can my scientific skills best be used to serve the people; to expose and correct the role of science and technology in wreaking genocide in wars, in oppressing individuals and minorities by acting as an agency of civil power, and in permitting malnutrition and disease both throughout the world at large, and even in rich societies?"

We are not facing a single once-and-for-all crisis in science but rather a chronic condition where the world, made one by massive science and technology, lurches either towards barbarism or progress. Survival itself demands revolutionary changes in society, and a precondition of this in a scientific society is the mobilization of the scientific community.

Where once revolutionary activity was argued for as a guarantor of progress, we now have to argue for its necessity as a guarantor against barbarism.

FOOTNOTES

1. E. B. Chain, "The Social Responsibility of the Scientist," *New Scientist*, October 22, 1970.

2. D. Joravsky, *Soviet Marxism and National Science*, London, Routledge & Kegan Paul, 1961.

3. J. Haberer, *Politics and the Community of Science*, New York, Van Nostrand-Reinhold, 1969.

4. J. D. Bernal, *The Social Function of Science*, London, Routledge & Kegan, Paul, 1939; also Cambridge, Massachusetts, MIT Press, 1967.

5. N. Moss, *Men Who Play God*, Harmondsworth, Middlesex, The Penguin Press, 1970.

6. H. Rose and S. Rose, *Science and Society*, London, The Penguin Press, 1969.

7. T. S. Kuhn, *The Structure of Scientific Revolutions*, Chicago, IL, University of Chicago Press, 1970.

8. D. McKay, "The Bankruptcy of Determinism," *New Scientist*, July 2, 1970.

Sociobiology, a Pseudo-Scientific Synthesis
—Barbara Chasin

This article pokes holes in the precarious underpinnings of Sociobiology, the newly invented science which attributes social behaviors like greed, altruism and sex roles to our genes.

Theories of human nature are, by virtue of their subject matter, political theories. Inevitably they make statements about the limits or potentialities of people and have serious social implications. Any notion of what kinds of beings we are will be used in struggles to preserve or change society. Theories which assert fixed, unchanging human qualities will be used to justify situations of inequality, while those stressing the human potential to develop and create will be used to open opportunities for greater numbers of people. History illustrates this point.

Writing about 2400 years ago, Plato discussed how to establish the ideal city-state. Society in the ideal state, he argued, would have to be hierarchical. He suggested that a myth be promulgated to maintain this structure. The people of the city should be instructed that:

> ...God, as he was fashioning you, put gold in those of you who are capable of ruling; hence they are deserving of most reverence. He put silver in the auxiliaries, and iron and copper in the farmers and the craftsmen.[1]

Plato, at least, was honest about his intentions; he recognized that he was creating a myth to support a society based on a strict class structure.

Modern writers couch their arguments in the jargon of science. During the 1960s—a time of great social unrest, of questioning of basic American institutions, and of growing interest in socialism as an alternative to the failures of capitalism—there appeared a spate of books with the theme that human beings are only another species of ape. Konrad Lorenz, Robert Ardrey, Desmond Morris, Lionel Tiger and Robin Fox, as well as less well known writers tried to convince their readers that animal studies are the key to an understanding of human social behavior.[2] War, violence of all sorts, private property, inequality, all of the major negative features of capitalism were said to be the natural and inevitable legacy of our primate origins.

A major contributor to these theories has been Edward Wilson, a Harvard biologist, who has given these theories of biological determinism a somewhat more sophisticated tone. His synthesis, which he's dubbed "socio-biology," seeks to unify disparate research into a coherent "scientific theory." Instead of basing social structure on the presence of gold, silver, or copper, Wilson argues that both the nuclear family and a sexually based division of labor are, like blue eyes and curly hair, genetically based. In his textbook, *Sociobiology: The New Synthesis*, he argues:

> The building block of all human societies is the nuclear family. The populace of an American industrial city, no less than a band of hunter-gatherers in the Australian desert is organized around this unit...During the day the women and children remain in the residential area while the men forage for game or its symbolic equivalent in the form of barter and money. The males cooperate in bands to hunt or deal with neighboring groups. If not actually blood relations, they tend at least to act as "bands of brothers.[3]

In an article in the *New York Times Magazine*, he continues:

> In hunter-gatherer societies, men hunt and women stay at home. This strong bias persists in most agricultural and industrial societies and on that ground alone appears to have a genetic origin. No solid evidence exists as to when the division of labor appeared in man's ancestors or how resistant to change it might be during the continuing

revolution for women's rights. My own guess is that the genetic bias is intense enough to cause a substantial division of labor in the most free and most egalitarian of future societies.[4]

From our earliest days, so the story goes, the man was the active, aggressive, subsistence providing person, while the little woman cleaned the cave, cooked the mastodon and reared the kiddies. A charming picture, but one which is just as mythical as Plato's story about the gold and silver.

By using the jargon of science, modern writers try to appear not as myth-makers but as hard-headed realists led willy-nilly by "scientific evidence" and "scientific reasoning" to the inevitable "scientific conclusion." Genetic theory, evolutionary biology, and anthropology are cited as the basis for pessimistic conclusions; yet, as we will see, underneath the jargon, there is little of valid substance. The research supposedly proving that our history is ruled by our genes is shoddy. Evidence which contradicts the theory of biological determinism is ignored or caricatured. Reasoning proceeds by the leaps and bounds of "analogies" between animals and humans. Conclusions are never tested against everyday experience. What results is a pseudo-science which provides an ideological prop not only for political ideologies, especially for sexism, but also for class and ethnic stratification.

What follows is an examination of some of the key links in this "new" science of "sociobiology." Despite the thousands of pages written to justify theories of biological determinism, three ideas underlie its arguments. One is the nature of aggression, another is on the origin of sex roles and sex role differences, and the third is the social inequities which seem to flow from them.

Sociobiology and Aggression

One of the major elements of sociobiological theory is that of the biological basis for aggression. In his *New York Times* article, Wilson cites a book by Eleanor Maccoby and Carol Jacklin, *The Psychology of Sex Differences*. In this book, the authors aim to "sift the evidence to determine which of the many beliefs about sex differences have a solid basis in fact and which do not."[5] They conclude that there are two areas in which biological differences between men and women result in different behavioral characteristics. One is spatial-visual skills. The other, important to the arguments of the sociobiologists, is aggression. These authors claim that males are more aggressive than females, and that this difference is genetically based.

Most biological determinists consistently use the notion of the greater aggressiveness of males to explain the "fact" that men are the doers, the makers of history. For example, Wilson argues that it is because of this aggression that even when men and women have "identical education and equal access to all professions, men are likely to play a disproportionate role in political life, business and science."[6] Because of this alleged link between aggression and sex, it is important to look into the nature of the arguments more closely.

In their characterization of the relation between sex and aggression, Maccoby and Jacklin cover the range of arguments used by biological determinists on this issue. For this reason and because Wilson uses Maccoby and Jacklin as evidence for his own assertions, a critique of their discussion is important.

Maccoby and Jacklin cite four kinds of evidence as pointing to the biological basis for male aggressiveness. They are:
1. a relation between aggression and levels of sex hormones;
2. similar sexually based differences in aggression can be found in man and subhuman primates;
3. sex differences in aggression are found early in life at a time when differential socialization cannot be occurring;
4. males are more aggressive than females in all human societies for which evidence can be found.
What evidence is cited for these forms of sex differences? And how does the data relate to the fundamental character of aggression?

The first line of evidence, that supporting the contention that there is a link between levels of sex hormones and aggressive behavior, comes from studies of rats and monkeys. Since humans consciously control their behavior to an extent unimaginable in a rodent or even in a monkey, it seems unsound to assume, where social interaction is concerned, that as animals go so do humans.

Looking at people, Maccoby and Jacklin do admit that, in fact, little is known about the relation beween sex hormones and behavior. They refer to two studies where people were involved. One is a report on testosterone (the primary male sex hormone) levels of 21 young men in prison. The men with the higher levels of this hormone had allegedly committed more violent and aggressive crimes during adolescence than men with lower levels. But to make this point, the authors of the original study had to resort to a curious definition of aggression. Aggression for them includes not only such acts as murder and assault, but *escape from institutions.*[7] The most generous thing you can say about this research is that it is inconclusive.

The other study they use has a certain ludicrous quality. Seventeen fetally androgenized girls were compared to their eleven normal sisters. "Fetally androgenized" means that these girls received excessive amounts of male hormones while they were fetuses. They were born with masculinized genitalia, which were later surgically corrected. But even after surgery, "their behavior continued to be masculinized in the following ways: they much more preferred to play with boys; they took little interest in weddings, dolls or live babies and preferred outdoor sports."[8] Despite such forms of "aggressive" behavior, these girls did not significantly initiate fighting more than their sisters.

We are being asked to believe that an interest in such things as dolls, weddings, live babies and sports are genetically related to our hormonal make-up. This is a highly dubious proposition. The girls, it should be remembered, were born with male genitalia and their parents were undoubtedly influenced and maybe even confused by this fact. The parents may, for example, have doubted that their daughters were "truly girls"; and parental doubts and anxieties have an uncanny way of being transmitted to the children. In any case, this study is hardly strong evidence for a genetic basis for "masculine aggression."

The arguments which link human aggression with primate behavior contain a great deal of oversimplification. Primates, even of the same species, differ remarkably from one another in their patterns of aggression, dominance, sex roles, etc. Baboons living on the plains do show the classical pattern of dominant male roles in such activities as making decisions about troop movements, having privileged access to food and sexual activity, and acting as the protectors of the young. (These animals are prominently featured in a film on sociobiology entitled "Doing What Comes Naturally.") But what sociobiologists Maccoby and Jacklin do not tell us is that forest baboons display little aggression and no male dominance hierarchies. When troops meet, which is rare, the encounters are friendly. When danger is perceived, the males run up the trees, leaving the females and young to deal with things on their own. The adult females are more likely to direct troop movements than are the males.[9] Thus, it is not at all clear that male primates are more dominant or aggressive than females.

But even if they were, we would not have an explanation of

human behavior. The behavior of primates does not automatically explain ours. Theories about primate behavior are themselves often based on theories of human behavior. The animal behaviorist uses his or her own experience as a human being as the basis for the study of "lower" animals. The analogy between humans and animals is equally forced in both directions.

Maccoby and Jacklin's third point—that sex based differences in aggressive behavior are found so early that socialization can't account for them—is an exceedingly tenuous argument. What is aggressive behavior in a newborn infant? Furthermore, there is evidence that differential behavior toward babies based on their sex occurs very soon after birth.[10]

Finally, there is the most important question of cross-cultural evidence. If we find that males are more aggressive than females in all human societies, there would be some reason to think that this is a sex-linked characteristic. Maccoby and Jacklin use only two studies. One has to do with playground behavior in the United States, Switzerland and Ethiopia. Boys more often hit or push each other without smiling in each of these societies. The other cross-cultural evidence they use is based on Whiting's and Pope's discussion of data drawn from six cultures. Maccoby and Jacklin themselves note that in this material there are few physical assaults of children upon one another and that sex differences didn't account for these. But boys were more likely to engage in "rough and tumble play" (which, interestingly, has now become synonomous with aggression, when the original definition was the intent to hurt), to trade more verbal insults among themselves, and to counterattack if physically or verbally assaulted. We don't need these studies to know that there are many societies where boys and men are more likely to engage in some type of aggressive behavior than are girls and women. Do limited studies like these show that sex-linked aggression is a universal trait among humans? Can they be used to prove that men are *innately* more aggressive than women?

In each of these four arguments there is a thread of dishonesty. Biological determinists consistently omit data which is inconsistent with their own statements. They thus relieve themselves of the burden of showing how their theory could be used to explain seemingly contradictory evidence.

Cross-cultural evidence, in fact, shows that there are societies where neither sex is aggressive. Preliminary reports on the Tassaday of the Philippines have noted the gentleness of both males and females and the lack of anything resembling fighting. There is no war, not even a word for war, nor is there a sexual division of labor. Such leadership as there is has at times been exercised by women.[11]

In other societies too it is hard to find examples of males being more aggressive than females.[12] Ruth Sidel's description of children in socialist China is worth citing here as well:

> The emphasis on the People's Liberation Army and on defending the motherland stands in sharp contrast, however, to the lack of aggression you see in the children in their day-to-day life. That we never saw a child push another child, never saw a child grab a toy from another child, never saw any hostile interaction between children or between adults and children truly amazed us. When we asked about aggression at a kindergarten at the worker's village in Shanghai, we were told by the kindergarten teacher, Lu Shiutsung, that aggression is not a problem, because the children have already "received collective training in nursery." She allowed that occasionally a child might be aggressive, but this can usually be handled through "education."[13]

Continuing, she notes:

> What is so amazing, of course, in walking the streets of Peking or Shanghai, or visiting a commune or urban neighborhood, is that we never saw aggression among the children. No doubt it exists, but we never witnessed it. At one park in Hangchow, one of us handed a piece of candy to a boy of about ten: he immediately passed it on to his baby sister. He was then handed a second piece, which he passed to his mother. He kept the third piece; he had no one else to give it to.

Felix Greene noted the same lack of aggression in children during his trip:

> "I have spent alot of time watching children in the streets— little tots all on their own. They are endlessly inventive in their games— a piece of wood or a bit of string will keep them happy for hours. They never fight. Why don't they? They never snatch—never 'That's mine.' "[14]

Sociobiology and Sex Roles

The ideas about sex role differences play a crucial part in the construction of the theory of sociobiology. Yet cross cultural evidence on sex roles is largely ignored or misrepresented in the works of the sociobiologists. There are societies, ones which might well have been the typical human grouping for millenia, in which there is little sexual division of labor. But even where some divisions exist, they are far different than those portrayed by Wilson et. al. Men and women may engage in different tasks, but it is not women's fate to be confined to a small area all day, puttering around the campfire doing household chores.

Colin Turnbull spent several years living with the Pygmies of the Ituri forest in the Congo. He writes:

> Between men and women there was... a certain degree of specialization, but little that could be considered exclusive...[16] The woman is not discriminated against, she has a full and important role to play. There is relatively little specialization according to sex. Even the hunt is a joint effort. A man is not ashamed to pick mushrooms and nuts if he finds them, or to wash and clean a baby. A woman is free to take part in the discussions of men if she has something relevant to say.[15]

Wilson does not make any reference to Turnbull's work with the Pygmies in his textbook, though he does refer to his studies of the Ik in Uganda, whose behavior is more congruent with his sociobiological theory.[18]

Similarly, Patricia Draper's account of the !Kung Bushmen of the Kalahari desert reveal that the women provide from 60 to 80 percent of the daily food supply. Their gathering requires them to go quite a distance from the camp, and eight to ten miles is not an unusual distance. The women are skilled in understanding the meaning of animal tracks and provide invaluable information for the male hunters. The women also have the "ability to distinguish among hundreds of edible and inedible species of plants at various stages in their life cycle."[17] Furthermore, men are in no sense the dominant authority figures. The women control the food they collect to a far greater extent than the men, whose kill is divided according to a rigid set of rules.

This egalitarian situation is changing as the !Kung are being resettled with agricultural people and their whole way of life is undermined. These changes are important to understand. Explanations in terms of genetic characteristics or hormonal balance do not enhance that understanding. In the new agricultural setting, the men acquire certain kinds of responsibilities and chores which are not matched by those of the women, the mode of child rearing changes, and so on. These are just examples of the variables that replace a biologically-based analysis.[18]

Biological determinists do not deal with any of this data. While Wilson claims, for example, that the Bushmen data supports his view of human nature, he virtually ignores the material collected by Patricia Draper, and is extremely selective in the use he makes of Richard Lee's findings. Lee is one of the world's foremost experts on the Bushmen and for all practical purposes Wilson ignores his work.

Why Do We Have Sociobiology?

Sociobiology's descriptions of what men and women do are wrong for the United States as well as for other societies. It is touted by the media not for its scientific value but for its political functions. This is readily apparent when Wilson and others like him attempt to convince people that women's place is really in the home.

Yet women are an essential part of the labor force, one of the most exploited parts of the labor force in fact. For their exploitation to continue unchallenged, women, and men, must accept the idea that women's work is something secondary to their lives and even something they should not really be doing.

Certainly some women spend most of their time in what Wilson calls the "residential area," but an ever increasing proportion do not. Only 20% of women were in the labor force in 1930; now 48% of women are working or looking for work.[19]

Capitalism needs women in the work force. Changes in the occupational structure have resulted in a lower proportion of industrial and manual jobs and a concomitant increase in office work, a change which necessitates hiring people who are relatively educated but also willing to work for low wages. Women fit the bill. Women in 1960 were 97% of the secretaries, 84% of the bookkeepers, 96% of the telephone operators, and 86% of the file clerks.[20] While the demand for women workers has increased, a drop in their husbands' real earnings has made it an economic necessity for even married women to work. In addition, the number of female-headed households has doubled between 1940 and 1975, and these women need to work as well.[21]

The work women do may be crucial but this is not reflected in their wages. According to Labor Department figures, the difference between an average man's annual earnings and those of the average working woman is now over $5,000. From 1955 to 1974 the gap between men's and women's earnings increased by an amazing 74%. The typical white man working year round full time in 1974 earned $12,343; the typical black man earned $9,082. Year round full time white women workers brought home an average of $7,025 and black women took home $6,611.[22]

These figures reveal in statistical fashion a major aspect of women's oppression. The women's liberation movement has attacked and is attacking the exploitation that women suffer. All agree, whatever else their differences, that the situation must be challenged and a new social order built. There is energy, strength and organization among women. Links have been built and continue to be forged with other political groups and struggles. Many women have developed a new confidence in themselves and their own abilities and are fighting the myriad forms that sexism takes.

The women's movement and its gains are under counter-attack. There is direct political repression to face. Under the Freedom of Information Act, the F.B.I. has released a 137 page report which chronicles its anti-feminist activities between 1969 and 1973. The Church Committee which is investigating government intelligence activities has produced evidence that the C.I.A., military intelligence and local police red squads have also operated against women's groups.[23]

The police agencies represent one kind of assault on women. Theories of biological determinism are yet another kind of weapon used to preserve inequalities. This is not to say that biological determinists such as Wilson have consciously decided to protect U.S. capitalism from the threat of women's liberation. But their ideas are used by the people who control the media, the publishing industry, and the scientific and social scientific establishments. Those who create these theories are rewarded; they are given money and prestige. Ambitious students and colleagues see which way the wind is blowing and add to the proliferation of books and articles that so misrepresent the real nature of human beings.

Sociobiology Justifies Inequality

While sociobiology is most explicit and elaborate in its attitudes toward women, it is also being used to "explain" class and ethnic differences.

Edward Wilson, for example, accepts as a well established fact the belief that social stratification and competition are universal features of human society. "The members of human societies," he writes, "sometimes cooperate closely in insectan fashion, but more frequently they compete for the limited resources allocated to their role sector. The best and most entrepreneurial of the role-actors usually gain a disproportionate share of the rewards while the least successful are displaced to other less desirable positions."[24] Wilson never refers to the anthropological and historical literature demonstrating the tremendous range of social organizations and forms that have existed among humans.

Even when he does draw upon some anthropological material, Wilson remains extremely biased. The !Kung Bushmen, discussed above, in their hunting and gathering society are characterized by a high degree of social equality and have developed mechanisms for maintaining this situation.[25] Wilson ignores this and asserts that among the !Kung "some are exceptionally able entrepreneurs and unostentatiously acquire a certain amount of wealth. !Kung men (sic) no less than men (sic) in advanced industrial societies, generally establish themselves by their mid-thirties or else accept a lesser status for life."[26] If Wilson believes that anthropologists such as Richard Lee have made gross errors in describing the Bushmen as egalitarian, he has a scientific obligation to show how he arrived at his own conclusions.

Wilson also implies that those who are the wealthiest in society are being rewarded for being the "best and most entrepreneurial of the role-actors." In his understanding of society, the cultural inheritance of wealth and position is irrelevant to the analysis of inequality. The only form of inheritance that is of interest to him is the inheritance of mental traits.

There are, in this formulation, "hereditary factors of human success." I.Q. is said to be such a factor and so are "creativity, entrepreneurship, drive and mental stamina."[27] Again, no evidence is presented to support his claim that these are inherited traits. There are numerous works showing that the arguments for the inheritance of I.Q. are spurious, in some instances even based on fabricated data, and at present, without any empirical basis.[28]

Once again, Wilson cannot even be bothered with alluding to

the controversy. Nowhere in his approximately 660 pages of double columned text does he present the criticisms of the genetic arguments and respond with his own counter evidence.

By taking the position that I.Q. and other mental traits are inherited, Wilson is giving credence to those who, like Arthur Jensen, have argued that blacks are inferior to whites and that compensatory education could not eliminate this inequality. Wilson's work goes even beyond this in providing support for racism. He states that the genes "maintain a certain amount of influence in at least the behavioral qualities that underlie variations between cultures."[29] We have already mentioned his contention that there are ethnic bases for such personality traits as introversion-extroversion, depression, etc. "Even a small portion of this variance invested in population differences might presuppose societies toward cultural differences." He contends that we should try to measure these differences—"in short, that there is a need for a discipline on anthropological genetics."[30] We might well ask Wilson why is such a subdiscipline needed?

Anthropologist Marvin Harris has written:

> We...bear witness to the extraordinary mutability of national character in relation to the evolution of socio-political systems. We see the Mozart-loving Germans transfigured into the beasts of Buchenwald, the authoritarian Japanese become the democrats of Asia, meek Balinese boys joining death squads,...American troops turning over Vietcong for torture, the staid Britishers dressed in mini-skirts, the racially democratic Portuguese in league with South African apartheid, promiscuous South Sea Islanders turned Protestant prudes and Appolonian Pueblo Indians who have become the drunken wards of their former enemies.[31]

There is a lot we need to know about the way the world works, but looking for explanations in the genetic differences between peoples seems more likely to obscure than clarify reality.

If Wilson's remarks have a certain ludicrous quality, they also contain dangerous implications. To postulate the need for anthropological genetics is to assert that the important differences among peoples are biological ones; in the process, Wilson resurrects a discredited nefarious racist notion. Will it be discovered that certain African tribes have a genetic disposition to work in the mines, that Latin American peasants are genetically adverse to work and should be replaced by machines, while Native Americans have genes predisposing them to alcoholism, unemployment and tuberculosis?

According to Wilson's analysis, not only are there biologically-based differences between peoples, but there is a genetic basis for the enmity and aggression that groups feel toward one another. Xenophobia, too, is a part of our biological heritage.[32] In recent years there has been a marked increase in hostility between ethnic groups. Nazis march, the Klan demonstrates, racist groups physically attack black children going to school, to the beach and so on. There have been a large number of incidents in which police have killed Chicanos, Puerto Ricans and blacks with the killers receiving light punishment, if any.

Racial violence has been one of the ugliest features of Western civilization. There have been useful analyses published that aid our understanding of this problem.[33] These stress the social and economic concomitants of racism, and the ways in which racist ideas are manipulated by powerful political groups. There are also periods and places in which racism is much diminished and even begins to disappear, as in Cuba. Sociobiology, however, tells us that fear of the foreign is a perennial feature of human life—as is aggression—so that we need look no further than our genes to understand it.

Sociobiology or Liberation?

But Wilson, like the other biological determinists, is wrong. The fault is in our society, the injustices; the inequalities do not lie in our genes. They are rooted in social institutions and class structures. All over the world people have challenged—with a growing success—sexism, racism, poverty, degradation and brutality. Cuba, Vietnam, China, Mozambique, Angola are not Utopias, but they are supporting a real effort to remove inequalities.

Whenever people join together to create a new order, those benefiting from the old try to crush the people and their vision. They use weapons that terrorize, maim, and kill; they use theories which aim to demoralize people, to convince them of their essential inferiority, and to reconcile them to the world as it is. But despite the damage that they cause, despite their capacity to hold back genuine progress, neither the armies, the napalm, the bombs, the C.I.A., nor the theories of biological determinism can stop the movement to build a new society.

FOOTNOTES

1. *The Republic of Plato,* A. D. Lindsay, translator, New York, E. P. Dutton, 1957, p. 124.

2. The particular books by these authors are: Konrad Lorenz, *On Aggression,* New York, Harcourt, 1966; Robert Ardrey, *African Genesis,* New York, Dell, 1967, first published in 1961; Desmond Morris, *The Naked Ape,* New York, Dell, 1969; Lionel Tiger and Robin Fox, *The Imperial Animal,* New York, Dell, 1971. Tom Alexander, writing "The Social Engineers Retreat Under Fire" in *Fortune* magazine, October, 1972, urged that more attention be paid to our genetic propensities. The arguments have been significantly extended beyond social measures and concerns. Many industries encourage genetic research to show that the health of their workers is not impaired by job hazards or industrial pollution but by the workers' own genetic propensities towards injuries. Useful criticisms of the biological approach have appeared in Ashley Montague, ed., *Man and Aggression,* New York, Oxford University Press, 1968; Alexander Allard, Jr., *The Human Imperative,* New York, Columbia University Press, 1972; and the Ann Arbor Science for the People Editorial Collective, *Biology as a Social Weapon,* Minneapolis, Burgess, 1977.

3. Edward O. Wilson, *Sociobiology: The New Synthesis,* Cambridge, Massachusetts, Harvard University Press, 1975, p. 553.

4. Edward O. Wilson, "Human Decency is Animal," The *New York Times Magazine,* October 12, 1975, p. 48.

5. Eleanor Maccoby and Carol Jacklin, *The Psychology of Sex Differences,* Stanford, California, Stanford University Press, 1974, p. vii.

6. Wilson, "Human Decency is Animal," p. 50.

7. Leo E. Kreuz and Robert M. Rose, "Assessment of Aggressive Behavior and Plasma Testosterone in a Young Criminal Population," *Psychosomatic Medicine,* 34, July-August, 1972, p. 321.

8. Maccoby and Jacklin, *The Psychology of Sex Differences,* p. 243.

9. David Pilbeam, "The Naked Ape: An Idea We Could Live Without," in David E. Hunter and Phillip Whitten, *Anthropology: Contemporary Perspectives,* Boston, Little, Brown, and Company, 1975, pp. 65-75.

10. See for example, Jeffrey Z. Rubin, Frank J. Provenzano, and Zella Luria, "The Eye of the Beholder: Parents' Views on Sex of Newborns," *American Journal of Orthopsychiatry,* 44, 1974, pp. 512-519; Carol Seavey, Phyllis Katz, and Sue Zalk, "Baby X: The Effects of Gender Labels on Adult Responses to Infants," *Sex Roles,* 1, June, 1975, pp. 103-109.

11. John Nance, *The Gentle Tasaday,* New York, Harcourt Brace Jovanovich, 1975.

12. Collin Turnbull, *The Forest People,* New York, Simon and Schuster, 1962.

13. Ruth Sidel, *Women and Child Care in China,* New York, Penguin, 1973, p. 150.

14. *Ibid.,* p. 114. Felix Greene, *China,* p. 54, quoted in Sidel, p. 114.

15. Turnbull, *op. cit.,* p. 154.

16. Wilson, *Sociobiology,* p. 554.

17. Patricia Draper, "!Kung Women: Contrasts in Sexual Egalitarianism in Foraging and Sedentary Contexts," in *Towards an Anthropology of Women*, Rayna Reiter, ed., New York, Monthly Review Press, 1975, pp. 77-109.

18. A useful book discussing the material basis for women's roles is Ernestine Friedl, *Women and Men: An Anthropologist's View*, New York, Holt, Rinehart and Winston, 1975.

19. Arlene Eisen, "More Women Work—The Double Shift," *The Guardian*, January 12, 1977.

20. Lisa Vogel, *Women Workers: Some Basic Statistics*, New England Free Press, Union Square, Somerville, Massachusetts.

21. Eisen, *op. cit.*

22. The *New York Times*, November 29, 1976.

23. Arlene Eisen, "The FBI's Tactics Against Women," The *Guardian*, February 23, 1977; Arlene Eisen, "FBI Targets Women's Movement," *Guardian*, February 16, 1973; see also Howard Husock, "The Feminist Papers: FBI Plays 'I Spy,' " and "More Tales from the Feminist Papers," *Boston Phoenix*, February 15 and 22, 1977.

24. Wilson, *Sociobiology*, p. 554.

25. Richard Lee, "Eating Christmas in the Kalahari," *Natural History Magazine*, December, 1969.

26. Wilson, *Sociobiology*, p. 549.

27. *Ibid.*, p. 550.

28. An excellent collection of articles on this subject is N. J. Block and Gerald Dworkin, eds., *The I.Q. Controversy*, New York, Pantheon Books, 1976.

29. Wilson, *Sociobiology*, p. 550.

30. *Ibid.*

31. Marvin Harris, *The Rise of Anthropological Theory*, New York, Thomas Y. Crowell Company, 1976.

32. Wilson, *Sociobiology*, p. 565.

33. For example, see: George Novack, *Genocide Against the American Indian*, New York, Pathfinder Press; C. Vann Woodward, *The Strange Career of Jim Crow*, New York, Oxford University Press, 1966; Jim Green and Allen Hunter, "Racism and Busing in Boston," *Radical America*, 8, November-December, 1974.

Genetics as a Social Weapon
—Gar Allen

This article draws attention to the similarities of the eugenics movement of 1920-1925 to the IQ/race debate of the 1970s. The former used so-called scientific data to limit opportunities of immigrant groups thought to be genetically inferior, and the latter uses similar kinds of theories to justify tracking of students in the educational system by race and class.

A raft of reports has appeared claiming a genetic basis for intelligence in human beings. These hereditarian explanations for intelligence have been given considerable publicity—by far more than given to opposing views. As a result, whether consciously or not, the U.S. scientific and general public has begun to absorb the mistaken notion that the biological evidence supports the idea that intelligence is largely inherited. Considering the large number of other scientific developments which could have been given such wide publicity during the same period, the frequency of articles on the heritability of I.Q. is somewhat surprising. It raises the question of why certain ideas in a field of science are investigated more at some times than others, and why they receive so much popular exposure.

Perhaps a good way to answer this question is to examine past cases of a similar nature. This article will compare the present I.Q. studies with the eugenic arguments of the early part of this century. Not only does that movement provide the scientific foundation for the present controversy, but we shall see that there are many similarities in the two historical periods which may help explain why the current rebirth has occurred.

Eugenics claims to apply genetic principles to the "improvement" of humankind. There are two general subdivisions to its efforts: Positive eugenics—increasing the reproduction of especially "fit" individuals, and Negative eugenics—reducing the breeding of particularly "unfit" types. At the turn of the century, the eugenics movement proposed both types of programs and had a wide influence. Between 1905 and 1920 eugenics courses were quite fashionable in colleges. A number of institutions devoted largely, or solely, to eugenic research and propaganda were founded in the same period. Two international congresses of eugenics were held, and a number of scholarly and propagandistic journals were published on the subject. The impact of eugenics was not, however, limited to academics. Eugenics and eugenicists exerted a considerable influence on popular opinion and on state and federal legislation. Twenty-four states passed sterilization laws for various social "misfits" (e.g. criminals, mentally retarded, or the insane). Some thirty states passed miscegenation laws restricting or outlawing interracial marriage. Perhaps the key triumph of the eugenics movement was the passage in 1924 of the Johnson Act by the Congress. This immigration law almost totally stopped immigration into the U.S. from Eastern European and Mediterranean countries. This act also brought the eugenic doctrines the most public exposure.

From its beginnings, the eugenics movement was closely associated with a sense of white Anglo-Saxon superiority and racism. Francis Galton, the founder of the movement, was an elitist and racist. He was first drawn to the study of human heredity and eugenics looking for a genetic source of his own family's "genius." (His cousin was Charles Darwin and his family tree was decorated with numerous illustrious ancestors.)

The eugenics movement in the U.S. started in 1904 when Charles B. Davenport persuaded the Carnegie Foundation to establish a Laboratory for Experimental Evolution at Cold Spring Harbor, of which he became the director, and at the same time leader of the U.S. eugenics movement. In 1907 he persuaded Mrs. E.H. Harriman, wife of the head of the Union Pacific Railroad, to financially support a Eugenics Records Office, also at Cold Springs Harbor. Here, Davenport and his colleagues made studies aimed at developing eugenic programs in the U.S.

Davenport shared Galton's belief in superior and inferior races (with the Anglo-Saxon at the pinnacle). Galton had remarked, "there exists a sentiment, for the most part quite unreasonable, against the gradual extinction of an inferior race." Davenport emphasized the possible ill effects of "race-crossing," especially between blacks and whites. Racism was a prominent element in Anglo-Saxon middle-class society at the time, and easily became part of hereditarian doctrine.

Before 1915, a number of prominent biologists supported and actively took part in the eugenics movement. Davenport himself was a respected geneticist and one of the early supporters of Mendel's theories of inheritance in the U.S. Other prominent biologists included E.G. Conklin, T.H. Morgan, H.S. Jennings, and W.E. Castle: all professors at elite American universities and members of the National Academy of Sciences. Castle wrote a popular textbook, *Genetics and Eugenics* which became a standard text for eugenics courses. Conklin edited a eugenics text and supported the eugenics movement in public lectures. These and many other less prominent biologists contributed large numbers of articles to the American eugenics movement's "scientific" publication, the *Journal of Heredity* which blended research, reporting, and propaganda on eugenics.

It is understandable that immediately after 1900 many geneticists wanted to see the newly discovered theory of Mendelian heredity applied to humans. Indeed, a few studies, such as those by Landsteiner on the A-B-O blood groups, and Garrod on metabolic disorders (alkaptoneuria and phenylketoneuria), provided good evidence for the existence of Mendelian inheritance in humans. However, these studies dealt with easily identified clinical traits whose inheritance could be checked by reference to clear-cut family pedigrees. Eugenically oriented geneticists such as Davenport and Castle, on the other hand, tried to show the inheritance of more complex traits in simple Mendelian terms. For example, Davenport tried to show that alcoholism, seafaringness, degeneracy, and feeble-mindedness were each due to single Mendelian genes, inherited in a dominant or recessive way. Similarly, Castle tried to argue by analogy that marriage between human races might produce the same type of misfit hybrid as a cross between a thoroughbred and a draft horse.[1] Intelligence was prominent among the traits that eugenecists tried to demonstrate as inherited. With the newly designed Binet test for intelligence as the standard of measurement, studies flowed forth showing connections between low test scores ("feeblemindedness") and delinquency, criminality, sexual promiscuity and degeneracy. Needless to say, the "evidence" for such claims was meager, based largely on assumptions, biases, analogies, and a variety of non-rigorous methods of "proof." Yet before 1915 many biologists still hoped that the study of human heredity would allow the elimination of some of the worst human diseases and the improvement of mankind's genetic potential.

As the popular side of the eugenics movement picked up steam after 1915, biologists began to withdraw their support. There were several reasons, which we can summarize briefly:

1. increasing evidence that few genetic traits were determined by single genes;
2. evidence that even genetically identical individuals showed variation, underscoring the importance of gene-environment interactions;
3. penetration of the idea of genetic equilibrium, which began to convince scientists of the difficulty of removing undesirable genes from a population;
4. increased skepticism about the methodology used by eugenic researchers.

This last problem was highlighted by the difficulties in measuring human intelligence. The first large scale study of I.Q. in the American public (done by the U.S. Army) showed that by current standards, half of the U.S. population was "feeble-minded"! Further, although the tests fulfilled racists' expectations in that on the whole blacks did worse than whites, northern blacks did better than southern whites! Biologists began to see the immense difficulties in trying to produce a simple genetic explanation of intelligence or intellectual differences, and indeed of even measuring such an ill-defined trait. Without planned matings, the study of human heredity even for well-defined traits seemed a long-term project. For vague traits such as intelligence it was clearly impossible. In general, biologists shifted their attention to studying laboratory animals. Scientific support for eugenics gradually dwindled away, although die-hards such as Davenport and Castle remained believers.

Despite the withdrawal of professional support, popular exposure to eugenics continued to grow. In 1914, 44 colleges taught eugenics. By 1928, the number had swelled to 376, or roughly ¾ of all colleges and universities—some 20,000 students! At the same time, popular books on the subject began to appear more frequently. Although these usually claimed to be "scientific," they often demonstrated a highly biased and racist tone. Perhaps the most popular of these works were Madison Grant's *The Passing of the Great Race*, published in 1916, and Lothrop Stoddard's *The Rising*

Tide of Color Against White Supremacy, published in 1920.[2] Both
lamented the increasing number of foreign immigrants in the
United States and the decline of "Nordic civilization" in the West.
Both supported their arguments with references to the works of
Davenport, Castle, and other geneticists who had suggested
biological ill-effects of race crossing. As Grant wrote:

> Whether we like to admit it or not, the result of the
> mixture of two races, in the long run, gives us a race
> reverting to the more ancient, generalized and lower
> type...The cross between a white man and a negro is a
> negro;...and the cross between any of the three European
> races and a Jew is a Jew.[3]

Race feeling might be called prejudice, Grant said, but it was a
"natural antipathy" which served to "maintain the purity of type."
Both books drew heavily not only on biological, but also anthropological and historical "evidence" to show that the white race, the
Anglo-Saxon and the Nordic, was the superior group on the human
evolutionary tree. While not all eugenics books were as overtly
racist, the more subtle texts contained most of the same implications. By the end of World War I, the eugenics movement had
taken on a distinctly pro-Nordic, anti-everything-else demeanor.

Such eugenics propaganda led to the passage of strongly racist
legislation. Perhaps the most important law passed was the
Immigration Restriction Act of 1924 (the Johnson Act). Except for
the war years, prior to 1921 the U.S. government had placed
virtually no restrictions on immigration to the United States.[4] Only
after the war did people begin to call for halting or greatly
restricting both the numbers and types of immigrants. The major
reasons for this were economic; it seemed necessary to stop
additions to an already glutted labor market. The immediate
response of Congress was the passage of a temporary Emergency
Act of 1921, which restricted immigration from any European
country to three percent of the foreign-born of that nationality
listed in the 1910 census. (This act, as well as the permanent
Immigration Restriction Act of 1924, applied only to European
immigration. Oriental immigration had been restricted by earlier
measures in the 1880s.) Since the Emergency Act was only
temporary, proponents of immigration restriction began work
immediately for a more permanent law. Between 1921 and 1924
biological (genetic) arguments became important in justifying a
campaign against all non-Nordic immigration. Eugenicists were
very active in this campaign.

The eugenicists claimed that the new immigrants were genetically inferior to the Nordic or Anglo-Saxon, and even to the older immigrants who had come to the United States in the 1850s and 1860s. Like the Social Darwinists several decades earlier, the eugenicists argued that a person's economic and social status showed his or her hereditary worth. The high levels of disease, illiteracy, poverty, and crime in immigrant neighborhoods proved to eugenicists that the non-Nordics were inferior and debased. Eugenicists also claimed without proof that heredity was far more important than environment in determining human behavior. They also fastened onto the genetic idea of "disharmonious crossings"— the idea that the children of interracial marriages will always be inferior to both parental races. (By 1920 most geneticists were well aware that mating between different "pure" strains produced more vigorous offspring than the continued mating within the "pure" strain, thus the idea of "disharmonious crossings" was an outmoded concept.) But the conclusion from these two "genetic" beliefs, according to eugenicists, was that the inferior qualities of immigrants could never be improved by the new American environment; and indeed, that dilution of the superior American blood (genes) by intermarrying with inferior immigrants would produce an inferior population. The eugenicists argued not for an end to immigration, but for *selective* immigration favoring the "better" races of Europe, meaning the Nordic and Anglo-Saxon.

To make the biological side of the argument most effective, eugenicists tried to summarize and "document" the genetic claims in a "scientific" way. This task was undertaken by Davenport's Eugenics Records Office. In April, 1920, Harry Laughlin was appointed as "expert eugenic agent" of the House Committee on Immigration and Naturalization. During the next three years he appeared in person several times before the committee, and in one of his testimonies, in 1922, concluded sweepingly:

> Making all logical allowances for environmental conditions, which may be unfavorable to the immigrant, the recent immigrants, as a whole, present a higher percentage of inborn socially inadequate qualities than do the older stocks.[5]

His evidence from the start was questionable, and his conclusions totally unjustified by the facts, yet the "Laughlin Report," made up of Laughlin's congressional testimonies, complete with "scientific evidence," became widely regarded as a scientific and unbiased presentation of fact. Laughlin's testimony, backed by the authority of the Eugenics Record Office and the Laboratory of Experimental Evolution at Cold Spring Harbor carried the weight of scientific authority with many congressional leaders.

The legislative debates over immigration restriction were furious throughout 1923 and early 1924. In the Senate hearings, biological arguments were minimal; but in the House, they became a major factor in getting the bill passed. Almost the only strong opponents in the House were either themselves representatives of minority groups (which at that time meant Jewish), or else came from northeastern states, where immigrant groups were well organized politically. The House Committee hearings were enormously biased, since "experts" called in to testify were hand-picked to present the eugenicists' Nordic and hereditarian line. At the insistance of Representative Celler of New York, the professional geneticist H.S. Jennings of Johns Hopkins was grudgingly asked to testify. Jennings was by this time entirely out of sympathy with the eugenics movement, but he was given only a few minutes to speak. He was told by the Chairman of the Committee not to present his arguments at that time in any detail, but to submit a written report later. In the end, the Immigration Restriction Act passed by large majorities in both the House and Senate.

After 1924 more scientists began to speak openly against the eugenic and racist propaganda which was being published in the name of "science" and "biology." Jennings was one of the first to speak out loudly—for he had seen the uses to which garbled biology could be put and the countless ill-effects it could have. Later Raymond Pearl, E.M. East, T.H. Morgan and W.E. Castle all joined in publicly repudiating the racist propaganda of the eugenicists on biological grounds. Unfortunately, the geneticists' efforts were too little too late. The major effects of the eugenics movement had already been achieved by the time geneticists began to criticize the movement publicly. The results included the restriction of immigration, especially from eastern and southern European countries—restriction which was not repealed until 1965. But more damaging perhaps than this were the attitudes of racism, superiority, and outright hate which the movement helped to intensify among the people of the U.S. Some of this racism might have been counteracted had scientists as a whole spoken up earlier and louder than they did against the eugenicists' biological arguments.

* * *

Graphic of Nick Thorkelson
Science for the People, May-June 1977

Why the Eugenics Movement

I do not wish to suggest that scientists should bear all or even a major part of the blame for the success of the eugenicists. Why, after all, did the eugenics movement get going in the first place? Who supported it and for what reasons? Was it simply the work of some crackpots and overly ambitious zealots? The sudden rise in the general popularity of hereditarian and racial views was not an accident or the result of good publicity on the part of a few fanatics either within or outside of the scientific community. It was at heart a social, political and economic phenomenon arising from the struggle between classes. What does this mean, and how can it help us understand more recent events such as the I.Q. and race issue?

First, we can ask: Who were the people in the eugenics movement between 1900 and 1925? Those non-scientists who founded, financed or in other ways supported the eugenics movement from the early 1900s onward were, almost to a person, wealthy businessmen, investors and other representatives of the financial and ruling elite of the United States at that time.

The ruling elite which in the early decades of this century initiated and provided financial support for the eugenics movement included: David Fairchild, President of the American Genetic Association (AGA—publishers of *Journal of Heredity* which became for a time the chief outlet for "scientific" studies of eugenics) and wealthy brother-in-law of the founder of National Geographic; Corcoran Tom, Treasurer of the AGA and vice-president of American Security and Trust Co., Washington; Mrs. E.H. Harriman (who personally supported the Eugenics Record Office at Cold Springs Harbor) whose husband, E.M. Harriman, was a railroad and telegraph magnate in the late nineteenth century; J.H. Kellogg, founder of Kellogg foods, who was the financial and ideological force behind the Race Betterment Foundation established in 1913 at Battle Creek, Michigan; Robert D.C. Ward, who established the Immigration Restriction League in 1894, and who was a member of the Saltonstall family of Boston, a Harvard graduate of the class of 1894 (along with his friend Charles B. Davenport), and later a Harvard professor; and Madison Grant, who was a conservative New York lawyer, privately wealthy, and deriving from an old aristocratic family.[6]

A major portion of the support for many philanthropic or social movements comes from wealthy interests, since the financial elites are usually the only ones having enough time and money to support the "humanitarian" causes. This was true to some extent with respect to the eugenic movement of the early 20th century. The question is whether their self-interest prompted them to give extra aid to the eugenics groups. In fact this appears to be the case. For example, Franz Boas, an eminent anthropologist and anti-eugenicist in the early decades of the century, attempted to raise funds for an African Museum in the United States. Boas appealed to the same financial elite, like the Rockefeller or Carnegie Foundations, which supported eugenics, only to be turned away flatly.[7] It is apparent that members of the ruling class supported first and foremost those movements which agreed with their ideology and hence served their own ends. They made rational (to them) choices about how to spend their philanthropic dollars.

In what way did the eugenics movement serve the class interests of the ruling elite? One answer can be found in the social history of the U.S. between 1890 and 1920. At the turn of the century the propertied, politically conservative classes enjoyed considerable position and influence.[8] But the Haymarket Riots, the Homestead Strike, the Pullman Strike, and the Populist Revolt, to mention only a few militant movements between 1880 and 1910, showed a rising unity and power within the working class. The years 1900-1929 saw an increase in the organization of labor and trade unionism, the rise of the Industrial Workers of the World (the IWW or "Wobblies"), the founding of the CIO, and the militant, revolutionary Seattle General Strike of 1919.[9] During the war, hysteria had increased the suspicion with which immigrants and foreigners were regarded. Immigrants were linked with the "Red Scare" of the 1920s, the increasing radicalization of workers, in general, and the IWW, in particular. Indeed, many of the more radical union leaders were immigrants who had found the land of milk and honey less utopian than they had expected. The U.S. rulers clearly saw the labor movement and other socialist-oriented mass organizations as endangering their (ruling) class interests. They used one of the classic techniques of those in power to maintain their position: "Divide and Conquer." The eugenics movement was one way—not the only way but certainly a very effective one—of implementing this strategy. Nothing works more effectively to keep people apart than to convince one group that others are biologically inferior and thus "not as good." This is what racism does, and the eugenics movement was a form of racism with "scientific" backing. The eugenics movement, in general, and immigration restriction, in particular, were responses of the ruling class to a growing popular demand for more control over society. It was not a conspiracy in the usual sense. There was no one powerful ruling class leader or group who laid out long-range plans. But class interests are such that members of the class generally know what movements serve their vested interests and what movements pose a threat.

The writings of members of the ruling class show they were aware of the threat which continued immigration posed. It was evident in social, political and economic ways. Lothrop Stoddard's book *The Rising Tide of Color Against White Supremacy* (1920) was inspired by the threat which the awakening of the peoples of Asia, Africa and Latin America would have on U.S. (and particularly Nordic) domination of the world. Madison Grant shows the aristocrat's reaction to increased immigration. He described New York as becoming a *"cloaca gentium* which will produce many amazing racial hybrids and some ethnic horrors..." The old stock American, he complained bitterly, *"is today being literally driven off the streets of New York City by the swarms of Polish Jews."*[10] This fear was genuine and not uncalled-for. While most immigrants were, in fact, conservative, overall the working class strongly favored a reordering of society. The era of the robber-baron had made it clear to American workers that anything they got they would have to fight for. Immigrants figured prominently among the radical leaders of the labor movement and the IWW. It was this which led the ruling class to aim the argument of biological inferiority at immigrants, for the ruling class found in the eugenics movement a strategy for disrupting workers' moves toward collective action.

How did the ruling class influence the spread of eugenic ideas and the passage of eugenic-inspired legisation? It exerted its considerable effects on public opinion and legislative action through several processes. The news media and publishing houses were by and large controlled by the same wealth which controlled the steel mills and coal companies. Publication became a useful channel for moulding public opinion. Through their control of these channels, the ruling class could introduce and defend ideas which supported their general interests. The ruling class also controlled the universities financially, and these institutions pushed ruling class ideas. This was not done by forcing faculty to teach ideas which they did not agree with, rather it was done by hiring and giving special support to those whose ideas were agreeable. Anti-eugenic ideas were debated in universities during the heyday of the hereditarian movement, but Castle's textbook was the most widely circulated and influenced thousands of students. There were few publications available to counteract that book's eugenic assumptions.

How does class analysis account for the line-up of forces during the debate on the Immigration Restriction Act of 1924? In fact, a quick look at the lobbying groups involved seems to contradict what we'd expect from class interests. For instance, in the classical sense it might be expected that industry would favor the bill, since industrial leaders, as members of the ruling class, would fear an increased population of genetically inferior and socially malevolent proletariat. On the other hand, it might be expected that labor leaders would have opposed the bill, since increased immigration meant increased numbers of workers in the unions and increased strength to fight the bosses. In fact, however, the matter was more complex. Testifying on behalf of the bill were various patriotic groups, fraternal societies, eugenics organizations, and organized labor. Opposing the bill were steamship companies, agriculture, immigrant aid societies and industry.[11] The positions of all except labor and industry are easy to understand. If we look closely, the positions of these two latter groups are neither unexpected, nor in contradiction to a class analysis.

Labor supported the bill because the job market in 1920 was rapidly becoming glutted with an oversupply of labor. In the north, particularly, this was due both to immigration and job reduction because of slowdown from wartime expansionism, and the migration of workers (mostly black) from the south. Partly because of the eugenicists' propaganda, and partly because the unemployment problem was a reality, labor tended to focus especially on the immigrant as the immediate threat. Industry was opposed to the bill because unemployment was desirable to drive wages downward; industry was ever on the prowl for ways to reduce labor costs. It is interesting to note that industry was *less* opposed to the 1924 bill than it had been to the 1921 emergency measure. In the interim, the popular fear of the racial effects of immigration had increased, businesses had continued to prosper even without the previously large supply of cheap labor, and large-scale black migration from the south had kept the labor pool from falling too low.[12] Thus, industry and labor both acted within the framework of their perceived class interests.

We must remember in applying a class analysis to historical events, that there are many different class interests, some of which may be predominant at one period, some at another. Many times these interests can be contradictory within each class. For example, the ruling class faced two conflicting self-interests in regard to immigration in the 1920s. Unrestricted immigration meant a larger labor pool and thus low wages (and high profit). At the same time, too many immigrants, or too great an increase in the size of the proletariat, meant increased danger of labor organizing and eventual social revolution. To the ruling class the advantage of increased cheap labor had to be balanced out against the disadvantage of labor discontent and unionization. A fear of the latter turned many ruling class members to support the eugenics movement.

Labor also had contradictory positions. Increased immigration provided more recruits to the rank of the proletariat: a necessary condition for effectively organizing against the bosses. On the other hand, with a restricted job market, increased immigration heightened competition for available jobs. A dislike of the racial hatred fostered by the eugenicists caused some workers (especially immigrants themselves) to oppose immigration restriction. The immediate problems of unemployment, however, caused labor leadership, along with other workers to favor the immigration act of 1924.

The Current Controversy

The hereditarian arguments over race and I.Q. in the 1970s have many similarities to the eugenics movement during the period 1900-1925. Both attempt to differentiate between superior and inferior characteristics allegedly associated more with one race or class than another; both have based their arguments on supposed biological traits (inherited differences); both have found support within the scientific community and have tried to derive prestige from "scientific data"; both have involved large elements of subjectivity and bias in the use of evidence; and both have been picked up by the ruling-class controlled media and have received far more publicity than their questionable conclusions would warrant; both have drawn favorable attention from political and governmental leaders of their day, and have had a variety of influences on public policy; thus, a study of the older eugenics movement can help us understand and respond to some of the deleterious effects that might arise from a general acceptance of a new brand of hereditarianism.

That the hereditary differences in I.Q. between races has already, as a policy, begun to enter the public domain, may be demonstrated by several examples. Shortly after Jensen's original article appeared in February of 1969, a southern congressman had the entire 123-page article read into the *Congressional Record*. Daniel Patrick Moynihan reviewed Jensen's studies to the Nixon cabinet, pointing out that Jensen's scientific credentials were exemplary. A Virginia court introduced Jensen's work as evidence in a desegregation case.[13] In recent months hereditarian thinking has been overtly reflected in a public statement by Dean Watkins, chairman of the Board of Regents of the University of California: "It is just possible that the reason some people are rich is because they are smarter than other people; and maybe they produce smarter children."[14] What is also significant is the timing of both the old and the new hereditarian movements. Both emerged in periods following considerable social upheaval: the labor movement and strike agitation in the 1890s and first decades of the present century; and the strong civil rights and anti-war organization of the period 1963-1970. Both movements sought to explain social inequalities and injustice by appealing to hereditary differences between the people on top and those on the bottom. Both such explanations are merely different brands of racism.

It would be rash to claim that the eugenics movement of the 1920s, the Nazi racism of the 1930s, or the hereditarian views of the 1970s could have been totally defeated had scientists spoken out at the time. Academics do not often have such power. But strong opposition from scientists would have made those earlier movements less easy to build, and would have forced their inherent racism to appear more strikingly. The same can be said for the hereditarian movement of the present. By understanding that movement we can more easily lay bare the fallacious conclusions which, as a brand of self-serving racism, are masquerading under the mantle of legitimate biology.

FOOTNOTES

1. W. E. Castle, *Genetics & Eugenics*, Cambridge, Massachusetts, Harvard University Press, 1916.

2. Mark Haller, *Eugenics: Hereditarian Attitudes in American Thought*, New Brunswick, New Jersey, Rutgers University Press, 1963.

3. Madison Grant, *The Passing of the Great Race*, New York, Charles Scribner's Sons, 1916.

4. Kenneth Ludmerer, *Genetics & American Society*, Baltimore, Johns Hopkins Press, 1972.

5. Harry H. Laughlin, "Analysis of America's Modern Melting Pot," Hearings before the House Committee on Immigration and Naturalization, 67th Congress, Third Session, Washington, D.C., U.S. Government Printing Office, 1922, p. 255.

6. Ludmerer, *op. cit.*

7. E. H. Beardsley, "The American Scientist as Social Activist: Franz Boas, Bert G. Wilder, and the Cause of Racial Justice, 1900-1915," *Isis,* 64, 1973, pp. 50-66.

8. Ludmerer, *op. cit.*

9. *Ibid.* ; Anon., "Seattle General Strike, 1919: Can We Do Better Next Time?" *Progressive Labor,* 9, July 1973, pp. 33-44; Harvey O'Connor, *Revolution in Seattle,* New York, Monthly Review Press, 1974.

10. Grant, *op. cit.*

11. Ludmerer, *op. cit.*

12. *Ibid.*

13. John Neary, "A Scientist's Variations on a Disturbing Racial Theme," *Life*, 63, June 12, 1970, pp. 50B, 65.

14. *The Daily Californian*, April 2, 1973.

GENERAL REFERENCES

Racism, Intelligence, and the Working Class, New York, The Progressive Labor Party, 1973.

Corporate Roots of American Science
—David Noble

This article describes how large corporations came to control whole sections of modern science via involvement in educational institutions like M.I.T. It also provides insight into the reasons why U.S. scientists are trained not to question basic assumptions about their work.

Modern technology has lost its magic. No longer do people stand in awe, thrilled by the onward rush of science, the promise of a new day. Instead, the new is suspect. It arouses our hostility as much as it used to excite our fancy. With each breakthrough there are recurrent fears and suspicion. How will the advance further pollute our lives; modern technology is not merely what it first appears to be. Behind the white coats, the disarming jargon, the elaborate instrumentation, and at the core of what has often seemed an automatic process, one finds what Dorothy found in Oz: modern technology is human after all. Particular people in particular places do particular things for particular purposes. Modern technology is social. Modern technology is political.

For some, this insight is a liberating revelation which renders technology more familiar, more comprehensible, and more susceptible to change (even by them). But such insight has been woefully long in coming. Previously, esoteric science, the automatic workings of the market mechanism, the monopolization of knowledge by capital and professionals, and the division of head and hand characteristic of the capitalist mode of production, all combined to lend credence to the view that technology was a force unto itself, self-propelled and impinging upon human affairs from "outside." One might well wonder, therefore, what we have missed in not having seen the obvious sooner. What follows is a brief exercise in hindsight: substituting history for metaphor, we lift the technological veil that has obscured human actions in the past. We begin to show how people, and not some remote, mysterious force called technology, gave shape to the technological society which we call home.

The Capitalization of Science

During the first few decades of the twentieth century, the mythology that technology is beyond human control was gaining wide popular currency. In fact, during this time, the actual process of modern technology was steadily being enlarged and brought under human control by private capital. The dramatic scientific revolution in material production and the profound transformation of social life which it entailed were in reality aspects of a particularly potent phase in the history of industrial capitalism. It was during this time that the now familiar patterns of work, science, and education characteristic of advanced technological societies were first established. As scientific knowledge increasingly became a crucial determinant of industrial advance and competitive strength, the large industrial firms which had been created to extend and dominate markets, stabilize prices, integrate productive activity, and provide ready returns for speculators and financiers, undertook also to secure control over science. (Indeed, some of them—General Electric and General Chemical, for example—were actually created for the express purpose of controlling patents).

In the vanguard of this new enterprise were the giant firms which early dominated the science-based electrical and chemical industries: General Electric, Westinghouse, AT&T, and DuPont. Rooted from the outset in the soil of science and thus unprecedented in their demand for scientific knowledge, these companies sought at once to stimulate and to regulate the growth of industrial science. Their efforts were carried forth on a number of fronts. Corporate reformers were themselves predominantly scientifically trained engineers who had become managers and executives in the new firms, professors, deans, and presidents in the new technical colleges, and officials of the various trade, technical, and professional societies. They sought control first over the *means* of scientific industry through the establishment and enforcement of industrial and scientific standards; second over the *products* of scientific industry through the monopoly of patents and reform of the patent system itself; third over the *process* of scientific invention and discovery through the organization of industrial and university research; and finally fourth over the *practitioners* of industrial science through transformation of public schooling, technical and higher education.

Gaining control over science was an enormous task. It entailed the invention, transformation, and revitalization of social institutions, the preparation, habituation, and mobilization of a society of people for wholly new forms of productive activity. And gearing the society for new modes of production—capitalist, corporate, and scientific—entailed the creation of new forms of social life, individual identity, relations between people, patterns of work, leisure, consumption, definitions of human potential, education, knowledge, the good—in short, the production of society itself. What follows is a very brief look at two critical aspects of this social transformation: the development of institutions for industrial research and the retooling of institutions of higher education.

Research

The decisive factor in the development of modern technology was the linking of the laboratory with the workshop, the search for truth about nature with the utilitarian and pecuniary objectives of "manufacture," and of science with the tradition-bound useful arts. Up until the last quarter of the 19th century this link was only minimal and haphazard. It rested largely on the efforts of well-heeled gentlemen who cultivated both an artistic and a pecuniary interest in it. But by the end of the century, those new firms which were dependent upon modern technology found it necessary to invent ways of establishing the connection on a regular basis, or of routinely bringing the laboratory and the workshop together in a manner that allowed at once for both the stimulation and supervision of "progress." Although firms in the mining, petroleum, electrical, steel, and chemical industries had occasionally hired university-based consultants during the latter half of the last century, it wasn't until the 1890s that they undertook to establish scientific research laboratories as an organized activity within the firm itself. The pioneers were the large, well-endowed corporations such as GE, AT&T and DuPont; it was here that the first "synthetic genii," as Philip Alger of GE called them, were formed: teams of assembled specialists "held together by bonds of sympathy and understanding, as well as by the company management." [1] Such laboratories quickly became enormous enterprises, employing hundreds of highly trained scientists, engineers and technicians and fostering, through careful supervision and a military organization of work, what Frank Jowett of the Bell Labs termed "cooperative effort under control."[2]

Unlike the large, heavily capitalized firms like GE and AT&T (both J.P. Morgan operations) smaller science-based companies could barely afford to set up their own laboratories or bear the risk of uncertain, long-term research. Thus, they relied upon independent contractors, such as Arthur D. Little, to do their research work for them and minimized both an individual company's risk and cost by establishing cooperative trade association laboratories, such as the N.E.L.A. labs, which served the manufacturers of electric lamps. In addition, they relied upon the service activities of new government agencies, particularly the National Bureau of Standards, set up in 1901 at the behest of industrial leaders and scientists.

Private contractors, trade association labs, and government agencies, however, could alone neither meet the growing demand for research nor satisfactorily link the world of science with that of industry. What was required was closer cooperation between the traditional domain of science where the bulk of research activity was being done, the universities, and the industries which aimed to put the results of that effort to profitable use. As Dugald Jackson, head of M.I.T.'s electrical engineering department and a leading utilities industry consultant, observed, "American industry finds it increasingly profitable to become interested in and to aid by means of money and counsel, research in universities...The more influential of the men of the technical industries have come to recognize the desirability of cooperation in the joint processes of education and industry."[3] Jackson, like his colleagues in the electrical manufacturing, telecommunications, chemical and chemical process industries, had a rather broad view of industrial society; there would be no need to try to duplicate the research work, the facilities, the personnel, of the university enterprise of science within private corporations if it were possible to "integrate universities [themselves] as research centers within the industrial structure."[4]

By 1920, various schemes for industry-university cooperation had been developed, all of which tied the universities firmly into the industrial arena and redefined the patterns and ends of the scientific efforts of faculty, staff, and students. Industrial fellowships were created in support of graduate study in science and to allow faculty more time for research. The most famous of these was the plan developed by Robert Kennedy Duncan of Kansas which became the foundation for the Mellon Institute in Pittsburgh. Extensive cooperative research programs were undertaken at universities throughout the country, primarily in the engineering schools and at the departmental level. The plant of the nation's colleges was expanded dramatically with the construction of new chemistry and physics and engineering buildings, at industrial expense. Engineering experiment stations, like the agricultural stations created by the Hatch Act, were established, primarily at state schools, to provide extension services for local industries (despite an unsuccessful attempt by industry leaders to secure federal support for such stations—their opposition could not condone public subsidy for private enterprise). Networks were established to facilitate the interchange of personnel and ideas between the schools and the industries: industrial advisory committees, industrial sabbaticals for professors, formal consulting arrangements, and the like. Increasingly, through such cooperative institutional ties, industries "put-out" their research tasks to universities, usually for a modest fee, and were thereby spared the overhead costs of facilities, staff, libraries, and training of research personnel.

The university as industrial service center was perhaps illustrated best and earliest by M.I.T. The physical chemistry laboratory set up in 1903 by A.A. Noyes and Willis Whitney (a founder of Cal Tech and first director of the GE labs, respectively) was, in M.I.T. President Henry Pritchett's words, "the first effort of any technical school in the country to offer research work distinctive from that of the colleges and directed toward engineering topics."[5] Pritchett's endorsement was hardly surprising. Two years earlier he had funded the industrially oriented National Bureau of Standards. In any case, the next decade saw a tremendous growth in industrial research at M.I.T., primarily in the departments of chemical engineering (presided over by William Walker, former partner of Arthur D. Little) and electrical engineering (presided over by Dugald Jackson). By 1920, these departmental efforts coalesced in the establishment of a centralized "division of industrial cooperation" headed by Walker. The new division was created to administer what was known as the "Tech Plan," by which any industrial firm could contract with the Institute for specific research work; by paying a fee for the service, the firm would receive not only the particular work specified but also access to staff, faculty in related areas, library facilities, information on the work done in Institute laboratories which might be relevant, bibliographic services, and information—personal and academic— about all present and former M.I.T. students and faculty who might be able to contribute to the research effort or to the general work of the firm. In the 1940s, the highly successful Division of Industrial Cooperation became the Division of Sponsored Research, its responsibilities broadened to include military and governmental, as well as industrially sponsored research.

What the centralized division of industrial cooperation did for the fragmented efforts of departments at M.I.T., the National Research Council did for the research activities of the nation's universities as a whole. Set up during World War I and funded primarily by such private agencies as the Industrial Engineering Foundation, the NRC assumed the task of coordinating the integration of universities within the industrial structure, of promoting research in science while at the same time fostering efforts along industrially-defined lines. The most active division of the NRC both during and after the war, was the Engineering Division, headed by such industry leaders as Frank Jewett, Elmer Sperry, and, by 1930, Dugald Jackson. The NRC provided invaluable assistance to burgeoning science-based industry, sponsoring research projects, conducting extensive surveys of research facilities in the government and the nation's colleges and universities, publishing bibliographies of research in progress, compiling personnel rosters of research institutes, university science and engineering faculty, graduate students, and recent PhD's, and even conducting tours for businessmen of major research facilities in industry and universities. The NRC, in short, spread the message and served the ends of the science-based industrial corporations.

Industrial sponsorship and direction of university research successfully shifted the burden of the major costs of science-based industry from the private to the public sector. But this was not all. Perhaps more important, it redefined the form and content of scientific research itself. This involved more than the general shift away from natural philosophy—the search for metaphysical truth through an understanding of Nature—to utilitarian science—the quest for intervention in, and power over, Nature, through knowledge of the fundamental relationships between matter and energy (a shift already well underway by the end of the nineteenth century). Now the shift toward utility assumed particular forms, those measured by the specific, historical needs of private industry, by particular firms intent upon increasing their profit-margins and their power. The industrial transformation of science affected not only what kinds of questions would be asked but also what particular questions would be asked, which problems would be investigated, what sorts of solutions would be sought, what conclusions would be drawn. Science had, indeed, been pressed into the service of capital.

Higher Education

If research was vital to science-based industry, so too was technical manpower. The first industrial concerns to employ college graduates (engineers) on a large scale were the electrical manufacturing companies, GE and Westinghouse; it was thus these firms, and others like them which emerged later, which developed the form of higher education demanded by modern corporate industry. With only slight exaggeration, the in-house training programs of these companies, the so-called "corporation schools," may be seen as the pilot programs, the experimental models of higher education as a whole in twentieth century America.

Until relatively late in the nineteenth century, the colleges were dominated by classicists and clerics, both of whom shared a Platonic disdain for the practical arts and its correlate, money-making enterprise. Thus, the colleges tended to remain removed from the steadily expanding realm of industry, with its noisy shops and less noisy counting-houses, and schools of technical education were forced to take root outside of, and oftentimes in opposition to, the established colleges. If the colleges happened to turn out men who ultimately entered industry, they did so largely by chance and from a distance. Between the end of the Civil War and the turn of the century, however, owing in large part to the pressure placed upon the established colleges by wealthy industrialists and the federal support of land-grant colleges in agriculture and the mechanical arts, there was a tremendous growth in the number of engineering schools, and thus a growing pool of technically-trained men for industry.

For a number of reasons, however, industrial managers found

technical graduates ill-prepared for immediate use to the company. For one thing, since the colleges could scarcely keep pace with the rapidly changing industrial state of the art, students were generally given instruction in obsolete methods with outdated equipment. At the same time, since engineering educators were preoccupied with enhancing their academic respectability, they tended to emphasize scientific theorizing and mathematics at the expense of practical training in industrially-applicable engineering. Thus they turned out graduates who were neither cultured gentlemen nor effective practitioners. Perhaps most important, graduates were imbued with the aristocratic arrogance of a university elite, the entrepreneurial spirit of laissez-faire capitalism, or the scientific zeal for untrammeled inquiry—traits which hardly suited them for efficient, loyal employment as subordinates in an authoritarian corporate enterprise. "The fundamental difficulty," Charles Scott of Westinghouse complained in 1907, "is lack of adaptations to new circumstances and conditions. We do not underrate knowledge and training, but we want the graduates to be of use...We want men who can see the situation and fit themselves to it. The possibilities and the outcome depend upon the ability of the man for harmonizing himself with his environment, and the more complete and efficient this adjustment, the more useful the life."[6] What is called for on the part of the graduates, another corporate educator concluded, is "self-forgetfulness."[7]

The in-house corporation schools created by GE, Westinghouse, AT&T, Kodak, DuPont, Dow, various Edison companies, and other leaders of science-based industry were designed to habituate college graduates to industrial employment, to give them additional technical training and the proper business point of view, to teach them how to follow orders. The importance of these schools in the training of generations of engineers should not be underestimated.

In electrical engineering, for example, the college graduate during the first three decades of this century had of necessity to become a "testman" at Schenectady or a "special apprentice" at Pittsburgh in order to complete his professional training. Along the way he usually learned to see the world as his superiors at GE or Westinghouse saw it. (Some in-house programs, like that at the GE plant in Lynn, trained graduates specifically for future managerial responsibility in the industry and as a matter of course brought them into contact with leaders in the field whose success they were encouraged to emulate.) In addition to their actual educational function, the corporation schools constituted an important phase in the evolution of modern personnel management, pioneering in methods of testing, rating, selecting and classifying graduates, of "scientifically" fitting the man to the job. (Thus, it ought not be surprising that the National Association of Corporation Schools (NACS), formed in 1913 to coordinate in-house educational activities nationwide, changed its name after World War I to the National Personnel Association and, again, to the American Management Association.) By the second decade of the century, however, the most pressing task the corporation educators faced, in their view, was that of putting the corporations themselves out of the education business, of gearing the colleges to do the job right the first time.

By enthusiastically promoting closer cooperation between industry and the colleges, the corporate educators sought to shift the burden of "correct" training back to the colleges and the taxpayers. Operating through such agencies as the NACS and the Society for the Promotion of Engineering Education (SPEE), they strove to transform the universities into efficient processing plants—"factories" as they usually referred to them—for the production, selection, and distribution of the human material required by industry, according to changing industrial specifications. SPEE played a crucial role. Organized in 1893 by engineering educators, it was the first national association of college educators devoted exclusively to educational matters. For the first decade and a half after its establishment, the SPEE members concerned themselves primarily with minor matters of pedagogy and the perennial problem of academic status. In 1907, however, the year Dugald Jackson became SPEE president, membership roles were swelled by the influx of "practicing engineers and businessmen" and the focus of discussion shifted dramatically to such questions as "adapting technical graduates to industry," "making graduates more efficient," and the like. From this point on—until the 1930s—the SPEE was the major forum for the corporate reform of higher education. "Each institution is in reality a factory turning out engineers," H.F.J. Porter of Westinghouse explained to SPEE members.[8] The challenge is how to manufacture a "uniform product." Porter's colleague at Westinghouse, Charles Scott (soon to become head of electrical engineering at Yale) concurred, summing up the task at a joint meeting of the engineering educators and industrialists.

> If producers and users of steel rails were in conference they would discuss the uses which rails are to serve,

classifying the kinds of service, considering wherein past products had failed, inquiring as to chemical analysis and metallurgical treatment. They would seek improvement in production and discrimination in use. But the more difficult problem of human material...has received less attention; how seldom do representatives of engineering industry and of engineering education meet together for conference. Yet they are users and producers of a vital product. Let us try to agree on what we want and then determine how to get it and how to use it. How many boys of differing kinds can be individually developed and fitted to varying needs?[9]

A major step forward along these lines was the cooperative education movement, begun in the engineering school of the University of Cincinnati in 1907 and pressed ahead enthusiastically by the NACS and the SPEE. "The aim of the course," Dean Herman Schneider boasted, "is not to make a so-called pure engineer; it is frankly intended to make an engineer for commercial production...This system will furnish to the manufacturer a man skilled both in theory and practice, and free from the defects concerning which so much complaint is made."[10] The cooperative course successfully brought the school into the shop; students spent alternating periods in the factory of a cooperating firm and in the classroom of the school. In this way, students were able to get the "proper" business point of view, the necessary habits of industrial discipline and corporate subservience while still in school. The movement spread rapidly throughout the country, at the prompting of both industrialists and corporate reformers among engineering educators. At M.I.T., for example, cooperative programs were begun in electrical engineering by Magnus Alexander, educational director of GE's Lynn plant and in chemical engineering by Arthur D. Little, the country's foremost industrial research consultant, while at the University of Pittsburgh a college-wide program was initiated by Dean Frederick Bishop, national secretary of the SPEE. By the 1920s variants on the cooperative plan were introduced at such schools as Northeastern, Tufts, Drexel, Case, Union College, Marquette, New York University, Antioch and Harvard, and included liberal arts students as well as undergraduate engineers.

While the cooperative education movement established closer industry-education interaction, other corporate reform innovations had the purpose of rationalizing the "processing plants" themselves. The corporate educators were ardent promoters of testing programs and efficient selection, rating and classifying processes much like those developed in the corporation schools and other areas of personnel management. Charles Mann, the author of the first national study of engineering education in the U.S. (sponsored by the SPEE and funded by the Carnegie Foundation) explained the primary purpose of introducing testing into the schools in an address to the NACS in 1914. "The one point that I want to bring out clearly to you," Mann stressed, "is that definite objective tests which define the type of ability which you wish to have developed are the most valuable, not only to yourselves as employers in selecting your help, but also as your most powerful means of controlling what is done in the school."[11]

The development of testing procedures for evaluating the aptitude of students, advanced considerably by the corporation schools, was paralleled by the creation of mechanisms for selecting and distributing the educational products. The first placement bureau in a U.S. university, for example, was established at Kansas State College by GE engineer Andrey A. Potter, who served as both dean of engineering and president of the local chamber of commerce. Later dean at Purdue, president of the SPEE, and a major force in engineering education, Potter had also developed the personality rating scales for evaluating students according to standards of "character," as well as more traditional academic measures, scales which were widely adopted both by educators and corporate managers (at Westinghouse for one).

The biggest push toward the rationalization of higher education came during the First World War, which also saw the unprecedented advance of intelligence testing and educational psychology. During the war, the nation's colleges came under the authority of the War Department Committee on Education and Special Training, designers and directors of the Student Army Training Corps. This committee, surprisingly enough, was composed of corporate educators from Westinghouse, Western Electric (who was also president of the NACS), and other firms as well as leaders from the SPEE, all of whom had donned uniforms for the duration. With the authority of the War Department behind them, these people were able to introduce many of their educational innovations with relative ease while conditioning a good many educators to produce according to specifications, industrial as well as military.

After the war, the corporate reform of higher education was

continued under other auspices: the National Research Council, the SPEE, and, perhaps most important, the new American Council on Education. The latter, dominated from the outset by War Department Committee members Samuel Capen and Charles Mann, both prime movers in corporate educational reform, quickly became the chief sponsor of the new "science of education" and promoter of testing in the schools (ACE testing programs coalesced eventually into the Educational Testing Service). It was through the ACE that the educational reform movement begun in the corporation schools of science-based industry and developed in the engineering schools entered the body and soul of higher education as a whole. In 1924, ACE Chairman Henry Suzzalo aptly summed up its mission.

> The American system of schools has a sanction in public efficiency as well as in equality of personal opportunity. [University educators] have an immediate responsibility to make the prospect more effective...Soon we must become as wise in pedagogical method as we have long been in scientific method. The processing of human beings through intellectual experiences is far more important socially than the processing of material things. Yet physical technology holds a place of respectability among us which human technology has not yet won.[12]

During the first half of the 20th century, and at the initiative of reformers from science-based industry, colleges and universities in the U.S. were retooled to fit the contours of a corporate, technological society; institutions of higher education were transformed into processing plants, integral parts of the industrial structure charged with the production of manpower and the habituation of students to the disciplines of loyal, efficient corporate service.

Perhaps no single individual better embodied this transformation than William Wickenden. An electrical engineer, Wickenden taught in the cooperative program at M.I.T. with Dugald Jackson before leaving the academy to become the first personnel director of the Bell Labs and then vice president for technical personnel for AT&T. Leaving AT&T in 1923, Wickenden became director of the most comprehensive study of engineering education in history (The SPEE Wickenden Report) and thereafter president of Case Institute. A giant in his profession and probably the foremost figure in 20th century engineering education, Wickenden summed up the meaning of his life's work a year before his death, in 1946. "The very word university," Wickenden observed, "comes from the Latin word for corporation and the college dormitory is simply a continuation of the plan of the guilds by which the master workmen not only trained their apprentices but took them into the household to live."

> That is where our circle began, but as it swung out on its wide arc, the world of education drew further and further away from the world of industry...The Sorbonne and Oxford scarcely knew of the world of science and for the world of industry they had only disdain. But the two circles went swinging on, bringing industry and education ever closer and closer, until together they are closing back once more at the point of origin where industry and education are one, where corporation and university again mean the same thing.[13]

****

As in Oz, the wizards of our technological society have been human, particular people working to achieve what they believed to be a rational, humane, "better" social order. As the brief histories above suggest, the creators of at least some aspects of our technological society were agents of both a revolution and a counter-revolution. On the one hand, corporate reformers from science-based industry were moved by "destiny," seeking to foster scientific progress and thereby reduce human toil and misery. On the other hand, they were moved by the specific historical needs of corporate capitalism, striving to channel scientific progress along lines which were compatible with the requirements of corporate stability and expansion. In their work, then, the contradiction between science and commerce, between technical rationality and market irrationality, the tension which Marx and Veblen thought would ultimately tear capitalism apart, collapsed and softened in practice. Modern technology—the people as well as the things— became a vehicle of corporate power, an extension of authority, a reinforcement of existing social relations. No wonder, then, that modern technology has lost its magic and predictions of liberation through science ring so hollow.

Footnotes 1-13 are all from David Noble's book *America by Design* (New York: Alfred Knopf, 1977).

How Poverty Breeds Overpopulation
—Barry Commoner

This article presents a classic case of how a belief in science to solve social ills can lead scientists to attempt solutions from a narrow technical perspective and to dismiss the social and political realities of the problem, thereby creating more oppressive social relations.

The world population problem is a bewildering mixture of the simple and the complex, the clear and the confused.

What is relatively simple and clear is that the population of the world is getting larger, and this process cannot go on indefinitely, because there are, after all, limits to the resources, such as food, that are needed to sustain human life. Like all living things, people have an inherent tendency to multiply geometrically—that is, the more people there are the more people they tend to produce. In contrast, the supply of food rises more slowly, for unlike people it does not increase in proportion to the existing rate of food production. This is, of course, the familiar Malthusian relationship and leads to the conclusion that the population is certain eventually to outgrow the food supply (and other needed resources), leading to famine and mass death unless some other countervailing force intervenes to limit population growth. One can argue about the details, but taken as a general summary of the population problem, the foregoing statement is one which no environmentalist can successfully dispute.

When we turn from merely stating the problem to analyzing and attempting to solve it, the issue becomes much more complex. The simple statement that there is a limit to the growth of the human population, imposed on it by the inherent limits of the earth's resources, is a useful but abstract idea. In order to reduce it to the level of reality at which the problem must be solved, what is required is that we find the cause of the discrepancy between population growth and the available resources. Current views on this question are neither simple nor unanimous.

One view is that the cause of the population problem is uncontrolled fertility, the countervailing force—the death rate—having been weakened by medical advances. According to this view, given the freedom to do so, people will inevitably produce children faster than the goods needed to support them. It follows, then, that the birthrate must be deliberately reduced to the point of "zero population growth."

The methods that have been proposed to achieve this kind of direct reduction of the birthrate vary considerably. Among the ones advanced in the past are: (a) providing people with effective contraception and access to abortion facilities and with education about the value of using them (i.e. family planning); (b) enforcing legal means to prevent couples from producing more than some standard number of children ("coercion"); (c) withholding of food from the people of starving developing countries which, having failed to limit their birthrate sufficiently, are deemed to be too far gone or too unworthy to be saved (the so-called lifeboat ethic).

It is appropriate here to illustrate these diverse approaches with examples. The family planning approach is so well known as to need no further exemplification. As to the second of these approaches, one might cite the following description of it by Kingsley Davis, a prominent demographer, which is quoted approvingly in a recent statement by "The Environmental Fund" that is directed against the family planning position: "If people want to control population, it can be done with knowledge already available...For instance, a nation seeking to stabilize its population could shut off immigration and permit each couple a maximum of two children, with possible license for a third. Accidental pregnancies beyond the limit would be interrupted by abortion. If a third child were born without a license, or a fourth, the mother would be sterilized."[1]

The author of the "lifeboat ethic" is Garrett Hardin, who stated in a recent paper that: "So long as nations multiply at different rates, survival requires that we adopt the ethic of the lifeboat. A lifeboat can hold only so many people. There are more than two billion wretched people in the world—ten times as many as in the United States. It is literally beyond our ability to save them all...Both international granaries and lax immigration policies must be rejected if we are to save something for our grandchildren. "2

Actually, this recent statement only cloaks, in the rubric of an "ethic," a more frankly political position taken earlier by Hardin: "Every day we (i.e. Americans) are a smaller minority. We are increasing at only one percent a year; the rest of the world increases twice as fast. By the year 2000, only one person in 24 will be an American, in one hundred years only one in 46...If the world is one great commons, in which all food is shared equally, then we are lost. Those who breed faster will replace the rest...In the absence of breeding control the policy of 'one mouth one meal' ultimately produces one totally miserable world. In a less than perfect world, the allocation of rights based on territory must be defended if a ruinous breeding race is to be avoided. It is unlikely that civilization and dignity can survive everywhere, but better in a few places than in none. Fortunate minorities must act as the trustees of a civilization that is threatened by uninformed good intentions."3

The Quality of Life

But there is another view of population which is much more complex. It is based on the evidence, amassed by demographers, that the birthrate is not only affected by biological factors, such as fertility and contraception, but by equally powerful *social* factors.

Demographers have delineated a complex network of interactions among these social factors. This shows that population growth is not the consequence of a simple arithmetic relationship between birth rate and death rate. Instead, there are circular relationships in which, as in an ecological cycle, every step is connected to several others.

Thus, while a reduced death rate does, of course, increase the rate of population growth, it can also have the opposite effect— since families usually respond to a reduced rate of infant mortality by opting for fewer children. This negative feedback modulates the effect of a decreased death rate on population size. Similarly, although a rising population increases the demand on resources and thereby worsens the population problem, it also stimulates economic activity. This, in turn, improves education levels. As a result the average age at marriage tends to increase, culminating in a reduced birthrate—which mitigates the pressure on resources.

In these processes, there is a powerful social force which, paradoxically, both reduces the death rate (and thereby stimulates population growth) and also leads people voluntarily to restrict the production of children (and thereby reduces population growth). That force, simply stated, is the quality of life—a high standard of living, a sense of well-being and of security in the future. When and how the two opposite effects of this force are felt differs with the stages in a country's economic development. In a pre-modern society, such as England before the industrial revolution or India before the advent of the English, both death rates and birthrates were high. But they were in balance and population size was stable. Then, as agricultural and industrial production began to increase and living conditions improved, the death rate began to fall. With the birthrate remaining high the population rapidly increased in size. However, later, as living standards continued to improve, the decline in death rate persisted but the birthrate began to decline as well, reducing the rate of population growth.

For example, at around 1800, Sweden had a high birthrate (about 33/1000), but since the death rate was equally high, the population was in balance. Then as agriculture, and, later, industrial production advanced, the death rate dropped until, by the mid-nineteenth century, it stood at about 20/1000. Since the birthrate remained constant during that period of time, there was a large excess of births over deaths and the population increased rapidly. Then, however, the birthrate began to drop, gradually narrowing the gap until in the mid-twentieth century it reached about 14/1000, when the death rate was about 10/1000.* Thus, under the influence of a constantly rising standard of living the population moved, with time, from a position of balance *at a high death rate* to a new position of near-balance *at a low death rate*. But in between the population increased considerably.

This process, *the demographic transition*, is clearly characteristic of all Western countries. In most of them, the birthrate does not begin to fall appreciably until the death rate is reduced below about 20/1000. However, then the drop in birthrate is rapid. A similar transition also appears to be underway in countries like India. Thus in the mid-nineteenth century, India had equally high birth and death rates (about 50/1000) and the population was in approximate

---

*This and subsequent demographic information is from: Agency for International Development, *Population Program Assistance*, December, 1971.

balance. Then, as living standards improved, the death rate dropped to its present level of about 15/1000 and the birthrate dropped, at first slowly and recently more rapidly, to its present level of 42/1000. India is at a critical point; now that the death rate has reached the turning point of about 20/1000, we can expect the birthrate to fall rapidly—provided that the death rate is further reduced by improved living conditions.

One indicator of the quality of life—infant mortality—is especially decisive in this process. And again there is the critical point—a rate of infant mortality below which the birthrate begins to drop sharply and, approaching the death rate, creates the conditions for a balanced population. The reason is that couples are interested in the number of *surviving* children and respond to a low rate of infant mortality by realizing that they no longer have to have more children to replace the ones that die. Birth control is, of course, a necessary adjunct to this process; but it can succeed— barring compulsion—only in the presence of a rising standard of living, which of itself generates the necessary motivation.

This process appears to be just as characteristic of developing countries as of developed ones. This can be seen by plotting the present birthrates against the present rates of infant mortality for all available national data. The highest rates of infant mortality are in African countries; they are in the range of 53-175/1000 live births and birthrates are about 27-52/1000. In those countries where infant mortality has improved somewhat (for example, in a number of Latin American and Asian countries) the drop in birthrate is slight, (to about 45/1000) until the infant mortality reaches about 80/1000. Then, as infant mortality drops from 80/1000 to about 25/1000 (the figure characteristic of most developed countries), the birthrate drops sharply from 45 to about 15-18/1000. Thus a rate of infant mortality of 80/1000 is a critical turning point which can lead to a very rapid decline in birthrate in response to a further reduction in infant mortality. The latter, in turn, is always very responsive to improved living conditions, especially with respect to nutrition. Consequently, there is a kind of critical standard of living which, if achieved, can lead to a rapid reduction in birthrate and an approach to a balanced population.

Thus, in human societies, there is a built-in control on population size: If the standard of living, which initiates the rise in population, *continues* to increase, the population eventually begins to level off. This self-regulating process begins with a population in balance, but at a high death rate and low standard of living. It then progresses toward a population which is larger, but once more in balance, at a low death rate and a high standard of living.

Demographic Parasites

The chief reason for the rapid rise in population in developing countries is that this basic condition has not been met. The explanation is a fact about developing countries which is often forgotten—that they were recently, and in the economic sense often still remain, colonies of more developed countries. In the colonial period, Western nations introduced improved living conditions (roads, communications, engineering, agricultural and medical services) as part of their campaign to increase the labor force needed to exploit the colony's natural resources. This increase in living standards initiated the first phase of the demographic transition.

But most of the resultant wealth did not remain in the colony. As a result, the second (or population-balancing) phase of the demographic transition could not take place. Instead the wealth produced in the colony was largely diverted to the advanced nation—where it helped *that* country achieve for itself the second phase of the demographic transition. Thus colonialism involves a kind of demographic parasitism. The second, population-balancing phase of the demographic transition in the advanced country is fed by the suppression of that same phase in the colony.

It has only been known that the accelerating curve of wealth and power of Western Europe, and later of the United States and Japan, has been heavily based on exploitation of resources taken from the less powerful nations: colonies, whether governed legally, or—as in the case of the U.S. control of certain Latin American countries—by extra-legal and economic means. The result has been a grossly inequitable rate of development among the nations of the world. As the wealth of the exploited nations was diverted to the more powerful ones, their power, and with it their capacity to exploit, increased. The gap between the wealth of nations grew as the rich were fed by the poor.

What is evident from the above considerations is that this process of international exploitation has had another very powerful but unanticipated effect: rapid growth of the population in the former colonies. An analysis by the demographer, Nathan Keyfitz, leads him to conclude that the growth of industrial capitalism in the Western nations in the period 1800-1950 resulted in the development of a one-billion excess in the world population, largely in the tropics. Thus the present world population crisis—the rapid growth of population in developing countries (the former colonies)—is the result not so much of policies promulgated by these countries but of a policy, colonial exploitation, forced on them by developed countries.

A Village in India

Given this background, what can be said about the various alternative methods of achieving a balanced world population? In India, there has been an interesting, if partially inadvertent, comparative test of two of the possible approaches: family planning programs and efforts (also on a family basis) to elevate the living standard. The results of this test show that while the family planning effort itself failed to reduce the birthrate, improved living standards succeeded.

In 1954, a Harvard team undertook the first major field study of birth control in India. The population of a number of test villages was provided with contraceptives and suitable educational programs; birthrates, death rates and health status in this population were compared with the comparable values in an equivalent population in control villages. The study covered the six-year period 1954-1960.

A follow-up in 1969 showed that the study was a failure. Although in the test population the crude birthrate dropped from 40 per 1,000 in 1957 to 35 per 1,000 in 1968, a similar reduction also occurred in the control population. The birth control effort had no measurable effect on birthrate.

We now know *why* the study failed, thanks to a remarkable book by Mahmood Mamdani.[4] He investigated in great detail the impact of the study on one of the test villages, Manupur. What Mamdani discovered is a total confirmation of the view that population control in a country like India depends on the economically-motivated desire to limit fertility. Talking with the Manupur villagers he discovered why, despite the study's statistics regarding ready "acceptance" of the offered contraceptives, the birthrate was not affected.

"One such 'acceptance' case was Asa Singh, a sometime land laborer who is now a watchman at the village high school. I questioned him as to whether he used the tablets or not: 'Certainly I did. You can read it in their books—from 1957 to 1960, I never failed.' Asa Singh, however, had a son who had been born sometime in 'late 1958 or 1959.' At our third meeting I pointed this out to him...Finally he looked at me and responded. 'Babuji, someday you'll understand. It is sometimes better to lie. It stops you from hurting people, does no harm, and might even help them.' The next day Asa Singh took me to a friend's house and I saw small rectangular boxes and bottles, one piled on top of the other, all arranged as a tiny sculpture in a corner of the room. This man had made a sculpture of birth control devices. Asa Singh said: 'Most of us threw the tablets away. But my brother here, he makes use of everything.' "

Such stories have been reported before and are often taken to indicate how much "ignorance" has to be overcome before birth control can be effective in countries like India. But Mamdani takes us much further into the problem, by finding out why the villagers preferred not to use the contraceptives. In one interview after another he discovered a simple, decisive fact: that in order to advance their economic condition, to take advantage of the opportunities newly created by the development of independent India, *children were essential*. Mamdani makes this very explicit:

To begin with, most families have either little or no savings, and they can earn too little to be able to finance the education of *any* children, even through high school. Another source of income must be found, and the only solution is, as one tailor told me, "to have enough children so that there are at least three or four sons in the family." Then each son can finish high school by spending part of the afternoon working...After high school, one son is sent on to college while the others work to save and pay the necessary fees...Once his education is completed, he will use his increased earnings to put his brother through college. He will not marry until the second brother has finished his college education and can carry the burden of educating the third brother...What is of interest is that, as the Khanna Study pointed out, it was the rise in the age of marriage—from 17.5 years in 1956 to 20 in 1969—and not the birth control program that was responsible for the decrease in the birthrate in the village from 40 per 1,000 in 1957 to 35 per 1,000 in 1968. While the birth control program was a failure, the net result of the technological and social change in Manupur was to bring down the birth rate.

Here, then, in the simple realities of the village of Manupur are the principles of the demographic transition at work. There *is* a way to control the rapid growth of populations in developing countries. It is to help them to develop—and more rapidly achieve—the level of welfare that everywhere in the world is the real motivation for a balanced population.

Enough To Go Around

Against this success, the proponents of the "lifeboat ethic" would argue that it is too slow, and they would take steps to *force* developing nations to reduce their birthrate even though the incentive for reduced fertility—the standard of living and its most meaningful index, infant mortality—is still far inferior to the levels which have motivated the demographic transition in the Western countries. And where, in their view, it is too late to save a poor, overpopulated country the proponents of this so-called "ethic" would withdraw support (in the manner of the hopelessly wounded in military "triage") and allow it to perish.

This argument is based (at least in the realm of logic) on the view, to quote Hardin, that "It is literally beyond our ability to save them all." Hardin's assertion, if not the resulting "ethic," reflects a commonly held view that there is simply insufficient food and other resources in the world to support the present world population at the standard of living required to motivate the demographic transition. It is commonly pointed out, for example, that the U.S. consumes about one-third of the world's resources to support only six percent of the world's population, the inference being that there are simply not enough resources in the world to permit the rest of the world to achieve the standard of living and low birthrate characteristic of the U.S.

The fault in this reasoning is readily apparent if one examines the actual relationship between the birthrates and living standards of different countries. The only available comparative measure of standard of living is GNP per capita. Neglecting for a moment the faults inherent in GNP as a measure of the quality of life, a plot of birthrate against GNP per capita is very revealing. The poorest countries (GNP per capita is less than $500 per year*) have the highest birthrates, 40-50 per 1,000 population per year. When GNP

---

* These and subsequent values are computed as U.S. 1969 dollars. The data relate to the 1969-70 period.

per capita per year exceeds $500 the birthrate drops sharply, reaching about 20/1000 at $750-$1000. Most of the nations in North America, Oceania, Europe, and the USSR have about the same low birthrates—15-18/1000—but their GNP's per capita per year range all the way from Greece ($941 per capita per year; birthrate 17/1000) through Japan ($1,626 per capita per year; birthrate 18/1000) to the richest country of all, the U.S. ($4,538 per capita per year; birthrate 18/1000). What this means is that in order to bring the birthrates of the poor countries down to the low levels characteristic of the rich ones, the poor countries do not need to become as affluent (at least as measured, poorly, by GNP per capita) as the U.S. Achieving a per capita GNP only, let us say, one-fifth of that of the U.S.—$900 per capita per year—these countries could, according to the above relationship, reach birthrates almost as low as that of the European and North American countries.

The world average value for birthrate is 34/1000, which is indicative of the overall rate of growth of the world population (the world average crude death rate is about 13/1000). However, the world average per capita GNP is about $803 per year—a level of affluence which is characteristic of a number of nations with birthrates of 20/1000. What this discrepancy tells us is that if the wealth of the world (at least as measured by GNP) were in fact evenly distributed among the people of the world, the entire world population should have a low birthrate—about 20/1000—which would approach that characteristic of most European and North American countries (15-18/1000).

Simply stated, the world has enough wealth to support the entire world population at a level that appears to convince most people that they need not have excessive numbers of children. The trouble is that the world's wealth is *not* evenly distributed, but sharply divided among moderately well-off and rich countries on the one hand and a much larger number of people that are very poor. The poor countries have high birthrates because they are extremely poor, and they are extremely poor because other countries are extremely rich.

The Roots of Hunger

In a sense the demographic transition is a means of translating the availability of a decent level of resources, especially food, into a voluntary reduction in birthrate. It is a striking fact that the efficiency with which such resources can be converted into a reduced birthrate is much higher in the developing countries than in the advanced ones. Thus an improvement in GNP per capita per year from let us say $682 (as in Uruguay) to $4,538 (U.S.) reduced the birthrate from 22/1000 to 18/1000. In contrast, according to the above relationships, if the GNP per capita per year characteristic of India (about $88) were increased to only about $750, the Indian birthrate should fall from its actual value of about 42/1000 to about 20/1000. To put the matter more simply, the per capita cost of bringing the standard of living of poor countries with rapidly growing populations to the level which—based on the behavior of peoples all over the world—would motivate voluntary reduction of fertility is very small, compared to the per capita wealth of developed countries.

Food plays a critical role in these relationships. Hunger is widespread in the world and those who believe that the world's resources are already insufficient to support the world population cite this fact as the most powerful evidence that the world is overpopulated. Conversely, those who are concerned with relieving hunger and preventing future famines often assert that the basic solution to the problem is to control the growth of the world population.

Once more it is revealing to examine actual data regarding the incidence of malnutrition. From a detailed study of nutritional levels[5] among various populations in India by Revelle & Frisch[5] we learn, for example, that in Madras State more than one-half the population consumes significantly less than the physiologically required number of calories and of protein in their diet. However, the *average* values for all residents of the state represent 99 percent of the calorie requirement and 98 percent of the protein requirement. What this means, of course, is that a significant part of the population receives *more* than the required dietary intake. About one-third of the population receives 106 percent of the required calories and 104 percent of the required protein; about 8 percent of the population receives 122 percent or more of the calorie requirement and 117 percent or more of the protein requirement. These dietary differences are determined by income. The more than one-half of the population that is significantly below the physiologically required diet earn less than $21 per capita per year, as compared with the state-wide average of $33.40.

What these data indicate is that hunger in Madras State, defined simply in terms of a significantly inadequate intake of calories and protein, is not the result of a biological factor—the inadequate production of food. Rather, in the strict sense, it results from the *social* factors that govern the *distribution* of available food among the population.

In the last year, newspaper stories of actual famines in various parts of the world have also supported the view that starvation is usually not caused by the insufficient production of food in the world, but by social factors that prevent the required distribution of food. Thus, in Ethiopia many people suffered from starvation because government officials failed to mobilize readily available supplies of foreign grain. In India, according to a recent *New York Times* report, inadequate food supplies were due in part from a government policy which "resulted in a booming black market, angry resentment among farmers and traders, and a breakdown in supplies." The report asserts further that "The central problem of India—rooted poverty—remains unchecked and seems to be getting worse. For the third year out of four per capita income is expected to drop. Nearly 80 percent of the children are malnourished...The economic torpor seems symptomatic of deeper problems. Cynicism is rampant: the Government's socialist slogans and calls for austerity are mocked in view of bribes and corruption, luxury construction and virtually open illegal contributions by businessmen to the Congress party."[6]

Given these observations and the overall fact that the amount of food crop produced in the world at present is sufficient to provide an adequate diet to about eight billion people—more than twice the world population—it appears to me that the present, tragically widespread hunger in the world cannot be regarded as evidence that the size of the world population has outrun the world's capacity to produce food. I have already pointed out that we can regard the rapid growth of population in developing countries and the grinding poverty which engenders it as the distant outcome of colonial exploitation—a policy imposed on the antecedents of the developing countries by the more advanced ones. This policy has forcefully determined both the distribution of the world's wealth and of its different populations, accumulating most of the wealth in the Western countries and most of the people in the remaining, largely tropical, ones.

Thus there is a grave imbalance between the world's wealth and the world's people. But the imbalance is not the supposed disparity between the world's *total* wealth and *total* population. Rather, it is due to the gross *distributive* imbalance among the nations of the world. What the problem calls for, I believe, is a process that now figures strongly in the thinking of the peoples of the Third World: a return of some of the world's wealth to the countries whose resources have borne so much of the burden of producing it—the developing nations.

Wealth Among Nations

There is no denying that this proposal would involve exceedingly difficult economic, social and political problems, especially for the rich countries. But the alternative solutions thus far advanced are at least as difficult and socially stressful.

A major source of confusion is that these diverse proposed solutions to the population problem, which differ so sharply in their moral postulates and their political effects, appear to have a common base in scientific fact. It is, after all, equally true, scientifically, that the birthrate can be reduced by promulgating contraceptive practices (providing they are used), by elevating living standards, or by withholding food from starving nations.

But what I find particularly disturbing is that behind this screen of confusion between scientific fact and political intent there has developed an escalating series of what can be only regarded, in my opinion, as inhumane, abhorrent political schemes put forward in the guise of science. First we had Paddock's "triage" proposal, which would condemn whole nations to death through some species of global "benign neglect." Then we have schemes for coercing people to curtail their fertility, by physical and legal means which are ominously left unspecified. Now we are told (for example, in the statement of "The Environmental Fund") that we must curtail rather than extend our efforts to feed the hungry peoples of the world. Where will it end? Is it conceivable that the proponents of coercive population control will be guided by one of Garrett Hardin's earlier, astonishing proposals:

> How can we help a foreign country to escape over-population? Clearly the worst thing we can do is send food...Atomic bombs would be kinder. For a few moments the misery would be acute, but it would soon come to an end for most of the people, leaving a very few survivors to suffer thereafter.[7]

There has been a long-standing alliance between pseudo-science and political repression; the Nazi's genetic theories, it will be recalled, were to be tested in the ovens at Dachau. This evil alliance feeds on confusion.

The present confusion can be removed by recognizing *all* of the current population proposals for what they are—not scientific observations but value judgements that reflect sharply differing ethical views and political intentions. The family planning approach, if applied as the exclusive solution to the problem, would put the burden of remedying a fault created by a social and political evil—colonialism—voluntarily on the individual victims of the evil. The so-called lifeboat-ethic would compound the original evil of colonialism by forcing its victims to forego the humane course toward a balanced population, improvement of living standards, or if they refuse, to abandon them to destruction, or even to thrust them toward it.

My own purely personal conclusion is, like all of these, not scientific but political: that the world population crisis, which is the ultimate outcome of the exploitation of poor nations by rich ones, ought to be remedied by returning to the poor countries enough of the wealth taken from them to give their peoples both the reason and the resources voluntarily to limit their own fertility.

In sum, I believe that if the root cause of the world's population crisis is poverty, then to end it we must abolish poverty. And if the cause of poverty is grossly unequal distribution of world's wealth, then to end poverty, and with it the population crisis, we must redistribute that wealth, among nations and within them.

FOOTNOTES

1. Quoted from the Environmental Fund's Statement "Declaration on Population and Food," original in *Daedalus*, Fall, 1973.

2. Presented in San Francisco at the 1974 annual meeting of the American Association for the Advancement of Science.

3. *Science*, vol. 172, p. 1297, 1971.

4. *The Myth of Population Control*, Monthly Review Press, New York, 1972.

5. Vol. III, "The World Food Problem," a report of the President's Science Advisory Committee, Washington, 1967.

6. *New York Times*, April 17, 1974.

7. "The Immorality of Being Softhearted," *Stanford Alumni Almanac*, Jan., 1969.

II
SCIENCE AND SOCIAL CONTROL

Introduction

The myth of the neutrality of science discussed in the previous section has far reaching consequences in our everyday lives and in the management of science. Having been convinced of the neutrality of science, most people find it impossible to accept the idea that social control is being exerted through science. The prevailing sentiment within and without the scientific community is that scientists are objective, that even when research is undertaken under corporate aegis the results will be beneficial to all humankind.

Under the guise of this neutrality, however, directions are being taken in basic and applied research that are all too often irresponsible and have inhumane consequences. There are many examples of this course of action. Technological advances in the area of contraception have ended up in sterilization abuse directed against poor and non-white women. Scientists who should be working to ensure safe working conditions have produced incomplete and misleading reports resulting in the perpetuation of unsafe conditions. Computers, used increasingly by business and government to perform a myriad of tasks and presented as the solution for the smooth running of modern society, have become an essential tool of the white minority of South Africa for maintaining repressive social control over the black population. And, in another example, many members of the science establishment have resisted the intervention and participation of the community-at-large when an important topic such as DNA recombinant research has been publicly discussed. The prevailing attitude is that common people do not know enough to be able to deal with the complex problems of our times. Scientists, seen as the experts, will find the right answers and indicate which directions to take. Even where situations are exploitative (of workers, women, or blacks), the scientific establishment lends its "objectivity" to the perpetuation of these conditions.

The articles in this section show how small groups of individuals representing narrow interests exercise control over science and thereby affect the lives of common people. The question of who controls science and what are the effects of that control are central to our understanding of modern society.

The U/California Operation of the Lawrence Livermore and Los Alamos Scientific Laboratories
—U.S. Nuclear Weapons Conversion Project

When weapons research is carried out under the supposed neutral and humane aegis of academia, who is responsible for the direction that research takes? In this article, we learn that the laboratories in question are responsible for the conception, design and testing of every nuclear warhead in the U.S. arsenal.

The University of California Nuclear Weapons Labs' Conversion Project (UNCNWLCP) is a broad coalition of groups and individuals from the University and from the community at large, which came together in October 1976 to challenge the University's operation of the two nuclear weapons research laboratories—Lawrence Livermore Laboratory (LLL) and Los Alamos Scientific Laboratory (LASL). The Project opposed UC's intention to renew, unchanged and with no meaningful public input, its five-year contracts with the federal government to administer the labs. We have since called on the University to review publicly its relationship with the laboratories; to facilitate a public process of discussion and education on the nuclear arms race and UC's role; to develop and work to implement plans for the complete conversion of the laboratories, from weapons work to life-supporting research; and to initiate an independent review of the public-health hazards presented by the laboratories' activities.

Our efforts were largely responsible for President David Saxon's appointment, in June 1977, of an eight-member administrative committee chaired by William Gerberding to review the University's operation of the two weapons labs. This Committee, known as the Gerberding Committee, had as its charge "to consider, from the University's perspective and in terms of the University's own welfare, if and under what circumstances it is appropriate for the University to continue managing these two laboratories."

We have followed very closely the work of this committee, from its birth to its final meetings. We have attended almost every open meeting (some of its sessions were closed to the public) and all public hearings held by the Committee, from July through November. In addition to submitting our own public testimony, we have talked with numerous individuals involved in the laboratories, including scientists, technicians and labs management. We have reviewed the history and current developments at the laboratories which are available in the public record. We issue this report in the hope that it will be circulated widely and will contribute to the discussion that will take place over the next few months about what the University should do with the weapons laboratories. We welcome your comments and your support.

The Work of the Laboratories: Nuclear Weapons

Although clear and undeniable, it is not widely known to the public that two laboratories operated by the University of California are together entirely responsible for the conception, design and testing of every nuclear warhead the U.S. has ever developed—from the "Little Boy" bomb dropped on Hiroshima to the hydrogen bomb, the warheads for current strategic missiles and new cruise missiles, and the myriad of "tactical" weapons like the neutron bomb.[1]

UC has operated the labs from their beginnings, Los Alamos since 1943 and Livermore since 1952, under contract with the federal government. The Department of Energy (DOE), which recently superseded the Energy Research and Development Commission (ERDA),* is now the federal agency which oversees the labs' work. DOE actually owns and funds the labs and provides UC with a "management fee" for its nominal administration. The fee, currently $3.8 million yearly, goes into the Regents' Nuclear Science Fund. Originally a reserve fund for nuclear research, the Nuclear Science Fund, now totalling $28 million, is used for general university construction and operation and not for offsetting the costs of labs management. The costs actually incurred for administering the labs are estimated by the University to be $3.5 million annually. This amount is included in the University's annual

---

*ERDA was created from a reorganization of the Atomic Energy Commission (AEC).

budget request to the state legislature and is reimbursed out of state tax revenues.

The two laboratories are the only places in the U.S. where nuclear warheads are developed. Actual production of the warheads takes place at other facilities under supervision of the DOE.

At each UC laboratory, direct nuclear weapons research consumes about 50% of the budget and work force. Another 20% is devoted to energy systems research having military applications (like the much-publicized work in laser fusion). Only 20% or less of the labs' work is devoted to basic energy research and other programs with little or no direct military applications.* The remaining small percentage of the workload is contracted from other agencies, such as the Department of Defense (DOD) or Nuclear Regulatory Commission (NRC).

In all, at least 70% of the labs' work is weapons-related—even though the advanced energy projects receive the most publicity, providing a convenient camouflage for the continued emphasis on advanced weapons development. LLL's director, Roger Batzel, clearly pointed out Livermore's priorities in the "Long Range Resource Projections" published in October 1976: "The laboratory's historic mission and continuing major responsibility is its nuclear weapons program." After explaining future programs which would strengthen the present weapons programs over the next five years, he went on to add that "present major energy and environmental programs are compatible with and have enhanced our work in weapons research."

Both the laser fusion and laser isotope separation programs are good examples of this compatibility. While laser fusion is generally publicized as "an energy hope for the future," its original and nearer-term purpose is a military application: nuclear weapons physics modelling and weapons effect simulation experiments.[2] The labs' hope is that the laser fusion program will be useful in bringing many, though not all, aspects of nuclear weapons testing into the laboratory. As Major General Edward Giller, former chief of national security for the Energy Research and Development

---

*The clearest indicator of the labs' lack of commitment to basic energy research is the number of jobs in each research area. If the figures from LASL and LLL are added together, for F/Y 1977, in the national security area, there were 7,308 jobs. In all advanced energy system areas (solar, geothermal, magnetic fusion, there were 1,235 jobs, less than 20% of the total for weapons-related work. By 1978, the situation worsens: 8,446 jobs in national security compared to 1,607 in advanced energy. This represents a gain of 1,138 weapon jobs and only 372 energy ones. (Figures from LLL and LASL Long-Range Resource Projections.)

Agency (ERDA) has described it, "Really this is a military program, and it always has been. It would be a very useful thing to have in a comprehensive test ban ... It would keep the weapons labs busy for 5 to 10 years anyway."[3]

As for laser isotope separation, the labs' researchers say it will provide an easier, cheaper method of enriching uranium fuel for nuclear power generation. But Barry Casper, writing in the January 1977 *Bulletin of the Atomic Scientists*, warns that laser enrichment could be a new path to unchecked, worldwide nuclear proliferation. Casper quotes a LASL scientist who says, "The whole world had better be a little bit uneasy, because it will be a whole lot easier to make bombs."[4]

In the future, the labs hope to strengthen their nuclear weapons work. Their long-range projections envision an expansion of this field, with the addition of one new weapons system to the U.S. arsenal each year for the next five years. Mary Gustavson, LLL's associate director for military systems, sums it up this way: "When you realize the wealth of developments that are possible in the nuclear arena, it is hard to foresee an end to our work."[5]

The Result of the Labs' Work: A Growing Threat of Nuclear War

The central question before us is whether the University should be involved in nuclear weapons research. In order to analyze this question, we must determine whether nuclear weapons research is beneficial to society or dangerous to its survival: that is, does such research bring us closer to nuclear annihilation? A critical look at the nature, purpose, and results of such research is necessary to arrive at an answer.

The basic assumption made by proponents of further research on nuclear weaponry is that continued gains in the national nuclear capability are necessary to *prevent* nuclear war. Only by increasing the quantity and sophistication of our weapons, the proponents argue—improving their range, accuracy and manoeuverability as well as total numbers of warheads—can we be sure they will not be used in warfare. Indeed, if this proposition is *not* true, in the context of today, there can be no point to weapons research: if it increases the chance of war, it is impossible to justify. As one LLL employee has stated, "In order to work here, you have to believe in nuclear deterrence."[6] Because the assumption that weapons research deters war is so fundamental to the labs' work, it is never brought into question. And because the University has never questioned the labs' work, the University has never questioned this assumption either. The University manages the labs as a "public service" without considering the most vital public concern: do the labs deter, or *increase*, the possibility of nuclear war?

One recognized worldwide authority, the Stockholm International Peace Research Institute (SIPRI), reported in its 1977 *Yearbook of Armaments and Disarmament* that "the probability of nuclear world war is steadily increasing."[7] Frank Barnaby, SIPRI Director, points out in this disturbing report that the worsening outlook is due to several interrelated factors: the growing worldwide stockpile of nuclear warheads and increasing sophistication of delivery systems (the U.S. foremost, with about 30,000 nuclear warheads in all); the increasing likelihood of nuclear proliferation due to the spread of nuclear-power technology among non-nuclear-weapons states; and finally, the escalating trade in armaments. Leaving the U.S. role as arms merchant out of consideration here, we can safely say that U.S. primacy in all nuclear developments is due, at least in part, to the technological advances pioneered by the two UC laboratories.

Two recent developments in official U.S. policy have also made the use of nuclear weapons more likely. One is the increased willingness on the part of the U.S. to fight "limited" nuclear wars, in which we would be the first to employ "tactical" nuclear weapons (such as the enhanced-radiation warhead or "neutron bomb"), in defense of Europe or Korea. By emphasizing reduced "collateral" damage to structures, etc., military planners hope to see such weapons considered more useable, blurring the distinction between conventional and nuclear weaponry.

Even more dangerous is the U.S. "counterforce" doctrine, requiring the capability to fight a limited, strategic nuclear war by hitting (Soviet) military targets, such as missile silos. The development of "counterforce" is inescapably destablilizing because it builds toward a "first strike" threat. The U.S. seems determined to lower the threshold of nuclear war, however, in spite of the consequences. James Schlesinger, our new Secretary of Energy (and as such, in charge of weapons development), defended the "counterforce" doctrine in 1974 when he was Secretary of Defense by saying: "I think we have to make the underlying calculation about nuclear war intellectually respectable."

Our primary reason for challenging UC's role in nuclear weapons research has been to bring home to the public the increasing insanity of the nuclear arms race, and to encourage actions which will lead eventually to the abolition of nuclear weapons. Our concern stems from a deep belief that most citizens do not realize, or understand, the great danger into which recent developments in nuclear weapons have led us.

It is in fact the very research of the laboratories that sets the pace and sophistication of the present arms race so that, far from preventing a nuclear war, the work of the labs greatly increases such a possibility. Herb York, the first director at Livermore, has summed up the ironic paradox of the arms race by noting that "nearly all the weapons which, in the hands of *others*, were (and are) threatening to our national security and indeed our very existence had been invented or perfected *by us* in the first place."[8]

Nuclear war is a real and growing threat, of which the labs are among the foremost promoters. The work done at LLL and LASL cannot be divorced conceptually from the most drastic consequences in which this work might result: the nuclear devastation of the earth and the destruction of human society.

How the UC Labs Shape U.S. Nuclear Policy

No control is exercised over the labs except by the military; there is very little effective review of the weapons program by responsible elected officials in either the legislative or executive branches of the federal government; and the University, by playing the silent partner to this arrangement, has committed a grave disservice to the people of this nation.

From the labs' earliest beginnings, first at Los Alamos and later at Livermore, the University of California has played a benevolent and protective role: on the one hand providing its good name, academic excellence and "objectivity" to attract both capable scientists and ample government funding; and on the other hand assuming no role at all in oversight or direction, but rather allowing the laboratories almost total freedom and autonomy. At the same time, the scientific and technical personnel at the labs, particularly those in management, have had a great deal to do with the direction of U.S. nuclear weapons development and resulting policies.

The lending of the UC name and prestige to the operation of the labs has created and perpetuated the myth that nuclear weapons research was under civilian control and influenced by the University in some positive way. Under the patronage of the Board of Regents, with the cooperation of the UC President's office and with the consent of the faculty, the University has been *nominal* administrator of the labs, under contract first with the AEC, then with ERDA, and now with the DOE. Edward Hammel, an assistant director at LASL and a member of the Gerberding Committee, recently made clear the exact nature of this relationship: "With respect to UC 'management' of LASL operations, UC plays the role of a 'benevolent absentee landlord.' The 'business' of LASL is for the most part with Washington. UC understands that and interferes in the program not at all." The major benefits to the labs from the University connection are, according to Hammel, "prestige" in that the UC name helps "in the recruitment and retention of scientific personnel" and "independence," in that the laboratory staff "enjoys a much greater degree of freedom in its interactions with government or industrial management."[9]

That this "freedom" which UC grants to the lab managers may not be such a good thing was one of the sharpest criticisms made in the 1970 report of the Zinner Committee: "We are particularly disturbed by the nominal leadership which the University provides. The laboratories enjoy a delightful autonomy within the protective shelter of the University, so delightful as to border on the licentious."[10] The Zinner Committee recommended, with overwhelming faculty approval, the continuation of the UC-labs relationship *only* if it were subject to substantial modification, designed so that the University would exercise from this time on some responsible guidance over the operation of the laboratories.

But in fact, the record of the past five years reveals an increasing isolation of the labs from meaningful interchange with the University. One of the major recommendations of the Zinner Committee was that the laboratories should prepare annual plans for their unclassified activities and subject these plans to review by campus authorities. Last year, in response to an inquiry from the UC President's office about what steps he had taken to implement this recommendation, LASL Director Harold Agnew replied, "While we will be happy to provide you with copies of these plans, it is not likely that it will be acceptable to ERDA to make any changes of substance."[11]

Thus the University provides an aura of academic legitimacy to the business of weapons development, effectively giving the laboratories' management a two-sided *carte blanche*: they are free of any supervision from within the University, yet the University name gives them independence from any other source of control.

This arrangement continued intact through the changes of AEC to ERDA and now to DOE. A general, with his military staff, selected by the Pentagon, directs the military applications of atomic energy from an office in a nominally civilian agency. General Alfred Starbird, who until recently was ERDA's administrator for national security, has made it clear that weapons work must remain the major task of the labs, countering a recent trend toward diversification of the labs' research program.[12]

There is substantial evidence of the very active political role that lab officials have played in shaping nuclear policy for the United States. Edward Teller and his associates at LLL lobbied strongly against the 1963 partial test ban treaty banning weapons tests in the atmosphere. In 1977, LLL Director Roger Batzel and LASL Director Harold Agnew testified in Washington before the Senate Foreign Relations Committee, in opposition to both the Threshold Test Ban Treaty awaiting ratification and the concept of the Comprehensive Test Ban actively supported by President Carter, which is still being negotiated between the U.S. and the USSR. While top ERDA officials backed the treaties, Agnew said they would give the USSR a "favored" status, while Batzel merely said their adoption would "preclude" advanced nuclear weapons research.[13]

On occasion—for example in November 1976—the LLL management team has not relied on the effectiveness of its inside connections in Washington but has taken its case directly to the people for support. On that occasion LLL directors called a press conference to boost their recommendations for an expanded budget appropriation for weapons development. Congressman Fortney H. Stark, commenting on the press conference later, said: "I am disturbed by statements attributed to weapons-minded nuclear scientists, because it is my conviction that unless the arms race and proliferation of nuclear weapons are stopped soon, the industrial nations of the world, including the U.S., are likely to be destroyed within fifteen years."[14] Congressman Stark added that the Livermore Labs are "an outstanding resource" and ought to be dedicated to constructive efforts on behalf of humanity.

One of the most revealing examples of how laboratory officials have gone about promoting new weapons is the story of the neutron bomb. LLL began work on an enhanced-radiation-warhead for missiles in the late 1950s. A revised version promoted by the Pentagon a decade later was refused funding by Congress in 1971. In 1973, ERDA and the Pentagon returned to lobby for an improved version. LASL Director Harold Agnew, testifying before a subcommittee of the Joint Committee on Atomic Energy, described efforts on behalf of the enhanced-radiation-warhead (whose name was censored from printed testimony) and expressed amazement that it had not been more acceptable to the military:

> I really don't know why people have not thought more on the use of these (deleted) weapons. It may be that people like to see tanks rolled over rather than just killing the occupants...I know we at Los Alamos have a small, but very elite group that meets with outside people in the defense community and in the various think tanks. They are working very aggressively, trying to influence the DOD to consider using these (deleted) weapons which could be very decisive on a battlefield, yet would limit collateral damage that is usually associated with nuclear weapons.[15]

Congress turned down the proposal in 1974; in January 1975, LLL undertook development of another artillery shell design; in August 1976 ERDA told the chairman of the Joint Committee on Atomic Energy it was planning limited production of neutron warheads; President Ford secretly approved its production shortly before leaving the White House in late 1976. In mid-1977, when already an accomplished fact, the "neutron bomb" finally became public knowledge.[16]

A number of directors and former directors of the laboratories are also involved in the Strategic Arms Limitation Talks (SALT), not only as technical experts but also directly as participants in negotiations. The present chairman of the General Advisory Committee to the U.S. Arms Control and Disarmament Agency (ACDA), the group that helps to shape President Carter's approach to arms control negotiations, is LASL Director Harold Agnew. It is hard to imagine a more glaring example of conflict of interest than that offered by allowing the promoter of the neutron bomb to enjoy this official relationship to the ACDA, and to advise the President on how he should respond in the SALT negotiations. Critics of the neutron bomb have said it will make arms reductions through SALT much more difficult to achieve.

The lack of informed supervision by Congress has been underscored by Senator Stuart Symington, member of the Joint Committee on Atomic Energy and a former Secretary of the Air Force. In 1973, Symington told his colleagues: "Not until I became a member of the Joint Committee and travelled to Europe with Sen. Pastore ...

in 1971, did I realize the military strength of the U.S. and become acquainted with the vast lethal power of our nuclear arsenal. I actually learned more about the true strength of the U.S. forces in Europe in those six days than I had in some 18 years on the Armed Services Committee. One cannot help but consider the implication incident to our defense and foreign policies if these facts were known by the appropriate committees of Congress, as well as...by the American people."[17]

In summary: the laboratories are directed by military, not civilian, officials; there is substantial interchange of upper-level personnel between the labs and the military establishment; laboratory managers take an aggressive role in promoting not only particular weapons systems, but also basic changes in military strategy and national policy, even those related to arms control; the degree of supervision exercised by our elected national leaders is at best superficial; and the amount of information provided to the general public is kept to an absolute minimum.

We conclude that our nation's nuclear policy has become the private property of the weapons technocrats, thanks in large part to the University's protection as "benevolent absentee landlord." The greater sins here are not what the UC has done in this connection, but what it has *not* done. It has not taken a responsible part in the management of the labs; it has not worked to decrease secrecy and disseminate knowledge about the weapons programs; it has not fostered research that might be critical of the military establishment or posed alternatives to present policies; nor has it encouraged education about the vital issues of the arms race.

A Public Service, or a Mantle of Legitimacy?

The rationale most often used to justify UC's operation of the weapons labs is that it is in some way "a public service." Since the University's traditional mission includes teaching, research *and* public service, operating these labs for the federal government is seen as a legitimate University function, like cancer research. Nuclear weapons research is "national policy" determined by Congress and the President, this argument goes, and is therefore a public service. But this "just following orders" rationale ignores several important facts and counter-arguments.

As we have shown, nuclear weapons policy is not determined solely by Congress and the President. To believe that this *is* the case, one must have a naive understanding of how national policy is contracting with the military, presents weapons policy options to Congress and the President, and then lobbies vigorously for them. These forces are so isolated from the public and so able to control the flow of information that the important policy decisions are often made with no substantial public knowledge or debate until long afterward.

A perfect example of this is the history of the so-called neutron bomb. After its original development by the labs in the early 60s, it went through many modifications, funded by Congressional approval of appropriations for further research and development under a blanket of secrecy. Finally Sen. Mark Hatfield and the *Washington Post*, reading between the lines in Congressional testimony and budget appropriations, alerted the public to the true state of affairs. We have already documented the labs' role as a powerful lobbying group for the neutron warhead.

We also believe it is a highly questionable definition of the term "public service" to create, develop and promote nuclear warheads— which could ultimately destroy the very public they are intended to "serve." If these weapons were ever unleashed, there would be no doubt that they were indeed not a public service. As we have shown, the possibility that these weapons might be used is a very real one that increases daily, thanks in large part to the work of these labs.

An equally disturbing aspect of the "public service" argument is the implicit acceptance, by the University, of secrecy and heavy security which prevail at the labs, preventing the press and public from obtaining much needed information. Ray Kidder, an associate division leader at LLL, says *"two-thirds* of the present effort at LLL is devoted to programs that are *primarily secret*, that is programs in which the most important results are classified as Secret Restricted Data" (emphasis Kidder's). He is also "troubled by University trusteeship of an activity that is primarily secret and shows every indication of remaining so."[18] The 1970 Zinner Committee report by UC faculty members went even further when it stated that "secrecy, generally speaking, is inconsistent with free inquiry to which the University is dedicated ... (and) is also inimical to the best interests of the nation."[19]

However, those with a vested interest in continuing the arms race have historically used the secrecy surrounding weapons policies to distort information, manipulate public opinion, and, by withholding key information, mislead the public into the belief that more and better weapons are the only route to "national security."

The only possible public service the University could have performed through its sponsorship of nuclear weapons research would be to facilitate public discussion and education about nuclear weapons policy and the arms race in order to shed more public light on such questions and foster the development of critical perspectives. In this area the University has done virtually nothing, while at the same time lending full support to the one-sided weapons promoters at the labs. In an effort to begin redressing this gross imbalance, we have repeatedly asked, over the past year, that the UC administration lend its name and resources to forums, classes, seminars or debates about nuclear weapons policy—only to be put off or rebuffed. The central UC administration would not even designate a representative to meet with us in public debate, but instead saw fit to arrest six people who had waited for two days in President David Saxon's outer office in the hope of persuading Saxon to change his mind about appointing such a representative. (A trial jury has just acquitted the six of all criminal charges, based on a finding that their presence in the office was justified by "legitimate business.")

In the words of Daniel Ellsberg, the University has not performed a "public service" in operating these nuclear weapons laboratories, but instead has lent "a mantle of legitimacy" to the development and promotion of nuclear weapons. By legitimizing the nuclear arms race and the need for secrecy, while failing to promote critical discussion, the UC has performed a profound public "disservice."

Environmental Considerations and Public Health

The need for secrecy, justified by "national security" considerations and a longstanding tradition of exempting federal property from jurisdiction by local regulatory agencies, has kept much of the public in the dark about the awesome hazards inherent in the labs' operations, especially at Livermore. We are concerned not only about the lack of any critical overview of these problems, but also about the legal and moral responsibility the labs have for the consequences of many years of nuclear weapons testing under their direction. Deaths and illnesses due to exposures to weapons test radiation and fallout may not occur until as much as 30 years after the event; the same interval may occur between inhalation of plutonium particles (as little as ten micrograms) and evidence of lung cancer. Recently an unusually high incidence of skin cancer has been identified among LLL employees, and undisclosed quantities of beryllium particles (another potent carcinogen) have been reported by guards at the non-nuclear test site near Tracy as the probable cause of the guards' striking incidence of lung disease.

The suppression of research findings on the effects of radiation on human beings has been documented by Dr. John Gofman, who as an Assistant Director at LLL, headed the Biomedical Division there and led a six-year study concluding that previously established radiation standards were inadequate. This finding was unacceptable to the AEC, and Dr. Gofman maintains he was consequently forced to resign. Similar stories involving health studies conducted at other facilities under AEC control are currently the subject of a Congressional investigation since all indicate that findings unacceptable to the nuclear industry were suppressed at a time when their widespread publication might have had important policy results.

In 1975, Dr. Gofman published a report showing that approximately one million cases of lung cancer, due solely to the plutonium content of weapons-test fallout, may be anticipated in the northern hemisphere, with 116,000 in the U.S. alone.[20] The next few years, marking the end of the latency period for plutonium exposure, may show whether Dr. Gofman's estimates are actually too conservative.

The quantity of plutonium used, stored, and transported to and from LLL is a closely guarded secret: only recently has it become known that the lab is licensed to have between 600 and 800 pounds of plutonium on the premises. Truck transport of this and other highly radioactive materials presents a continuous, but little-known environmental threat. Air transport of plutonium warhead components between LLL, the Nevada test site and Rocky Flats, Colorado, went on undetected and in secret for some 25 years. The plutonium was carried in small turboprop planes without crash-proof containers, and took off and landed at a small municipal airport. The secret flights were discovered by the press and public in 1977, and under pressure from Congressman Fortney Stark, were reduced in number by ERDA/DOE, which also claims that a "crash-proof" container will soon be available.

The most all-encompassing threat to public safety implicit in the quantity of plutonium at LLL is the ever-present possibility of a major earthquake. Thirteen known, geologically active faults pass near or under the lab site. According to a 1976 environmental impact statement, future earthquakes with a potential Richter magnitude greater than that which the lab structures could resist may be anticipated.[21] Many critical facilities could be damaged severely enough to allow the escape of radioactive liquids, gases and solid particles; in this category are buildings containing hot cells, glove boxes, filter systems, pipes and waste retention tanks. The Radioactive Liquid Waste Treatment Plant in Building 514 could, in the event of a serious quake, "release untreated, highly radioactive wastes into the Livermore sewage system." Five buildings described as having "the most potential for adverse impact" produce a major portion of the total radioactive wastes. The scenario contained in the report for a "maximum credible accident" posits the release of 250,000 curies of radioactive fission products, enough to contaminate nearby agricultural areas and the food chain for up to 1000 years.

There is evidence that LLL managers have chosen to disregard data on the magnitude of future earthquakes, published first in 1974,[22] in favor of less alarming estimates which appeared two years earlier. The two reports disagree on the definition of a "Safe Shutdown Earthquake," including the amount of expected ground acceleration which becomes the yardstick for upgrading structural safety. Ground acceleration estimates have been revised upward at least twice since the first LLL structures were built. "Secrecy" of research has been used by LLL management to prevent state and county building inspectors and structural engineers from making any objective safety evaluations.

Though the exemption of federal facilities from local air-pollution control efforts ended in August 1977, inspectors from the Bay Area Air Pollution Control District are still unable to verify the suspected beryllium contamination at Site 300. This situation is now being investigated by the Environmental Protection Agency (EPA).

EPA has also been investigating the ocean dumping ground for radioactive wastes generated by various nuclear research labs between 1946 and 1970. Some of the nearly 60,000 barrels of waste came from Livermore; about 25% have broken open on the ocean floor some 40-50 miles from San Francisco, and "detectable" amounts of plutonium and other long-lived radioactive material, leaked from the barrels, have been identified. In view of its own waste-disposal problems, past and present, the LLL has ironically contracted to do a study of safe disposal of radioactive wastes for the Nuclear Regulatory Agency.

We find that a University which sponsors major projects in

medical research, including cancer therapy, is an inappropriate agency to sponsor, and legitimize, secret research involving such immense public health hazards, most of them highly carcinogenic. The labs' resistance to inspection and regulation by the appropriate local and state agencies is grossly inconsistent with the University's "public service" charge.

The Labs' Conversion Potential

It should not be forgotten (though laboratory officials never stress it) that there is enormous potential at the laboratories for a shift to basic energy research. A number of small projects encompassing a wide range of energy alternatives are now being pursued there. These include comprehensive energy studies, geothermal development, natural gas stimulation, solar experiments, wind current studies, flywheel applications, and alternative methods of automotive propulsion. The necessary chemists, physicists, engineers, and environmental scientists are already at the laboratories, all working, for the most part, on some aspect of the weapons program.

There is an enormous need in the state of California for additional research and development of energy alternatives: solar, tidal, wind, geothermal, and other non-polluting energy sources. The labs could be instrumental in meeting this need. Funding for this type of work should be sought as aggressively as the labs managers have sought funding for contracts in weaponry. And the arbitrary limits now placed on non-weapons research should be lifted so that all these areas can be actively and fully explored. The labs have recruited work forces uniquely combining a variety of disciplines and skills and housed them in superbly equipped surroundings. Many members of these work forces are willing and anxious to turn from weaponry to more constructive and satisfying research alternatives.

This willingness, and the potential availability of funding for such work, were described to the Gerberding Committee at its August 9 public hearing in Livermore by Terry Rossow, an LLL engineer and then president of the Livermore chapter of the Society of Professional Scientists and Engineers (SPSE). Rossow stated unequivocally that funding for substantially more energy research at the labs did in fact exist, but that it was rejected by ERDA and the labs managers due to the arbitrary ceilings on such work which now prevail.

The University could begin playing a more positive role with regard to the labs by lending prompt support to those employees who sincerely want to see more funding for energy research accepted by the laboratories' managements.

Conclusions and Recommendations

There are two critical problems confronting the University and the community at large with respect to the Lawrence Livermore and Los Alamos laboratories. The first of these is the nature of work on advanced nuclear weapons, which is accelerating the international arms race and threatening human survival. The second is the nature of the administration of the laboratories, which permits them to influence the formation of U.S. nuclear policy while shielding them from public scrutiny and participation.

With respect to the first issue, our overriding concern is to eliminate nuclear weapons from the world. An important step the U.S. can take to deescalate the international arms race and move toward that goal, without endangering national security, is to cease researching, developing and testing advanced nuclear weapons systems.* We strongly believe that the University has an important role to play in supporting such U.S. initiatives.

On the second issue, we believe that democratic control—the participation by the University, the surrounding community, and elected representatives—in decisions affecting the labs—is crucial. Such participation, by experts and non-experts alike, is the best guarantee we have against the further abuse of power by the weaponeers. The importance of the matter of weapons development to the future of the human race demands the broadest public scrutiny of this kind of research from now on.

Many people feel that these problems point to only one conclusion: that the labs should be severed from the University of California. This has been the subject of extensive discussion among members of the UC Nuclear Weapons Labs' Conversion Project. Those who favor severance support their view with both moral and pragmatic arguments: that the University is an institution of learning and has no business making weapons; that it may not have the power to make any real changes in labs' policy; or that it is not the University's role to determine national policy and so it should simply get out of the picture.

---

* A similar proposal was drafted by the Union of Concerned Scientists, signed by some 13,000 of its members, and presented to President Carter on December 20, 1977.

Yet, after much consideration of this course of action, we stand firm in our conclusion *not* to endorse the simple separation of UC from LLL and LASL as a first option. "Purifying the air" around the campuses will contribute nothing toward a solution of the problems posed by the labs' operations; it could leave them unaffected, free to continue present policies under new management. We already know what happened in the past when concern over "war research" resulted in the severance of the Stanford Research Institute from Stanford University and the Draper Labs from Massachusetts Institute of Technology. In both cases, the research labs emerged independent, unscathed, and perhaps even stronger than before.

Instead, we hold the University strictly responsible for its past actions which helped to make the weapons laboratories the public threat they have become; and we insist that the University undertake the task of trying—no matter against what odds—to work vigorously for solutions of the problems we have described. For the University simply to "wash its hands" of the labs would be irresponsible. We believe that for the University which wishes to bolster a tradition of true public service, the way is clear to respond, sincerely and positively, to the recommendations which follow.

From the beginning, we have called for an end to all nuclear weapons-related work at LLL and LASL, and for the conversion of the labs to life-supporting, constructive fields of research including (but not limited to) safe and renewable energy sources. We again call for that conversion now. To take the first steps on the road to conversion is the only way for the University to cleanse itself of its past long-standing involvement with weapons research, and henceforward bring its deeds into conformity with its professed humanitarian ideals.

We are quite aware that the University cannot, on its own, convert the labs since they are federally owned and funded. Clearly any such decision would be subject to intense opposition by labs managers, the military establishment, and some members of Congress. Nevertheless, the University *has* the power to raise the questions, to initiate the process and to challenge the labs and the government to respond. The University's *commitment* to conversion can be demonstrated by the following concrete steps, all fully within its power to undertake now:

1. The University should publicly declare its desire for an immediate halt to nuclear weapons research at LLL and LASL, and for their complete conversion from weapons work to constructive purposes.

It should announce a commitment to developing conversion plans and seeing them implemented.

2. To bring the labs under public control, the University should create an Administrative Board, which will:

a) be composed of individuals representing a broad spectrum of views, chosen for their concern with the work of the laboratories and their standing in the community;

b) be charged to act as a counterweight to Pentagon influence at the labs by working: to eliminate secrecy, to provide the most complete information to the public about the labs' work, and to support and give a voice to those inside the laboratories, as well as outside, who oppose continued weapons research;

c) be empowered to review programs and policies of the laboratories, receive comments on laboratory activities from employees as well as from other citizens, make recommendations for changes (including changes of management personnel) and participate in all contacts between laboratory officials and representatives of other government agencies to insure against one-sided presentations;

d) establish a Conversion Planning Committee, broadly representative and sufficiently funded to enable it to make detailed studies and concrete proposals for the conversion of the laboratories, including provisions for the employees' job security.

3. The University should establish an educational program that would facilitate open, comprehensive public discussion among students, faculty, and the community on all its campuses and at the laboratories. The subject of such discussion would be U.S. nuclear weapons policy, research done at the labs, and alternatives to the status quo in both areas. This could involve classes, seminars, debates, forums, and creation of departments of Peace Studies, to be funded from the Regents' Nuclear Fund, which has been built up from fees the University received for managing the two laboratories.

4. Because serious questions have been raised about health and environmental hazards posed by types of work and the presence of highly radioactive materials at the labs, the University should lead the way in demanding a thorough review of all such questions by an agency or agencies totally independent of present lab operations or any military connections (for example, the General Accounting Office of the federal government).

5. If the University finds itself unable to implement these proposals, *then and only then* do we urge that it exercise its option to

terminate its operation of the laboratories within two years, but abide by the sense of these proposals by:

a) publicly stating its desire to see nuclear weapons research ended;

b) working to establish some other form of public control over the laboratories;

c) proceeding to establish an independent University committee to develop proposals for non-military scientific research which could utilize the laboratories' facilities.

It is understood that if these proposals were adopted, they would represent a serious challenge to the entrenched power of the military-industrial-scientific complex. In order for the University to make significant headway against the expected opposition to bringing the weapons laboratories under more democratic control and working for their conversion as outlined above, political support must be mobilized from a wide constituency outside the University complex. The first test, of course, will be whether the University is willing to accept this challenging task.

FOOTNOTES

1. Samuel H. Day, Jr., "The Nuclear Weapons Labs," *Bulletin of the Atomic Scientists*, 33, April, 1977, p. 22.

2. Lawrence Livermore Laboratory, *Long Range Resource Projections*, Livermore, California, October, 1976.

3. Robert Gillette, "Laser Fusion: An Energy Option, But Weapons Simulation is First," *Science*, 188, April 4, 1975, p. 30.

4. Barry Casper, "Laser Enrichment: A New Path to Proliferation?" *Bulletin of the Atomic Scientists*, January, 1977, p. 28.

5. Lawrence Livermore Laboratory, *Newsline*, September/October, 1976, p. 9.

6. "Popping the Strangelove Myth," LLL *Newsline*, September/October, 1976, p. 16.

7. Frank Barnaby, "The Mounting Prospects of Nuclear War," *Bulletin of the Atomic Scientists*, June, 1977. Drawn from *World Armaments and Disarmament*, SIPRI Yearbook, 1977.

8. Samuel H. Day, Jr., "The Nuclear Weapons Labs," *Bulletin of the Atomic Scientists*, 33, April, 1977, pp. 21-32.

9. From a memo circulated by Hammel in July, 1977 at Los Alamos Scientific Laboratory, and subsequently given by him to the Gerberding Committee.

10. Zinner Committee Report, 1970, p. 21.

11. Item C, "Items for Discussion," Regents' meeting of February 17, 1977.

12. John Walsh, "ERDA Laboratories: Los Alamos Attracts Some Special Attention," and "Critics Seek 'Conversion' of Labs," *Science*, May 13, 1977, pp. 744-746.

13. News story, *Contra Costa Times*, September 9, 1977, p. 6.

14. News story, *Livermore Independence*, December 13, 1976.

15. "Military Applications of Nuclear Technology," Hearings before the Subcommittee on Military Applications, Joint Committee on Atomic Energy, 93rd Congress, First Session, April 16, 1973 (Part I).

16. News story, *San Francisco Chronicle*, July 8, 1977, originating in the *Washington Post*.

17. Subcommittee on Military Applications, *op. cit.*

18. Letter from Kidder to William Gerberding, August 23, 1977.

19. John W. Gofman, "Estimated Production of Lung Cancer by Plutonium from Worldwide Fallout," *CNR Reports*, 2, July 10, 1975.

20. Draft Environmental Impact Statement of LLL and Scandia Laboratory, October, 1976.

21. "A Geological and Seismological Investigation of the LLL Site," June 3, 1974. Incorporated into the DEIS as Appendix 2A.

Sterilization Abuse
—Helen Rodriguez-Trias

This article describes how sterilization procedures are used in the U.S. today to control poor, Puerto Rican, Black and Native American women. It shows how racist and sexist ideology has led to a practice that can be recognized as genocide.

Recent events have shown that sterilization is a procedure freely chosen by some people in family planning but demanded of others against their will. Consider, for example, what has recently happened in India. Mrs. Indira Ghandi's government was defeated—an event many attribute to the mass forced sterilization program it sponsored—but before the final ousting, at least 300 Indians died in riots protesting the assault of forced sterilizations on both men and women. The Indian sterilization experience showed the world that some population control programs mean ugly coercion.

A second item is right here at home: large numbers of native American women have been sterilized by the Indian Health Services, a United States government agency. A recent report from the General Accounting Office, produced at the request of Senator James Abourezk from South Dakota, reveals 3,406 sterilizations of American Indian women between the ages of 15 and 44. These were performed in Aberdeen, Albuquerque, Oklahoma City, and Phoenix between 1973 and 1976.[1] Evidence that the basic elements of informed consent were not communicated to the patients lends credence to Dr. Connie Redbird Uri's many public statements charging the United States government with having a deliberate genocide policy against her people, who number under a million.[2] Although there has been a moratorium since April 1974 on government financing of sterilization of women under 21 years of age, there were 13 violations by the Indian Health Service in two years.[3] To date, there has been no action against the violators.

Sterilization is a time-honored procedure in the United States. The first of the laws empowering the state to sterilize unwilling and unwitting people was passed in 1907 by the Indiana Legislature. The Act was intended to prevent procreation of "confirmed criminals, idiots, rapists, and imbeciles" who were confined to state institutions. The law was clear in its tenet that heredity plays an important part in the transmission of crime, idiocy, and imbecility.[4] After World War I, a model federal law was proposed by Dr. Harry Hamilton Laughlin, Superintendent of the Eugenics Record Office, and copies were widely distributed in large quantities to governors, legislators, newspaper and magazine editors, clergymen, and teachers. According to the model law, the following ten groups were labeled "socially inadequate" and were therefore subject to sterilization: (1) feeble-minded; (2) insane (including the psychopathic); (3) criminalistic (including the delinquent and wayward); (4) epileptic; (5) inebriate (including drug habitues); (6) diseased (including the tuberculous, the syphilitic, the leprous, and others with chronic, infectious, and legally segregable diseases); (7) blind (including those with seriously impaired vision); (8) deaf (including those with seriously impaired hearing); (9) deformed (including the crippled); and (10) dependent (including orphans, "ne'er-do-wells," the homeless, tramps, and paupers.)[5] Laws such as this, known as the eugenics laws, were passed in 30 states and as of 1972 were still on the books in 16.[6]

It is shocking to learn that between 1907 and 1964, more than 63,000 people were sterilized under these eugenics laws in the United States and one of its colonies, Puerto Rico.[7] Practices sanctioned inside institutions often become commonly accepted practices in the larger community. It is therefore important to keep in mind this long history of legally sanctioned forced sterilization as a framework for understanding current hospital practices.

Labeling people mentally retarded, insane, criminal, or indigent is an act we must examine closely. This sort of labeling is a peril in itself, but when it is used as grounds for sterilization, it is doubly dangerous. The groups considered undesirable may change, but they always include people who work for wages or are unemployed; they are inevitably the most exploited, and, therefore, poor. In the United States, the labeling process has additional racial overtones, because most Third World people are in the least remunerated strata of the working class and are definitely poor. It is a cruel irony that people with preventable diseases due almost solely to poverty are included in groups seen fit for sterilization.

Under the eugenics laws, many black women had been sterilized without challenge. The challenge came only when in 1924 Carrie Buck, a poor, white 18-year-old woman institutionalized for mental retardation, was threatened. Although judged retarded, Buck had completed six grades of school in five years. She had defied the norms by bearing an illegitimate child and was about to be sterilized, when members of a religious group in Virginia challenged the law all the way to the Supreme Court. Justice Oliver Wendell Holmes handed down his well-known *Buck vs. Bell* opinion in favor of her sterilization in which he stated: "the principle that sustains compulsory vaccination is broad enough to cover cutting the fallopian tubes." He concluded that "three generations of imbeciles are enough."[8]

Implicit in Justice Holmes' opinion was the belief that Carrie Buck's alleged mental retardation was hereditary. Today mental retardation is often determined on such questionable evidence as inability to cope with the school system, the discredited I.Q. tests, or even evidence of cultural differences.

Perhaps an even greater impact of these infamous laws was the social legislation they inspired. At least 10 states have proposed compulsory sterilization of people on welfare.[9] No state has passed such legislation, but the very existence of such proposals should make us question the prevailing social climate. In a country plagued by chronic unemployment, such proposals reveal virulent feelings toward women who cannot earn a living because they must care for children, the elderly, or others.

Physicians play an important role in implementing the view that poor people have no right to decide on the number of their children. A survey of obstetricians showed that although only 6 percent favored sterilization for their private patients, 14 percent favored it for their welfare patients. For welfare mothers who had borne illegitimate children, 97 percent of the physicians favored sterilization.[10] Similarly, a number of polls of the public at large shows that the idea of sterilization of welfare recipients is very much accepted. In a 1965 Gallup poll, about 20 percent of the people surveyed favored compulsory sterilization for women on welfare.[11]

We are witnessing a resurgence of the Malthusian ideas which proclaimed the poor unfit to receive the knowledge and hygienic measures which might decrease their mortality.[12] The more sophisticated modern version calls for a decrease in the social, medical, educational, and other resources allotted to poor people and for an offer of sterilization instead. In lieu of social changes to provide a decent living for every American, the population planners choose to curtail population. In the words of Dr. Curtis Wood, past president of the Association for Voluntary Sterilization,

> People pollute, and too many people crowded too close together may cause many of our social and economic problems. As physicians, we have obligations to the society of which we are a part. The welfare mess, as it has been called, cries out for solutions; one of these is fertility control.[13]

The use of the phrase "fertility control" is itself deceptive. In reality, it means only one thing, permanent control—that is, sterilization. Therefore, it does not surprise us that a 1973 survey revealed that 43 percent of women sterilized in federally financed family programs were black.[14]

Hysterectomy, now the most frequent major operation, done four times as frequently in the United States as in Sweden, is an indication of still another way of sterilizing women without their consent.[15] Black women on welfare suffer the most abuse. According to the *New York Times:*

> In New York and other major cities, a hysterectomy which renders a patient sterile costs up to $800, while a tubal ligation (the tying off of the fallopian tubes), which does the same thing, pays only $250 to the surgeon, increasing the motivation to do the more expensive operation. Medicare, Medicaid, and other health plans for both the poor and the affluent will reimburse a surgeon up to 90 percent for the costs of any sterilization procedure, and sometimes will allow nothing for abortion. As a consequence, "hysterilizations"—so common among some groups of indigent blacks that they are referred to as "Mississippi appendectomies"—are increasingly popular among surgeons despite the risks.[16]

Several lawsuits since 1973 around the country provide evidence of both the widespread nature of abuse as well as of the rising redress on the part of the people.

Most notorious is the case of the two sisters, Mary Alice, then 14, and Minnie Lee Relf, who was 12 at the time of their sterilizations in Montgomery, Alabama in June 1973. As described in court by their mother, two representatives of the federally financed Montgomery Community Action Agency called on her requesting consent to give the children some birth control shots. Believing the agency had the best interests of her daughters' health in mind, she consented by putting an X on paper.[17]

Judge Gerhard Gesell, who heard the case, declared:

> Although Congress has been insistent that all family planning programs function purely on a voluntary basis there is uncontroverted evidence in the record that minors and other incompetents have been sterilized with federal funds and that an indefinite number of poor people have been improperly coerced into accepting a sterilization operation under the threat that various federally supported welfare benefits would be withdrawn unless they submitted to irreversible sterilization.[18]

In another case, a number of women from Aiken, South Carolina, sued Dr. Clovis Pierce, a white former Army physician, for his coercive tactics in obtaining consent, tactics which included threats to refuse to deliver their babies. In 1973, black women were objects of 16 of the 18 sterilizations paid for by Medicaid and performed by that physician.[19]

Norma Jean Serena, a Native American mother of three children, will be the first to raise sterilization abuse as a civil rights issue. She charges that in 1970 health and welfare officials in Armstrong County, Pennsylvania, conspired to have her sterilized when her youngest child was delivered.[20]

Ten Mexican-American women are currently suing the Los Angeles County Hospital for obtaining consent in English when they spoke only Spanish. Some were in labor at the time, others even under anesthesia. A few reported being told such things as "Sign here if you don't want to feel these pains anymore" while a piece of paper was waved before their eyes.[21]

Largely as a result of the pressure mounted when the Relf case came to light, the Department of Health Education and Welfare (HEW) decided to write guidelines on sterilization procedures

during 1974. In effect, these established a moratorium on sterilizations of people under 21 years of age, and on those who for other
reasons could not legally consent. In addition, the guidelines
stipulated that there must be a 72-hour waiting period between the
granting of consent and the carrying out of the sterilization. They
also required an informed consent process including a written
statement to the effect that people would not lose welfare benefits
if they refused to be sterilized, and they included the right to refuse
sterilization later, even after granting initial consent.

Although HEW promulgated the guidelines early in 1974, a
study conducted in 1975 by the Health Research Group, a
renowned Washington-based organization,[22] and later corrobo-
rated by Elissa Krauss of the American Civil Liberties Union,
showed that only about 6 percent of the teaching hospitals were in
compliance with these guidelines. Many of the hospitals provided
only the broadest of consent forms without proper explanations.[23]
A still more recent study by the Center for Disease Control, an
HEW agency, revealed that widespread non-compliance continued
to be the rule. The study attributed the fact to ignorance of the
guidelines.[24]

Early in 1975 those of us who were concerned about the issue
of abuse formed a committee which we called The Committee to
End Sterilization Abuse (CESA). We were faced with some hard
realities: First, HEW can only regulate for federally funded procedures, and although it is true that the primary targets of sterilization abuse have been women on welfare, there are still many other
vulnerable groups who are not welfare recipients, including the
recently unemployed, undocumented workers, and workers whose
earnings are just barely above the poverty line. Second, it seemed
obvious that without a national monitoring system, it is impossible
to determine what is happening to whom or whether guidelines are
being followed. Third, those who control information often
manipulate people's behavior. For example, the inclusion of hysterectomy as one form of sterilization in an HEW informative
pamphlet[25] tends to grant legitimacy to that mutilating operation
in the eyes of the reader. Finally, the need for strong enforcement
mechanisms became clear. There is no way that well-established
actions and practices can be uprooted without the use of some
measure of enforcement, particularly when the practices are
profitable and socially sanctioned.

These facts, coupled with the rise in number of sterilizations
observed in the New York City hospital system, particularly in

those hospitals serving black and Puerto Rican communities, prompted a number of concerned people from the Health and Hospitals Corporation, the New York agency responsible for the municipal hospitals, and from citizens' groups to form an *ad hoc* Advisory Committee on Sterilization Guidelines early in 1975. The Committee to End Sterilization Abuse, Healthright, Health Policy Advisory Center, the Center for Constitutional Rights, the community boards of the hospitals, and many other organizations and individuals were represented on this new committee. Most of the members were women involved in patient advocacy and who at the same time represented New York's various ethnic communities.

Our goal was to write new guidelines for the municipal hospitals. We met initially to ascertain the facts and to analyze the processes by which abuse takes place. Then we compiled the information in a report.[26] We identified existing weaknesses in the HEW regulations by using women's experiences as the touchstone for the drafting of stronger guidelines.

Many consents are obtained around the time of abortion or childbirth. The philosophy behind this practice is exemplified in the words of one doctor who said, "Unless we get those tubes tied before they go home, some of them will change their minds by the time they come back to the clinic."[27] The waiting period of only 72 hours after consent had been obtained at a time of great stress, allowed no opportunity for the woman to discuss the matter with friends, family, or neighbors to assure herself that she really wanted a sterilization. The time of abortion was particularly hazardous, because many teaching hospitals offer abortions as a "package deal" together with sterilization.[28] What kind of information was given to women was also key since both the vocabulary and the amount of information can clarify or confound. This reflects another weakness inherent in both the structure of the health system and the doctor-patient relationship: the coercive nature of medical advice given as it is in a patriarchal setting.

Our coalition of concerned groups drafted guidelines to remedy these weaknesses. They called for a 30-day waiting period; and interdiction of consent at time of delivery, abortion, or of hospitalization for any major illnes or procedure; the requirement that full counseling on birth control be available so that alternatives are offered; the stipulation that the idea for sterilization should not originate with the doctor; and the provision that informational materials must be in the language best understood by the woman.

The guidelines also stated that if she wished, the woman could bring a patient advocate of her choosing to participate at any stage of the process. She could also have a witness of her choice sign the consent form. Perhaps the most important point we made was that a woman should express in her own words, in writing on the consent form, her understanding of what the sterilization entailed, particularly its permanence.

We were unprepared for the ferocity of the opposition to our guidelines. Our files, replete with angry letters from obstetricians, organizations involved in family planning and population control, and other groups, attest to the length and difficulty of our struggle. The chiefs of the obstetrical services in the municipal hospitals marshaled many objections, especially to the extended waiting period and the prohibition of consent around the time of abortion or childbirth. Some based their arguments on dramatic stories about the "habitual aborter" and the "grand multipara." The "habitual aborter" was described as a young woman who is using repeated abortions rather than contraceptive methods; the "grand multipara" was a woman who could only consent to sterilization while in hospital for childbirth, this being the only time she sought services. Our response was to continue to bring testimony of how abuse takes place and to negotiate on the provisions of the guidelines until they were acceptable to our committee, its constituency, and to the obstetricians and other staff. It was a massive outreach effort that gained the support of community groups, boards of hospitals, health organizations, and legal groups. And it was this broad-based support, backed by several thousand letters and petitions, three or four demonstrations, hundreds of speeches and dozens of meetings that finally overcame still strong opposition in medical circles.

New York City's Health and Hospital Corporation had once more been responsive to public wishes, illustrating that even imperfect institutions can respond, provided consumers find the channels through which they can fight for change.

The guidelines became effective November 1, 1975. Barely three months later, six professors of obstetrics and gynecology representing six major medical schools filed suit opposing the guidelines issued by HEW, New York State, and the New York City Municipal Hospital System. They claimed the guidelines interfered with the rights of two women specifically: one a mentally retarded nineteen-year old, the other a woman about to have a third cesarean section, both of whom requested sterilization. In their own behalf, the doctors claimed infringement of their right to free speech, since they were mandated to discuss sterilization only in the context of other methods of birth control.[29] The obstetricians had carried their protest to the court, since their objections had been overruled by a vigilant public.

During the same period, a 19-year-old black woman, detained on a criminal charge at Rikers Island Prison, arrived for an abortion at King's County Hospital, largest of the municipals, and the one that had publicly refused to follow the regulations. She stated that she had been asked whether she desired pregnancies in the immediate future. When she said no, she was offered an operation for contraceptive purposes, which, she was told, could be reversed "when she became a normal citizen." She consented. Her uterus was removed during the operative procedure. This young woman is currently suing for gross malpractice. It was painful to see the need for enforcement of the guidelines through her suffering. The observance of just *one* single stipulation would have prevented that tragedy—consent cannot be obtained during admission for abortion.

We began to see that there were some critical problems to be solved. First, how could we implement and enforce the new regulations? Second, how could we establish a monitoring system to know about sterilizations on an ongoing basis? Third, how could we apply the guidelines to private hospitals, not just the municipal hospitals? At present, the medical schools contract with the municipal system to deliver medical services. Doctors work at both public and private hospitals and can carry out their programs at either place. Often doctors prefer the private setting as long as the fees for the services are forthcoming. We realized that guidelines could be circumvented simply by admitting Medicaid recipients or other insured patients to their private hospitals.

A fourth concern was the definition of "elective" in any surgical procedure. By specifying that the guidelines applied only to "elective" procedures, large loopholes were left for the "medically indicated." Doctors define "medical indications" on the basis of their experience or preference. Entrenched in their positions of power, doctors often resent any questioning. Attempts by patients to enter the sacrosanct areas of "medical indications" are invariably vigorously repelled.

A fifth and extremely important problem, mentioned previously, was the control of information. It is possible to sell a procedure by giving distorted accounts of its benefits and downplaying the risks. We strongly maintained that the mental, physical,

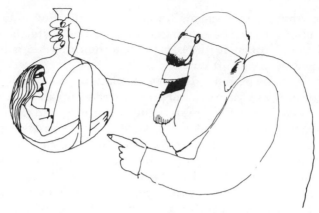

(*SftP*, July 1973)

and social hazards of sterilization should be discussed in informational materials. We felt these should be different from those of HEW, whose off-hand information can easily mislead.

Finally, the concern over the abuse of hysterectomy was still paramount. Since excessive hysterectomies are performed not only for reasons of money, custom, poor medical practice, and hostility to women, but also as an occult manner of sterilizing women without their informed consent, we recognized the need for guidelines on hysterectomies also.

These problems are as yet unresolved. Our approaches have been to continue to organize coalitions of people from within and without the hospitals in order to monitor what is happening and to continue to press in whatever ways are possible to have some impact upon these practices. For although one group of people managed to write and pass the guidelines, many more people are needed to see that they are honored.

The lessons from these battles have been invaluable. We have learned that we *can* organize coalitions of community groups and health workers, and that these coalitions can be effective in sharing information as well as applying pressure.

More important to me has been the experiences shared with women and community groups. We managed to identify some of the ways in which racist ideology keeps us from acknowledging our common oppression as women. Within the women's movement we sometimes found a denial of the experiences of others in statements such as "I had a hard time getting a sterilization five years ago. I can't see the need for a waiting period." And certainly it is true that in the not too distant past many middle-class women were denied sterilization by physicians. The issue became clear only when women understood that the same people who would deny a middle-class woman her request were the ones who were sterilizing working-class whites, blacks, Puerto Ricans, Native Americans, and Mexicans without ever bothering to obtain consent.

We examined social class attitudes of superiority which can lead to an acceptance of coercion of "others" such as welfare recipients, and we dealt with them in open discussion. We learned to identify our friends from the ranks of women, Third World people, health workers, and church groups. We likewise identified our opponents from the ranks of gynecologists, board members of the organizations dedicated to population control—which promulgate the "people pollute" ideology—and of those who favor a coercive society which oppresses people.

We also learned that there are many organizations which mask their ideology of population control by providing needed services in the areas of health, education, and family planning. These organizations are often linked to the large corporations and to a small number of private foundations in the United States.

In the process of study we analyzed the case of Puerto Rico. There, during the last 30 years the government with United States funding has sterilized over one-third of the women of childbearing age.[30] This was achieved by providing sterilizations free at a time when women were joining the workforce in large numbers. The lack of family support services, of legal and safe abortions, of alternative methods of contraception, and of full information about the permanency of sterilization have all combined to produce those startling numbers.

An analysis of the complex situation of Puerto Rican women showed us that there are many coercive factors in society which easily lead to sterilization abuse. Freedom of choice requires that there be real alternatives. We have deepened our understanding of the connections between the current denial of abortion rights to poor women, the dearth of child-care facilities, and the cuts in welfare and sterilization abuse.

We are now confident that we will halt sterilization abuse in New York City and that our example will serve as a model to groups of like-minded people who are springing up across the country to combat the same problem in their communities.

Epilogue

Since November 10, 1976, the date of this lecture, many important developments have taken place. My note of optimism on the effectiveness of coalitions to win protective legislation was justified. As consciousness developed, a movement grew to support further legislation in New York City. Public Law Number 37 was passed by the City Council with a vote of 38-0 in April 1977. The Law embodies the principles of the Guidelines on Sterilization of the New York City Health and Hospital Corporation and applies them to all New York City health facilities, both public and private. The law regulates sterilization of both women and men.

Provisions in summary are:

1. Informed consent in the language spoken or read by the person.
2. Extensive counseling to include information as to alternatives.
3. A prohibition of consent at time of delivery or any other time of stress, and of overt or veiled pressures on welfare patients.
4. The right to choose a patient advocate throughout the counseling or any other aspect of the process.
5. A 30-day waiting period between consent and procedure.

Support was broad and varied and included such groups as the Committee for Human Rights; Community Boards of Methodist and St. Luke's Hospitals; Committee of the United Neighborhood Houses of New York; New York City Coalition for Community Health; Committee to End Sterilization Abuse; New York Civil Liberties Union; Physicians Forum; Women United for Action; National Black Feminists Organization; New York National Organization for Women. Several other groups endorsed it in principle.

The opposition was primarily from the ranks of organized medicine and from the organizations with population control programs. The reasons often had to do with their specific interests. These are best described by Carter Burden, the City Councilman who introduced the legislation, when he commented on the role of one of these organizations in the following words:

> The principal, and certainly most effective, lobbyist against this bill has been Planned Parenthood, a distinguished and dedicated organization which we all have reason to respect. Many of its principals are old and admired personal friends, people I know too well to have any doubt about the sincerity of their concern or the honor of their motives. I do have some concern, however,

and I feel it should be said publicly, that Planned Parenthood's exaggerated fear about this bill has some relationship to the fact that they too will be directly affected by it. Some months ago, Planned Parenthood applied to the Health Systems Agency for authorization to perform female sterilizations in their outpatient clinics. That application was rejected by a special review committee because Planned Parenthood refused to adhere to certain criteria set down by the committee—including a 30-day waiting period.[31]

On the negative side are some currently unsolved problems. The Hyde Amendment, which withdraws federal funding of abortions, was upheld by the Supreme Court in June 1977. Besides the incalculable hardship this measure presents for women on welfare, who are condemned to bear unwanted children or to risk illegal abortion, which may lead to death, it opens the door to the sale of the "package deal" of abortion-sterilization. The federal government is willing to pay as much as 90 percent of a sterilization procedure, so for those institutions eager to get this fee, it becomes an additional incentive to coerce women who are desperate for an abortion by pressuring them to consent to sterilization. There is no doubt that the struggle for abortion rights is completely linked to the struggle against sterilization abuse.

FOOTNOTES

1. U.S. General Accounting Office, Report to Hon. James G. Abourezk, B 164031 (5), November, 1976, p. 3.

2. "Uri Charges I.H.S. with Genocide Policy," Hospital Tribune, 11, August, 1977.

3. U.S.G.A.O., op. cit., p. 4.

4. Allan Chase, The Legacy of Malthus: The Social Costs of the New Scientific Racism, New York, Alfred A. Knopf, 1977, pp. 15-16.

5. H. H. Laughlin, Eugenics Sterilization in the United States, Chicago, Psychopathic Laboratory of the Municipal Court, 1922, pp. 446-447, quoted in Chase, op. cit., p. 134.

6. Gena Corea, The Hidden Malpractice: How American Medicine Treats Women as Patients and Professionals, New York, William Morrow, 1977, p. 128.

7. Chase, op. cit., p. 16.

8. *Ibid.*, p. 6.

9. James E. Allen, "An Appearance of Genocide: A Review of Governmental Family Planning Program Policies," *Perspectives in Biology and Medicine*, 20, Winter, 1977, pp. 300-307.

10. "Physician's Attitudes: MDs Assume Poor Women Can't Remember to Take the Pill," *Family Planning Digest*, January, 1972, p. 3.

11. Chase, *op. cit.*, p. 2.

12. *Ibid.*, p. 6.

13. H. Curtis Wood, Jr., "The Changing Trends in Voluntary Sterilization," *Contemporary Obstetrics and Gynecology*, 1, 1973, pp. 31-39.

14. Denton Vaughn and Gerald Sparer, "Ethnic Group and Welfare Status of Women Sterilized in Federally Funded Family Planning Programs," *Family Planning Perspectives*, 6, Fall, 1974, p. 224.

15. Joann Rodgers, "Rush to Surgery," The *New York Times Magazine*, September 21, 1975, p. 34.

16. *Ibid.*,p. 40.

17. Jack Slater, "Sterilization: Newest Threat to the Poor," *Ebony*, October, 1973, p. 150.

18. Relf vs. Weinberger, 372 Federal Supplement 1196, 1199, D.D.C., 1974.

19. Slater, *op. cit.*, p. 152.

20. Joan Kelly, "Sterilization and Civil Rights," *Rights*, (publication of the National Emergency Civil Liberties Committee), September/October, 1977.

21. Claudia Dreifus, "Sterilizing the Poor," *The Progressive*, December, 1975, p. 13.

22. Robert E. McGarraugh, Jr., "Sterilization Without Consent: Hospital Violations of HEW Regulations: A Report by the Public Citizens' Health Research Group," January, 1975. Available from the Public Citizen's Health Research Group, 2000 P Street, Washington, D.C.

23. Elissa Krauss, "Hospital Survey on Sterilization Policies: Reproductive Freedom Project," *ACLU Reports*, March, 1975.

24. Carl W. Tyler, Jr., "An Assessment of Policy Compliance with the Federal Control of Sterilization," June, 1975. Available from the Center for Disease Control, Atlanta, Georgia.

25. U.S. Department of Health, Education, and Welfare, "Your Sterilization Operation, Hysterectomy," Washington, D.C., U.S. Government Printing Office, 1976.

26. New York City Health and Hospitals Corporation, "Why Sterilization Guidelines are Needed," 1975. Available from the Office of Quality Assurance, 125 Worth Street, New York, New York 10013.

27. Bernard Rosenfield, Sidney Wolfe, and Robert McGarren, "A Health Research Project Study on Surgical Sterilization: Present Abuses and Proposed Regulations," Washington, D.C., Health Research Group, October, 1973, p. 22.

28. *Ibid.*

29. Gordon W. Douglas, M.D., et al. and John L. S. Hollowman, Jr., et al., Civil Action File No. 76, CW 6, U.S. District Court, January 5, 1976.

30. Jose Vázquez-Calzada, "La Esterilización Femenina en Puerto Rico," *Revista de Ciencias Sociales,* 17, no. 3, San Juan, Puerto Rico, September, 1973, pp. 281-308.

31. Carter Burden, Testimony upon Introduction of Bill No. 1105, April 18, 1977, now Public Law No. 37. Available from the New York City Council, City Hall, New York.

GENERAL REFERENCES

Clara Eugenia Aranda, et al., *La Mujer: Explotación, Lucha, Liberación,* Mexico D.F., Editorial Nuestro Tiempo, S.A., 1976. Available from Avenida Copilco 300, Locales 6 y 7, Mexico 20, D.F.

Barbara Caress, "Sterilization," Health Pac Bulletin 62, January/February, 1975.

Barbara Caress, "Sterilization Guidelines," Health Pac Bulletin 65, July/August, 1975.

Linda Gordon, *Woman's Body: Woman's Right,* New York, Grossman/Viking, 1976.

Terry L. McCoy, et al., *The Dynamics of Population Policy in Latin America,* Cambridge, Massachusetts, Ballinger, 1974.

Bonnie Mass, *Population Target,* Brampton, Ontario, Canada, Charters, 1976.

Barbara Seaman and Gideon Seaman, *Women and the Crisis in Sex Hormones,* New York, Rawson Associates, 1977.

Asbestos, Science for Sale
—David Kotelchuck

This article discusses how industry has consistently obscured and controlled scientific evidence regarding asbestos and its relation to respiratory diseases, and how industrial scientists have manipulated data from studies to place the asbestos industry in a non-culpable light.

For almost a decade exposures of worker deaths due to asbestos have commanded newspaper headlines. In 1972 the U.S. government held hearings on a new asbestos standard for the workplace. Yet today the human cost of asbestos exposure remains a public scandal.

Despite this recent publicity the dangers of asbestos were discovered not in the 1960s, but back at the turn of the century. The first worker death due to asbestos exposure was diagnosed by a London physician in 1900.[1] His report lay interred in government records for over two decades.

What the general public did not know, the asbestos industry and the workers certainly did. In 1918 U.S. and Canadian insurance companies stopped selling personal life insurance policies to asbestos workers.[2] Also, many workers discovered the hazards of the job soon after being hired and quickly left.

Asbestos disease escaped notice by doctors, in part because its main effect was to exacerbate existing cases of tuberculosis or reactivate dormant cases. But perhaps a more basic reason was that the number of deaths was small, since the number of workers throughout the world was only a few thousand. The asbestos industry was still in its infancy in the early 1900s, with world production of asbestos in 1920 at only 200,000 tons, five percent of present production.

The industry began its rapid growth during the post-World War I construction and automobile booms of the 1920s. With this growth, inevitably, came an upsurge in worker deaths. The medical profession rediscovered asbestos disease in 1924, when Dr. W. E. Cooke reported in the British Medical Journal on the death of a 33-year-old woman from dust inhalation in an asbestos factory.[3] By the end of the 1920s British doctors had reported a total of 12 cases. What's more, in some instances asbestos disease was found at autopsy with no sign of tuberculosis, unequivocally implicating asbestos itself as the cause.

The "new" disease, called asbestosis, is caused by scar tissue forming around asbestos fibers trapped in the lungs. Its earlier symptoms appear mild—a slight persistent cough and shortness of breath upon exertion—usually developing about ten years after first exposure to the dust. If exposure continues, the disease can eventually lead to serious lung damage and death.

In the United States the first asbestos death was reported in 1930. By 1935 a total of 28 asbestos cases had been reported in Great Britain and the United States.[4] Industry had ignored all reports of asbestos disease in the past, but with the number of cases mounting it could no longer do so.

Corporate Strategy

During the 1930s, Johns-Manville, giant of the U.S. asbestos industry, began developing a strategy whose purpose was to insure the company's economic survival and profits. Begun in the midst of a major depression when competition was cutthroat, this strategy served the company well for more than 30 years.

The strategy developed on several fronts:

a. Become indispensable by building the company as rapidly as possible and weaving asbestos into the matrix of the economy.

b. Buy science, that is, fund medical research that would discredit reports of asbestos hazards.

c. Withhold information from workers, as much as possible.

d. Keep labor unions out of the plant.

(a) Becoming indispensable

The first imperative—to grow as rapidly as possible—was of

course common to all industry, and in this respect the asbestos industry succeeded phenomenally well. The engine of growth was the rapid development of literally thousands of new uses for the so-called magic mineral. For example, before World War I, transite (asbestos-reinforced concrete) water pipe had not yet been developed; today it is the single major use of asbestos. Asbestos insulation for ships came into widespread use during the ship-building boom of World War II, endangering several million shipyard workers. Today the estimated 3000 industrial uses for asbestos include products as varied as insulation for Apollo space rockets, roof shingles, siding, brake lining, clutch facing, linoleum, electric wire casing, draperies, rugs, floor tiles, ironing board covers, potholders and fireproof clothing.

With this boom, almost all of it taking place after extensive reports of asbestos hazards, Johns-Manville sales grew from $40 million in 1925 to $685 million in 1971, making it among the hundred largest U.S. corporations. Today the U.S. asbestos-manufacturing industry alone employs 50,000 people, asbestos-insulation workers in the building trades number 40,000 and an estimated 5 million people work daily with asbestos-containing products. As a result of this enormous expansion it is almost impossible, in terms of present political realities, to phase out nonessential asbestos production.

(b) Buying science

The second prong of industry's strategy was to buy scientific results that would refute the many case studies of asbestos deaths. In 1929 the Metropolitan Life Insurance Company was commissioned by the asbestos industry to conduct a study on asbestosis. Under the direction of Dr. A. J. Lanza,* Assistant Medical Director of Met Life, medical examinations were conducted on a total of 126 asbestos workers selected at random from five plants and mines in the U.S. and Canada, mostly Johns-Manville facilities. Sixty-seven of the 126 workers examined were classified by the insurance company doctors as positive cases of asbestosis, 39 as doubtful and only 20 as completely free of any sign of asbestosis. On their face

---

*Eventually Dr. Lanza became Director of the Bureau of Occupational Safety and Health (BOSH), the toothless federal predecessor to the present Occupational Safety and Health Administration (OSHA). Later he became chairman of the Department of Industrial Medicine at New York University Medical Center, and today the A. J. Lanza Institute for Environmental Medicine at NYU stands as his memorial.

these figures represent an epidemic of disease. Calculated as percentages, the findings showed 84 percent with some signs of disease (those classified as "positive" plus those classified as "doubtful") and only 16 percent with no signs of asbestosis at all. However, the authors did not publish these percentages. They simply listed the number of workers in each category and hurried on without comment. Short of suppressing the data, they could scarcely have done worse.

In addition to minimizing the incidence of disease, the authors also played down its severity. They dismissed workers' complaints of coughing and shortness of breath, typical early symptoms of asbestosis, with the responses "too much emphasis should not be placed on statements of subjective symptoms."

The U.S. government served as handmaiden to industry in this case by publishing the Met Life study as a Public Health Report of the U.S. Public Health Service. This gave the study the imprimatur of the federal government despite its origins in industry, its industry funding and its appalling pro-industry bias.[5]

Johns-Manville's other venture into medicine was its funding of animal studies at the Saranac Research Laboratory in upstate New York beginning in 1929. Although this work was continued for the next 25 years, it was of such poor quality that the National Institute for Occupational Safety and Health (NOISH) later deemed it of no use in setting an asbestos standard.[6] Nevertheless, industry was able to cite the work as evidence of its "long concern" about asbestos hazards.

In 1935 another asbestos-related disease appeared. Two doctors from the Medical College of South Carolina reported a possible link between asbestos and lung cancer.[7] By 1942 nine other case studies followed, showing that asbestosis victims suffer a high incidence of lung cancer.[8] Two scientists from Saranac, Arthur Vorwald and John Carr, dismissed the conclusions because, they argued, asbestosis victims might be especially susceptible to lung cancer.[9] What was clearly called for was a large-scale, plant-wide study, a so-called epidemiological study, in which workers employed at some particular date were followed for a period of years and all cases of disease recorded. But the hitch was that the asbestos companies had custody of the personnel records on which a study would necessarily be based, and they did not want the study to be conducted. In fact, it was not until 21 years later that the study was performed. In the interim the Vorwald-Carr paper was industry's "proof" that no link existed between asbestos and lung cancer.

(c) Keeping workers uninformed

Industry, which was resonsible for workers' health problems, withheld information about the hazards from their employees. For example, until recently the Johns-Manville medical staff denied workers access to their medical records. Furthermore, they refused to tell workers the results of physical examinations. Company spokespersons admit that until a few years ago the company did not tell workers that their respiratory problems were linked to asbestos. Joseph Kiewleski, an asbestosis victim, indicated that he was transferred without an explanation from a machinist's to a janitor's job after he had undergone a company physical examination. He found out from his own doctor years later that the reason for the job change was to remove him from the source of the exposure.

Moreover, company doctors in their cursory examinations of workers missed the most blatant diseases. In 1971, Daniel Maciborski was diagnosed to have cancer at the age of 49, a few weeks after he had been given a clean bill of health by the company. He died seven months later.

The company also tried to attribute occupational diseases to other causes. According to Dr. Maxwell Borow, a local doctor, "They claimed the workers had pneumoconiosis from mining coal in Pennsylvania." But ironically, after World War II the Manville N.J. plant had an influx of young veterans who had never mined coal and in fact had left Pennsylvania in part to avoid the black-lung disease that plagued their fathers.

(d) Asbestos-industry unions

The only part of the industry's strategy that was not successful in the period from 1930 to 1960 was its attempt to keep unions out of its plants. During this time most of the Johns-Manville plants were organized. But instead of having one or a few industrial unions at these plants, 26 different international unions were organized there, an almost certain guarantee that each would have a weak bargaining position witht the company.

After World War II

Industry's basic strategy, unchanged since the 1930s, began to unravel in the 1950s as a result of new medical reports of asbestos hazards. Individual case studies further linking asbestos and lung cancer kept accumulating.[10] Finally in 1955 a member of England's prestigious Medical Research Council analyzed government data on asbestos-industry deaths and found an unusually high rate of lung cancer among the workers.

In what for them was a lightning-fast response, the Quebec Asbestos Mining Association (QAMA) commissioned a study in the following year on lung cancer among Quebec asbestos miners.[11] This was 21 years after the first reports linking asbestos and lung cancer. What industry badly wanted was a whitewash job—and it got one.

The study was conducted under a QAMA grant by the Industrial Hygiene Foundation (IHF, now called the Industrial Health Foundation). IHF, located in Pittsburgh, performs occupational-health studies for corporations. It is openly pro-management and is supported almost entirely by major U.S. industries.

As with the studies of asbestosis, the contrast is striking between the enormous size and scope of the IHF experiment and that of the non-industry case studies—a fact that lent credibility to the former. The IHF investigation was an extensive epidemiological study of 6,000 asbestos miners from Quebec with five or more years of exposure.

All of this sounds impressive until one examines the IHF report itself.[11] Among numerous errors in method was one central, scientifically inexcusable flaw—the investigators, Daniel Braun and T. David Truan, virtually ignored the 20-year time lag between exposure to an agent known to cause lung cancer and the first visible signs of disease (the so-called latent period). They studied a relatively young group of workers, two-thirds of whom were between 20 and 44 years of age. Only 30 percent of the workers had been employed for 20 or more years, the estimated latent period for lung cancer. With so many young people in the study, too young to have the disease although they might well be destined to develop it, Braun and Truan of course did not find a statistically significant increase in lung cancer among the miners. As became obvious later, they had drowned out a clear danger in a sea of misleading data.

The practice of looking at a workforce with limited asbestos exposure is not an isolated error in a particular experiment; it is a hallmark of epidemiological studies funded by the asbestos industry. Even in the 1970s, researchers funded by industry continue to conduct studies on young workers despite scores of experiments by non-industry scientists showing that the various asbestos diseases take anywhere from 10 to 30 years to develop.

By 1960, medical research on asbestos was at a watershed. A total of 63 papers on the subject had been published in the U.S. and

Canada and Great Britain. The 52 papers not sponsored by industry, mostly case histories and reviews of case histories by hospital and medical school staff, indicated asbestos as a cause of asbestosis and lung cancer. The 11 papers sponsored by the asbestos industry presented polar opposite conclusions. They denied that asbestos caused lung cancer and minimized the seriousness of asbestosis. The difference was dramatic—and obviously dependent on the doctor's perspective, whether treating the victim of disease or serving as agent for its perpetrator.

The Lid Blows

In the early 1960s the research picture changed dramatically as a result of three separate studies. In 1960 a new malady was added to the lexicon of asbestos diseases: mesothelioma, a rare and invariably fatal cancer of the lining of the chest or abdominal cavity.[12]

In 1963 a study of lung sections from 500 autopsies of urban dwellers in Cape Town, South Africa—people who did not work in the asbestos industry and who died from a variety of causes—showed that 26 percent had asbestos bodies in their lungs.[13] These characteristic bodies are formed in the lungs around asbestos fibers and are commonly observed in the lungs of asbestosis victims. Similar studies, with similar results, were later found in major cities throughout the world. The findings received widespread publicity and raised the specter of asbestos as a modern environmental hazard affecting all citizens.

To top this off, in the early 1960s, Dr. Irving Selikoff and his associates at Mt. Sinai Medical Center in New York broke industry's hegemony over medical and personnel information by using the welfare and retirement records of the asbestos insulators' union as the basis for conducting an epidemiological study. Now for the first time in the U.S., scientists not beholden to industry conducted large-scale definitive studies on groups of asbestos workers. Beginning in 1964 the investigators reported an unusually high incidence of lung cancer and mesothelioma among asbestos-insulation workers, with time lags of 20 and 30 years, respectively, between exposure and disease.[14] By focusing on workers who were first exposed 20 or more years earlier, the studies highlighted its hazards. Together with the South African studies they made the "magic mineral" front-page news throughout the world.

The asbestos industry responded to these reports by spending $8.5 million on research and development in 1972, a large fraction of which went to outside medical research centers.[15] In contrast, the National Institute for Occupational Safety and Health (NIOSH) spent a mere $260,000 on asbestos research grants that year.[16]

As a result, an industry that had only managed to generate 11 research papers on asbestos in the three decades before 1960 has come up with 33 in little more than a decade since then. The recent studies are just as self-interested as ever. Industry has stopped denying that asbestos causes lung cancer, mesothelioma and asbestosis. But research proposals that industry thought would (a) minimize the problem or (b) shift the blame have been given' unstinting support.

(a) Minimizing the problem

A major epidemiological study was published in 1971 by J. Corbett McDonald and his associates at the Department of Epidemiology and Public Health at McGill University in Montreal. It was funded through a grant from the Institute of Occupational and Environmental Health of the Quebec Asbestos Mining Association.[17] The subjects were 11,000 miners in the two largest asbestos mines in Quebec.

Like the earlier IHF study on asbestos miners, this one looks quite impressive until it is examined carefully. Then we find as before that the workforce studied has had relatively limited exposure, and that many other serious methodological errors were made.

Let us consider the duration of exposure of the workforce. The research data shows that many of the miners included in the study worked in the mines for only a short time and then left. One-third of the miners in the study had worked less than a year in the mines, two-thirds had worked less than 10 years. So it is not surprising that their mortality was not much different from that of the general population. The authors go even further. They begin their comments on the results with the observation that workers in the asbestos-mining industry have "a lower mortality than the population of Quebec of the same age."

What's more important, the authors largely ignore the latent period between exposure and disease for lung cancer. They do not categorize workers by number of years since first exposure, which would highlight any latency effect. Workers with recent exposures, more recent than the 20-year latent period for lung cancer, are included in the study and may be placed in the same categories as those who have been exposed many years earlier.

In contrast, Selikoff and associates at Mt. Sinai in their earliest experiments only looked at workers with 20 or more years of work experience since first exposure.[18] Thus they focused their attention on precisely that group of workers most likely to develop disease, and thereby found evidence of serious hazards.

In fact, studies not supported by industry have consistently found asbestos to be a serious health hazard. While Braun and Truan, and McDonald found no increase in mortality rate due to asbestos or only small increases up to 20 percent, studies not financed by industry reported an increase in mortality rate among asbestos workers of from 200 percent to 9,000 percent above that of the general population.

(b) Shifting the blame

Industry has also gained time by shifting the blame. Pro-industry scientists have concocted one theory after another to prove that asbestos workers and their families were not dying from asbestos but from some impurity, some contaminant or some unusual type of asbestos.

One of the early theories, by Dr. Paul Gross of the Industrial Hygiene Foundation, was that trace metals were contaminating asbestos and causing the diseases attributed to asbestos. This work was supported by industry for six years until Gross and supporters finally had to admit that the theory was incorrect.

Another theory was that certain types of asbestos fiber are dangerous, while others are safe. Ninety-five percent of the asbestos used in the U.S. and Canada is of one type, chrysotile. Since the bad-fiber theory has its origins in industry-sponsored research, it comes as no surprise that fiber types other than chrysotile have been blamed for asbestos disease.

Probably the ultimate diversionary tactic was the theory propounded by Gibbs of McGill University and funded by the Quebec Asbestos Mining Association, that the polyethylene bags in which asbestos is stored produce oils that contaminate the asbestos and might cause the cancer associated with asbestos.

Whether or not industry has lost these battles, the eventual outcome of each is less important than the fact that each salvo has tied up scientific resources, defined research issues and bought time. In the case of almost every industry proposal, some non-industry scientists have had to conduct experiments in rebuttal, using up some of the meager resources in the process.

Sitting on the Victims

While industry was mounting its medical and scientific counterattack, it had to deal with asbestos victims and their families. Many of the victims' dependents filed suits against the company. To keep things quiet, Johns-Manville usually settled out of court. The average settlement in the mid-1960s was $10,000. In recent years the company has increased the settlement for mesothelioma victims. It now pays the deceased's hospital bills, as well as half the victim's salary for the rest of the surviving spouse's life. Asbestosis victims have fared even more poorly. In 1970 the awards for Johns-Manville's asbestosis victims averaged $2,175.

Recently, workers and their families have begun to institute large damage suits against individual companies. In California an asbestos worker won $351,000 in damages from a company physician who withheld information that he had developed asbestosis. This year in Paterson, New Jersey, the families of a number of workers who died of asbestos exposure sued the Raybestos-Manhattan company and its suppliers (Johns-Manville among others) for damages of $326 million. While suits are filed after the fact of disease and death, the plaintiffs have often expressed the hope that the suits' financial impact may be great enough to cause a major cleanup throughout the asbestos industry. However, U.S. courts have been notoriously unfriendly to labor in the past and seem a weak reed to lean on.

While victims and their families were trying to deal individually with the company, union locals such as the one in the Manville plant were slow to take any initiative on asbestos hazards. In 1970, for the first time in ten years, the union struck, crippling the plant for almost six months. The major concerns were bread and butter issues, but a vocal minority of younger workers began to raise questions about their health. When the strike was settled, the company agreed to permit workers access to their x-rays. As a "preventive measure," J-M also consented to establish a joint union-management environmental control committee. Union officials publicly proclaimed the committee a great victory.

But the company quite independently of this union-management committee had begun a major cleanup of its plants in the late 1960s, presumably in expectation of stiffer government regulations in the near future. Throughout all of its plants Johns-Manville lowered dust levels by eliminating many intermediate steps in the production process, enclosing or bettering ventilation in some areas, and improving housekeeping procedures. In the textile division of the Manville plant, for example, steps have been eliminated from carting, spinning and warping, according to officials who conducted a recent plant tour. Manville executives are elated. "By eliminating steps we don't need, we also save money," one engineer boasted. And, he might have added, the company cuts labor costs and improves productivity.

What happens to those whose jobs are eliminated? They are "absorbed in other parts of the plant," according to Wilbur Ruff, community Relations Director at the Manville plant. But that's not the whole story. J-M has cut its Manville work force in recent years mostly by attrition—that is, by not replacing many retirees and others who leave the plant. In the six-year period during which J-M was reducing dust levels, the nonsalaried work force at the Manville plant dropped almost 45 percent, from 3,200 to 1,800 employees, according to Ruff. The working people of Manville have exchanged jobs for improved health conditions at the plant.

The 1972 Asbestos Hearings

But however much the company was in control of events within its plants, constant publicity about scientific studies that demonstrated asbestos hazards took its toll.

Following passage of the federal Occupational Safety and Health Act in 1970, major attention was focused on asbestos hazards. OSHA held hearings to establish a new asbestos standard in 1972. At the hearings Dr. George Wright of St. Luke's Hospital in Cleveland, Johns-Manville's chief science advisor, was able to call on five studies supporting J-M's contention that the standard of five asbestos fibers per cubic centimeter should be maintained, not lowered. Of the five studies, four had been funded by the asbestos industry.

These studies helped put a "scientific" cover over industry's interests. Industry could not prevent the asbestos standard from being lowered to two fibers per cubic centimeter, but it contributed to a four-year delay in its effective date. Thus, the corporations had won precious time to regain their initiative in the struggle. For workers too, the time lost was critical. Dr. Selikoff estimates that this delay eventually will take as many as 50,000 lives.

But even when the fiber limit comes down, the battle is not over yet, not by a long shot. The 1972 NIOSH report on asbestos bases its two-fiber recommendation primarily on the British standard. This standard is now under question in England because the experiment it was based on appears to have underestimated the extent of disease. Also, whatever level is set, there is no known safe level of exposure for any cancer-causing agent, according to officials at the U.S. National Cancer Institute. Thus, public discussion about setting a legal exposure level in plants is largely based on a false premise—that a safe level of exposure exists.

Johns-Manville After 1972

Since the 1972 hearings, industry's decades-old strategy has changed. For example, instead of forging ahead with development of new uses for asbestos, Johns-Manville has for the first time seriously decided to diversify. It is planning to develop a major outdoor-recreation center in Colorado, and it is, with consummate audacity, selling environmental-control products and services to other companies based on the experience in its own plants. In fact, environmental controls have been extolled by the *Wall Street Transcript* as one of the company's "hottest growth areas." The corporation is also studying the use of fiberglass as a substitute for asbestos.

While Johns-Manville's stocks have gone down in recent years, company sales went up and its profits are steady. In 1972, *Value Line Survey* called Johns-Manville "the picture of financial health." To be sure, U.S. sales of asbestos are down, but Johns-Manville has been pushing its foreign sales and these have more than made up for domestic losses.

Not only have foreign sales increased, but the asbestos industry has also been exporting jobs especially those in the dusty asbestos-textiles trades. Since 1968 asbestos textile imports from countries with weak or nonexistent occupational-health laws have increased from 0.1 percent to a whopping 50 percent of the total U.S. imports. Mexico's asbestos-textile exports to the U.S. rose from a mere 180 pounds in 1969 to 1.2 million pounds in 1973. During the same period Taiwan's exports rose from 0 to 1.1 million pounds and Brazil's from 0 to 0.5 million pounds. Twenty-one of Mexico's 23 asbestos-processing plants have been built since 1965, a number of them by U.S. firms.[19] Thus, industry has taken operations that would be difficult and expensive to clean up and has, with full knowledge of the consequences, exported them abroad to maim and kill foreign workers.

Why have workers and medical scientists friendly to them

been unable to turn the tide against the asbestos companies, who seem to remain in control of the situation?

In the case of the workers, the reasons are quite clear. Short of closing down asbestos plants, the best solution would be automation of the production process, thereby removing workers from the exposure. While this appears technically feasible, in the present society it would be a disaster. Virtually all production workers, such as those at Manville, would lose their jobs and would be left to their own resources to find new ones. As one worker commented, "I am 52. I been workin' at J-M 27 years. Who would hire me? Where else could I go?"

A planned and people-oriented system could find alternative jobs for displaced workers. Then automation could be, in the fullest sense of the term, life-saving. But our society does not have this commitment. Instead, it discards people when they are no longer economically useful—as it has done to miners and aerospace workers, for example. Thus workers continue to be forced into the no-win "choice" between their jobs and their lives. No wonder that they have been afraid to push for strong health and safety measures.

Scientists: pro-worker and pro-industry

Worker-oriented scientists have played an important role in the asbestos struggle, but their failures were critical ones that might have changed the situation decisively. For example, had scientists and doctors who first found evidence of asbestos disease brought it directly to workers in plants, workers might have been much more willing to take on the company despite great odds, and the new information would have armed them in their effort. In fact, in the late 1920s and early 1930s, when the asbestos industry was still small, if medical people had not limited themselves to operating within narrow professional roles, the expansion of the asbestos industry might have been nipped in the bud and thousands of lives might have been saved.*

---

* A positive example of the importance of bringing medical information to workers is provided by the coal-miners' black-lung struggles. During the early 1960s scientists and medical people travelled to union and community meetings throughout Appalachia reporting to workers results of their studies and discussing with them the dangers of black lung. Later analyses have show that these scientists played an important role in the development of black-lung struggles.

The interaction between workers and scientists need not be one-sided. Looking at the history of research on asbestos disease, it becomes clear that a decisive scientific turning point took place in the early 1960s when the asbestos unions turned over their retirement and death-benefit records to Dr. Selikoff and thereby, for the first time, allowed non-industry scientists to examine the records of all workers in an industry. This provided critical new information on asbestos health hazards and helped overcome objections that earlier studies had focused on individual asbestos victims.

Today industry still supports individuals whose scientific practice demonstrates built-in biases useful to industry. These biases usually are the result of scientific and social values rather than dishonesty or conspiracy on the part of scientists. A critical examination of industry-funded asbestos research does not reveal overt falsification of data. In fact in many of the large-scale industry experiments (for example the Metropolitan Life study in 1935 and the McDonald studies in the 1970s) data indicating asbestos dangers is circumspectly presented in the reports themselves.

It is clear that the critical difference between industry- and worker-oriented research does not lie in the experimental methods employed, but in the questions that scientists try to answer and in the assumptions made when they analyze and present their data. If a scientist suspects that workers are often harmed on the job, he or she will adopt this as an implicit hypothesis and will focus attention on older, heavily exposed workers who are more likely to show signs of disease. Data will be presented that is designed to illuminate the hazard to this group of workers, and summary and conclusions will typically begin with a statement about the most serious hazard uncovered by the study.

On the other hand, a scientist who designs a study with the assumption that workers are not often harmed on the job is more likely to study a much larger, more heterogeneous group of workers in a given plant or industry. In this case, data will be presented that lumps workers together into a single group, a step which tends to bury the effect of unhealthy subgroups within the larger group. Summary and conclusions will usually open with a statement about the similarity in the mortality pattern of the entire group of workers to that of the general population. A comparison of the papers of Selikoff and McDonald, for example, illuminates these differences clearly.

Pro-worker scientists have much to learn from studying and understanding these differences. Some of our main tasks are to learn how to frame research questions in a pro-worker format and then proceed to definitely answer them. Formal scientific and technical education can help us a great deal with the latter task. But we must learn for ourselves how to ask the right questions in occupational health, since we are unlikely to acquire this skill at the university. At a later stage of the research process we must take the responsibility of sharing our findings with the workers whose health is affected.

Throughout, we must acknowledge and be aware of economic and political realities. If we do not, we may, as in the case of asbestos research, win the battles to discover truth and lose the war to save human lives.

Postcript (1978): In 1974 Johns-Manville hired Dr. Paul Kotin, former Director of the National Institute for Environmental Health Sciences (a member institute of the prestigious National Institutes of Health), to be its new corporate medical director. This was a personnel coup for the company, for Kotin was widely known and respected as a scientific figure and is one of the most prominent scientists in many years to leave academia and/or government service for a corporate medical position.

Since then J-M has finally admitted that asbestos workers suffer asbestosis, lung cancer and mesothelioma. But, in the spirit of its past, J-M and Kotin insist that with lower dust levels in the company's plants, asbestosis is no longer a threat to workers' health. Time and workers' lives will tell if this is so.

As for lung cancer, the corporate position, as expressed by Kotin at the 1977 Denver conference of the American Academy of Occupational Medicine (AAOM), is that "among asbestos workers, smoking causes lung cancer." To be sure, most asbestos workers who have died of lung cancer were cigarette smokers. But to conclude at this time that smoking alone *causes* lung cancer among asbestos workers is, at best, scientifically premature—and, of course, socially self-serving. Johns-Manville's "solution" is to ban smoking in its plants wherever state law allows, Kotin announced at the AAOM conference. By now, further studies substantiate a higher risk of lung cancer for non-smoking asbestos workers compared to the general non-smoking population, although the rate of lung cancer for the non-smoking workers is much lower than for workers who smoke.

Mesothelioma is being explained away much the way cancer is, by jumping to conclusions prematurely and probably incorrectly. Recent medical studies have shown, not surprisingly, that there is a gradation in the lethal effects of different crystalline types of asbestos fiber in causing mesothelioma, and that crocidolite (so-called blue asbestos), found mainly in South Africa, is the most dangerous type. From this the company argues that crocidolite *alone* causes mesothelioma, and that mesothelioma deaths in the U.S. were caused by the temporary use of South African asbestos in the U.S. during World War II and will soon stop so long as Canadian chrysotile asbestos continues to be used in this country as at present. Meanwhile an increasing number of mesothelioma deaths without any recorded crocidolite exposure continues to be documented.

While moving aggressively on the medical front, Johns-Manville has also been attacking workers from another direction by blaming them for the diseases to which they have fallen victim. J-M has been pushing hard recently to revive the concept of "hyper-susceptibility," that workers develop disease because of their unusual susceptibility due to genetic defects or social deprivation during early childhood. At the 1977 AAOM conference, Kotin said that many asbestos workers and coal miners were "clearly cases of hypersusceptibility." This by the representative of an industry where among manufacturing workers almost half of all workers die of asbestos-related diseases!

So the struggle to protect the health and lives of asbestos workers continues. The front is constantly shifting as new scientific and medical information develops. What remains constant is industry's effort to protect its own interests—and profits—despite the vast human costs to asbestos workers.

FOOTNOTES
1. H. M. Murray in *Charing Cross Hospital Gazette*, London, 1900; later published in *Report of the Departmental Committee on Compensation for Industrial Disease*, London, H. M. Stationary Office, 1907, p. 127.
2. National Institute of Occupational Safety and Health, *Occupational Exposure to Asbestos: Criteria for a Recommended Standard*, Washington, D.C., U.S. Government Printing Office, 1972, p. III-4.
3. W. E. Cooke, "Fibrosis of the Lungs Due to the Inhalation of Asbestos Dust," *British Journal of Medicine*, 2, 1924, p. 147.

4. D. S. Egbert, "Pulmonary Asbestosis," *American Review of Tuberculosis*, 32, 1935, p. 25.

5. A. J. Lanza, et al., "Effects of Inhalation of Asbestos Dust on the Lungs of Asbestos Workers," *U.S. Public Health Reports*, 50, No. 1, 1935.

6. National Institute of Occupational Safety and Health, *op. cit.*, p. III-12.

7. K. M. Lynch and W. A. Smith, "Pulmonary Asbestosis," *American Journal of Cancer*, 24, 1935, p. 56.

8. H. B. Holleb and A. Angrist, "Bronchiogenic Carcinoma in Association with Pulmonary Asbestosis," *American Journal of Pathology*, 18, 1942, p. 123.

9. A. J. Vorwald and J. W. Karr, "Pneumoconiosis and Pulmonary Carcinoma," *American Journal of Pathology*, 14, 1938, p. 49.

10. E. R. A. Merewether, *Annual Report of the Chief Inspector of Factories*, London, H. M. Stationary Office, 1947, and S. R. Gloyne, "Pneumoconiosis," *Lancet*, 1, 1951, p. 810.

11. D. C. Braun and T. D. Traun, "An Epidemiological Study of Lung Cancer in Asbestos Miners," *Archives of Industrial Health*, 17, 1958, p. 634.

12. J. C. Wagner, et al., "Diffuse Pleural Mesothelioma and Asbestos Exposure," *British Journal of Industrial Medicine*, 17, 1960, p. 260.

13. J. G. Thomas, et al., "Asbestos as a Modern Urban Hazard," *South African Medical Journal*, 37, 1963, p. 77.

14. Reviewed in I. J. Selikoff, et al., "Cancer of Insulation Workers in the United States," Lyons Conference on Asbestos, 1972.

15. Johns-Manville Corporation, Report to the U.S. Securities and Exchange Commission, form 10-K, Washington, D. C., 1973, p. 6.

16. NIOSH Contract and Research Agreements, Washington, U.S. Public Health Service, September, 1972, pp. i-vi.

17. J. C. McDonald, et al., "Mortality in Chrysotile Asbestos Mines and Mills of Quebec," *Archives of Environmental Health*, 22, 1971, p. 677.

18. Reported in *Lifelines: Oil, Chemical, and Atomic Workers Union News*, April, 1975. For a recent confirmation and expansion of this report, see remarks of Hon. David Obey, (D-Wisconsin) in the *Congressional Record*, June 29, 1978, pp. E3559-61.

19. Castleman, Barry, "The Exportation of Hazardous Industries to Developing Countries," in *International Journal of Health Services*, Vol. 9, no. 4, 1979, Baywood Publishing Co., Long Island City, NY, pp. 569-606.

Recombinant DNA
—Jeremy Rifkin

This article points out that despite the questions, debate and controversy surrounding the safety of recombinant DNA research, industries have already begun production of substances via this process. It makes clear the relationship between science and entrepreneurship in the hottest field of scientific research today.

On September 22, 1976, President Ford sent a memorandum to the heads of all major federal departments and agencies. Although it received no attention at the time, the President's memo may be remembered years from now as the keynote to one of the most significant developments of the 20th century. The memo concerned the formation of a new Interagency Committee of the federal government. Its mission: "TO REVIEW FEDERAL POLICY ON THE CONDUCT OF RESEARCH INVOLVING THE CREATION OF NEW FORMS OF LIFE."

That Interagency Committee has since convened at the National Institutes of Health (NIH) in Bethesda, Maryland, on November 4 and November 23, for a total of five hours and 30 minutes. There were no TV crews present, no photographer to shoot pictures of the proceedings for the record book. UPI and AP didn't even list the meeting on their daily calendars of important events to cover in Washington. Both sessions were conducted behind closed doors.

With these meetings, just three years after molecular biologists had succeeded in separating and recombining the DNA molecules that carry the genetic code for all living beings, unlocking for the first time the secret of creating life itself, the United States officially entered the Organic Age.

Up to now, human beings have been engaged in a constant battle against the elements. We have used our wits to harness the resources of the external world for our own survival. Ours was a finite reality. At best, we could manipulate and exploit parts of our universe for our own ends; at worst, we could destroy those parts of the outside environment that threatened our well-being. In short, our limits were established by what already existed. Within the past 30 years we have approached the outer limits of that finite world of matter and energy with the splitting of the atom, our entry into the Nuclear Age.

Yet now, even as we still grapple with the nuclear demon, a new phenomenon has emerged, having to do with the world of life itself. With the unlocking of the secrets of DNA, we will eventually be able to change the cellular structure of living beings and to create entirely new species. Biologists are already doing it with microorganisms. The Nuclear Age was the age of the physicist; the Organic Age is the age of the biologist.

At this moment, microbiologists are at work in more than 180 separate laboratories across the country, busily spending more than 20 million dollars in government grants in pursuit of the creation of new forms of life. They are experimenting with so-called recombinant DNA. By now most newspaper readers have heard of the controversy surrounding DNA research at universities. But, sheltered from the glare of publicity that bathes every new debate at Harvard or Stanford, something much more ominous is happening. Today seven major drug companies are engaged in, or about to begin, recombinant-DNA research. The companies will soon apply for patents on the new forms of life they are developing. In time this research will translate into an unparalleled commercial bonanza for the pharmaceutical, chemical and agricultural companies as they introduce literally dozens of new-lifeform products into the market place.

General Electric is already out in front with the announcement that it has applied for a patent on a tiny microorganism that can eat up oil spills.

While the commercial prospects for this new technology have whetted corporate appetites, the potential dangers in its further development and application—although some of them are still years off—pose perhaps the single greatest challenge to life that humankind has ever faced.

That challenge is made more awesome by the fact that virtually any amateur biologist can obtain the enzymes necessary to experiment with new life forms. Miles Laboratory, which markets the enzymes, admits that most of its enzyme sales are done through the mail, and that there are no "guarantees of what the customer will do" once that person receives the biological materials.

How does one even begin to look at a technology that could eventually lead to the creation of new plants, animals and even the alteration of the human species?

And then there is a more immediate question before us as we enter the Organic Age: should our present corporate system be used as the developing and marketing process when life itself is the product?

Recombinant DNA is a recently developed technique that recombines DNA segments (the basic material determining the hereditary characteristics of all life) from two different organisms. Scientists became able to do this when they found that DNA segments had "sticky" ends that, under proper laboratory conditions, could be fastened to another organism's DNA segments. Thus is formed the genetic basis for new living and multiplying organisms that do not exist in the natural evolutionary order.

Although most scientists agree that recombinant DNA is one of the most important scientific breakthroughs of modern history, they violently disagree as to whether the potential benefits of even the most restricted experimentation outweigh the grave potential dangers to human life and the environment.

Paul Berg, a prominent recombinant-DNA researcher at the Stanford University School of Medicine, believes experimentation in recombinants could result in the creation of major new food crops that can obtain nitrogen from the atmosphere rather than from fertilizer; a new form of medicine, gene therapy, to treat crippling genetic diseases; and such things as cheap and efficient production of vitamins, antibiotics and hormones.

On the other hand, scientists like Liebe Cavalieri of the Sloan-Kettering Institute for Cancer Research argue for a complete moratorium on recombinant-DNA research until the long-range implications are fully discussed. Cavalieri points out that such research "involves many unknown factors beyond the control of the scientist." According to Cavalieri, "it is necessary to create vast numbers of cells with unknown genetic alterations in order to obtain a cell containing a specific recombinant DNA." He continues: "The probability of creating a dangerous genetic agent in the process is real, and there is no way to test for the danger. The scientist does not know what he has done until he has analyzed the newly created cell—at which point it may be too late." At that point, science fiction's most horrible scenarios become fact.

Cavalieri and his colleagues are deeply concerned over the possibility that a new Andromeda-type virus, for which there is no known immunization, might accidentally be developed in a laboratory somewhere and spread a deadly epidemic across the planet, killing hundreds of millions of people. They also fear that a new, highly resistant plant might be developed that could wipe out all other vegetation and animal life in its path. Dr. Robert Sinsheimer, who chairs the Biology Division of the California Institute of Technology, warns that "the invention and introduction of new self-reproducing living forms may well be irreversible." Sinsheimer asks: "How do we prevent grievous missteps, inherently unretraceable?"

Sinsheimer got the first tentative answer to his question last summer (1974) when the City Council of Cambridge, Massachusetts, voted to prohibit work on recombinant DNA at Harvard and M.I.T. pending further public investigation. This unparalled public restriction of scientific research focused, for the first time, the attention of the national media on the question of the creation of new life forms.

But even as scientists, public-interest groups, social commentators and the media continue to rehash the implications of the Cambridge incident, another development, almost totally ignored in the press, may be far more significant than any event that has taken place in academia: the entry into the field of DNA research by the private corporation.

On September 23, 1976, a little-noticed *Washington Post* story reported: "U.S. health officials acknowledged that the government does not know what companies are trying to create revolutionary new forms of life or the whereabouts of their laboratories." Dr. Bernard Talbot of the Office of the Director of NIH told us that "as of now, there is no federal agency that is looking at research being done by private industry in recombinant DNA....We have no registry [of companies involved in this field]."

It is probably true that NIH and other federal agencies are unaware of the specific *nature* of the research going on in private industry. However, governmental authorities *do* know what companies are involved in this research and where their plants are located, but they have not been willing to make this information public.

At present, seven major pharmaceutical companies are now engaged in or about to be engaged in secret recombinant-DNA research (see the list in the box on p. 150). Nine other corporations involved in drugs, chemicals and agricultural products are now looking into the potential application of recombinant DNA. They are: Cetus Corporation, CIBA-Geigy Corporation, DuPont, Dow, W.R. Grace & Company, Monsanto, French Laboratories, Wyeth Laboratories and Searle Laboratories.

The almost air-tight secrecy surrounding this particular research, says *Medical World News*, is "reminiscent of the atmosphere surrounding biological-warfare research a few years ago."

Tom Craig, the public-relations representative for Abbott, said that his firm has no intention of informing the general public about Abbott's activities, "because it's often difficult to obtain an understanding of what is being done. It creates more alarm than is justified."

At Upjohn, public relations chief Joe Haywood even tried, at first, to deny that the firm had any role in recombinant-DNA research. Only when confronted with hard evidence did he finally admit to Upjohn's involvement. At Roche Institute, an assistant to the director who identified himself as Dr. Bartle refused any comment when asked when the new maximum-control facilities would be operational and how many researchers would be involved. Similar responses were invoked right down the line in interview after interview with various company officials across the country.

It's not surprising, then, that a check with public officials in Rochester, Kalamazoo, South Bend, and Nutley revealed that none were aware of secret research into recombinant DNA going on in laboratories in their communities. In Kalamazoo, Michigan, Mayor Francis Hamilton pointed out that, while the Upjohn laboratory was "within three blocks of where I'm sitting," he had not been informed by the company that it was involved in recombinant-DNA research. In New Jersey, Dennis Helms, in the Attorney General's office (who is already in charge of an investigation into recombinant DNA in his state), was asked if he knew of any firms doing any P-4 level recombinant-DNA research. ("P-4" refers to maximum-risk research controls.) Helms said it was his understanding from people in the industry that NIH, in Bethesda, Maryland, has the only P-4 facility in the country. Helms was then

CORPORATIONS IN DNA RESEARCH

Company	Location of Laboratory	State of Research
Miles Laboratories (Dr. Robert Erickson, Dept. of Science Information and Communication Services)	Rochester, NY, and South Bend, IN (Research under contract to universities)	Ongoing
Eli Lilly & Company (Dr. Cornelius W. Pettinga, Executive Vice President	Indianapolis, IN	Ongoing
Hoffman-LaRoche (Dr. Sidney Udenfriend, Director, Roche Institute)	Nutley, NJ	Ongoing
The Upjohn Company (Dr. Joe Grady, Section Head, Infectious Diseases)	Kalamazoo, NJ	Ongoing
Merck, Sharpe & Dohme Research Laboratories (Dr. Jerome Birnbaum, Executive Director, Basic Biological Sciences)	Rathway, NJ	Tooling up
Pfizer, Inc. (Dr. John DeZeeuew, Research Scientist)	Groton, CT	Tooling up
Abbott Laboratories (Dr. Lacy Overby, Director Experimental Biology)	North Chicago, IL	Tooling up

asked if the Hoffman-LaRoche Company had informed the Attorney General's office that they were constructing a P-4 facility in Nutley, New Jersey. The answer was no.

Even though NIH (the agency responsible for overseeing the Inter-agency Committee) continues to assert that it has no "official" knowledge of research going on in the private sector, its director, Dr. Donald Frederickson, initiated a meeting nine months ago (on June 2, 1976) with representatives of 20 U.S. corporations to ascertain their interests and needs regarding research into recombinant DNA. At a meeting held at NIH headquarters on December 3, 1976, Frederickson himself said: "It is essential there be a way the industrial technology of this country can take advantage of this."

This cozy behind-the-scenes relationship between industry and government officials isn't hard to understand once one looks into the backgrounds of the officials involved. A number of the consultants to or members of the NIH group that drew up the government's regulatory guidelines on DNA research have industry ties—among them Dr. Ernest Jaworski of the Monsanto Company and Dr. Louis G. Nickell of W.R. Grace & Company. More important still, of the fifteen members of the key Interagency Committee for whom we were able to obtain background biographies, seven had previously been employed with major U.S. corporations. Two of these had served with major pharmaceutical companies now involved in recombinant-DNA work. Oswald Ganley, the State Department representative to the committee, was previously employed as Assistant Director of International Relations at Merck, Sharpe and Dohme Laboratories; and Department of Transportation representative William D. Owens was at one time a director of a subsidiary of the Searle Corporation.

This industry-weighted committee on recombinant DNA was scheduled to present its recommendations on the "creation of new life forms" to the President in mid-January, although it may be weeks or months before the report's contents are made public. The report is expected to cover several major items:

• **How Will Dangers Be Contained?** When the federal government started doing research some years ago on infectious diseases afflicting cattle and other animals, it set up an Animal Disease Center on an uninhabited island off the easternmost tip of Long Island. The lab is windowless. A barbed-wire fence surrounds the entire island. Under the fence, three feet of buried concrete prevents rodents and insects from burrowing through to the facility. Every one of the

Center's workers gets an elaborate security check by the FBI before being allowed access to the Center. Employees wear sterilized garments and work under air pressure lower than outside, and all are required to shower between each experiment. Security guards protect the premises from unauthorized trespassers. Researchers and government officials are taking every precaution possible to make sure no germs that could infect U.S. farm animals escape the island.

By contrast, NIH's own maximum-security DNA-research facility is a mobile trailer parked off a side street outside the agency's office in Bethesda, Maryland, a residential suburb of Washington. It is protected only by a simple seven-foot cyclone fence. According to NIH officials, the mobile trailer is not yet operational because they are still fixing its leaky roof. Some secretaries and other office workers at NIH are quite frightened about the trailer parked outside their offices. One of them called the environmental organization Friends of the Earth, pleading for help. "The employees are all talking about this P-4 trailer and we're scared to death," she said. "Can you do something? Anything."

The set of guidelines the government has drawn up for recombinant DNA research, although elaborate, are nowhere near so tight as the security precautions at the Animal Disease Center. True, there are the so-called physical-containment provisions, requiring laboratory air to be kept under low pressure, researchers to take showers, and so forth. There are also a series of biological-containment requirements, which mandate that certain experiments be done with weakened strains of bacteria that theoretically could not survive outside the laboratory.

Nonetheless, these labs are usually located in populated areas, and the government does not have the staff to police them and to make sure the physical- and biological-containment requirements are met.

Most recombinant DNA experiments are done with E. coli bacteria, which exists in the intestinal tracts of all human beings. The chief danger involved here is that a research accident could produce a particularly virulent virus that causes a disease for which there is no immunization. A lab technician who accidentally breathed or swallowed a few of its particles could then begin rapidly spreading the virus to others—perhaps eventually to a whole population. "Only one accident is needed," biologist Cavalieri says, "to endanger the future of mankind."

Sinsheimer of Cal Tech believes that the government guidelines are insufficient. In a letter to NIH this year he says, "I cannot believe that under these proposed guidelines the organisms can be contained.... The organisms will inevitably escape and they will enter into the various ecological niches known to be inhabited by *E. coli*." The consequences, Sinsheimer says, "are highly predictable and highly likely dangerous."

• **Who Will Regulate the New Life Forms?** The most astonishing thing about commercial recombinant DNA research today is that nobody knows which government agencies have the authority to regulate it. Perhaps it is the Center for Disease Control, which oversees the interstate shipments of hazardous biological agents; perhaps the Food and Drug Administration, when companies begin using recombinant DNA techniques to create drugs or hormones for human use; perhaps the Patent Office, when companies apply for patents on these products; perhaps the Environmental Protection Agency, under the new Toxic Substances Control Act. Then of course there's the National Institutes of Health, which has drawn up the research guidelines on the subject but which has no power to enforce them.

Industry loves this situation, of course. The confusing welter of bureaucracies makes it much easier for the corporations to go ahead and do what they want. The gaps are huge. When Dr. Robert Elder of the FDA was asked if his agency would be informed if, for example, test animals in a commercial lab began mysteriously dying from unknown diseases after being injected with a new recombinant-DNA-type drug, he said that there would be "no requirement that [the company] inform us," and that *he knew of no other agencies that would be privy to such information.*

Given vast loopholes like this, it is still more remarkable to learn, as we go to press, that a significant faction of the Interagency Committee is urging that federal guidelines on recombinant-DNA research be made *voluntary* and that the industry be left to police itself. In the unlikely event that the the committee takes a tougher stance and recommends, for instance, the creation of a new superagency with enough money and muscle to closely police all DNA research, look for the industry to resist. There will be cries that the government is interfering with free scientific inquiry: the drug companies will fight back with all the lobbying power at their command. If the committee compromises and urges a distribution of regulatory authority among various agencies, government regulation may remain almost as diffuse and ineffective as it is now.

• **Who Will Own and Profit from New Life Forms?** Not a single person on the government's Interagency Committee who was interviewed even questioned the right of commercial firms to patent processes for creating new forms of life. Dr. Delbert S. Barth, the Environmental Protection Agency representative to the Interagency Committee, summed up the prevailing sentiment of his fellow committee members on the question: "This is a moral and ethical question—and I don't have a strong opinion."

And because the members of this group either are pro-business or do not have a strong opinion on the moral or ethical questions involved, they will in all likelihood recommend that private corporations like Miles, Upjohn and Abbott be entrusted with the authority to create and market new forms of life, for profit.

The only question regarding commercial patents being addressed by the committee is a technical one: how to protect the secrecy of research going on in commercial labs so that competitors will not steal trade secrets before the firm can patent a new life-form process.

Under the existing NIH guidelines covering university DNA research, scientists must disclose all their plans in advance. (Two universities—Stanford and the University of California—have applied for patents on their DNA-recombination processes.) Industry leaders say that these provisions of the guidelines would be unacceptable because, in the words of Dr. Jerome Birnbaum, director of Merck, Sharpe and Dohme Laboratories, "if you disclose your research plans, you lose the right to a patent." The Interagency Committee is expected to go along with the industry's demands to keep its research secret by establishing some kind of provision whereby only a select few government officials will be privy to the specific nature of the research going on in the corporate laboratories. An awesome thought, since it's only a matter of time before molecular biologists are able to create new plant and animal forms or alter the genetic characteristics of the human species.

A few researchers—despite the opposition of most scientists— are already talking about just that. One of them is Dr. James Bonner, Professor of Biology at the California Institute of Technology, who has written: "We can control the [genetic] changes to produce better individuals. This is even more important now that we see limitations being placed on the number of children that may be born into the world. There is a moral obligation to see that these children are free of genetic defects, and we may even have to proceed to the logical conclusion that these children should be provided with the best genetic material we can obtain. Man has done this with all of his domestic animals and plants. It seems likely that he will do it also with himself."

Dr. Bonner, whose words on the subject appear in a book published by the National Aeronautics and Space Administration, goes on to say: "The logical outcome of activities in modifying the genetic make-up of man is to reach the stage where couples will want their children to have the best possible genes. Sexual procreation will be virtually ended. One suggestion has been to remove genetic material from each individual immediately after birth and then promptly sterilize that individual. During the individual's lifetime, record would be kept of accomplishments and characteristics. After the individual's death, a committee decides if the accomplishments are worthy of procreation into other individuals. If so, genetic material would be removed from the depository and stimulated to clone a new individual. If the committee decides the genetic material is unworthy of procreation it is destroyed...The question is indeed not a moral one but a temporal one—when do we start?"

Up to now, recombinant-DNA research has been seen as largely a health and safety question. With the possibility of hazardous viruses escaping from labs, it is indeed. But even this question is bound to seem secondary as the broader implications of recombinant DNA begin to be understood by the general public. When the U.S. begins to ask itself whether individual scientists and a handful of government bureaucrats and private companies have the right to rearrange the evolutionary order and create new forms of life, the recombinant-DNA question will emerge as a focus of national attention.

Mayor Peter Memeth of South Bend, Indiana, one city in which recombinant-DNA research is going on, touches a central nerve when he says, "if they had all that trouble in Tennessee with the apes, then they haven't seen anything yet, I guess." Considering the fury that engulfed the Scopes trial, the issue of artificially creating and controlling new life forms may well reach very deep. The controversy it raises will make the debate over abortion seem a mere brush fire by comparison.

Dr. Harry Hollis, director of the Committee on Family and Special Moral Concerns of the Christian Life Commission of the Southern Baptist Convention, and a spokesperson for President

Carter's own denomination, says: "I feel very strongly that Huxley's warnings have a bearing for us today. After Dachau and Watergate, we shouldn't take lightly what human beings are capable of inflicting on each other. This is not just science fiction. Genetic engineering for the worst of reasons is a possibility in this world in which we live." The issue of who has the authority to develop and produce new forms of life is perhaps the single most important question any society has ever had to grapple with.

With the dawn of the Organic Age upon us, there is no longer any question of going back. The question now is how we proceed, and how we prevent ourselves from embarking on an inexorable corporate course towards Huxley's Brave New World.

Novus ordo seclorum. The new age now begins.

Dealing with the Experts
—Scott Thatcher and Bob Park

This article describes how representatives of Cambridge MA questionned and discussed DNA research that was to take place in their city. And it shows that this issue of DNA research, which may have major consequences for all people's survival, is best not left in the hands of a small group of technical "experts."

Molecular Biology Against the Wall

The proliferation of possibilities in recombinant DNA research has brought new excitement to molecular biology. Besides new vistas in "pure" research, remarkable applications and grim hazards have appeared on the horizon. Previously farfetched scenarios for genetic engineering seem much less distant.[1] The commercial aspects have aroused the curiosity not only of drug companies but of industry in general. Molecular biologists were invited to give briefings on Wall Street. A skirmish recently broke out in the Commerce Department when an official proposed accelerated patent procedures for recombinant DNA techniques. (So far GE holds three patents and both Stanford and University of California have applications pending.)

Simultaneously an unprecedented open debate has mushroomed on the control of this research. Numerous cities and towns, likely future hosts to recombinant DNA research, have joined the debate. For the first time, molecular biology has received local front-page coverage. No longer is the research a matter for "self-regulation" by scientists through the good offices of the National Institutes of Health (NIH) which funds most biomedical research. The issue has been catapulted to top-level policy-making, involving the Secretary of Health, Education and Welfare (HEW), the Commissioner of the Food and Drug Administration (FDA), and an interagency task force which has recommended comprehensive legislation. Bills are now being formulated in Congress and in State legislatures.

Harvesting the Culture of Elite Science

In recent years most working people have acquired a critical sense of the role of science and technology despite a tradition of science mystification and deference to authority. Many now recognize that unemployment, pollution, and disease are another side of the grand hype that science means automatic progress; they see that most of those white-coated experts are owned by business or government. Technology's record has fostered this disillusionment, witness: PCBs, kepone, SST, Tris, nuclear power, occupational hazards, etc.

In 1974, when molecular biologists called for a moratorium on certain potentially dangerous experiments and asked that scientists discuss among themselves safeguards for this research, the news spread readily far beyond science to a quite interested public. Popular skepticism has been further stimulated by the disagreement increasingly visible among the experts themselves. But perhaps it was the prospect of their actually engineering genetics-- whether of humans, plants, or microbes-- which finally cancelled the blank check of elite science.

Open Debate on Usually Closed Issues

Debate on recombinant DNA research, both in and out of the science establishment, reveals that a Pandora's box has been pried open; social control of science is a live issue. Specific questions arise in three areas: the ostensible benefits, probable uses, and unintentional hazards. However, we can go further and ask what underlies the disagreement among experts themselves and then ask how government policy toward science could become the province of the people?

One benefit recombinant DNA technology promises is a breakthrough in world food production. A new, specially engineered species of plants, it is claimed, would significantly reduce world hunger. This invites examination of the past effects of the Green Revolution's increased yields from selected hybrid varieties of rice, corn, and wheat. The results have not included the feeding of the hungry.[2] Other benefits are implied in the predictions of new drug sources and supertherapies for intractable diseases. These "promises" invite examination of three critical areas: the economic and social origins of most disease and health problems, medical research priorities in general, and the high technology, "technical fix" approach to health care.

While, conceivably, new therapies will be able to correct some of the known non-controversial genetic defects, there are many other conditions—virtually any characteristic with a claimed genetic predisposition—where the "correction" would amount to a form of genetic repression of individuals. Who decides when human variability becomes a genetic "defect?"[3] We need to spell out the implications--present and future--of our orientation that emphasizes genetic fixes over treatment of society as a whole. The results of our approach include declining social services, increased channeling of individuals (IQ in education, occupational hazards, vulnerability in employment), and ultimately, suppression of deviance, dissent, unrest, and other "maladaptive" behavior.

While the ultimate uses of recombinant DNA technology are probably the gravest threat, it is over the immediate hazards of doing the research that technical disagreements among the experts are most apparent.* The debate centers around the adequacy of containment for experimental organisms as well as the pretense that molecular biologists (or anyone else) know enough to guess at the broader ecological or evolutionary threats. How can supposedly objective experts** be in such disagreement? We think perceptions of "objective" reality are dependent on philosophical and ideological premises as well as on other immediate and material factors in people's lives. A large part of the "benefit to risk" estimate is speculative and thus is especially open to subjective valuation. For

---

* The concern arises from the use of the bacterium E. coli as a host because it is a normal inhabitant of the human gastro-intestinal tract (but occasionally causes a serious disease). E. coli is used because it is the best known bacterium. But hybrid versions, created unknowingly when random samples of foreign DNA are spliced into its chromosome, could create a whole new class of disease-causing organisms.
**Definition of "expert": a person with extensive personal experience, both in theory and practice, in some area of technical knowledge, not necessarily certified by an academic degree. Being an expert, however, does not mean knowing the "truth" on a technical matter within one's expertise or better understanding the social implications.

example, how one assesses benefits from recombinant DNA work is contingent upon one's view of the social role of technology; predicting hazards depends on one's technological optimism.

Another source of subjectivity derives from one's own contribution to, or interest in, technology. For many in science, the value of their work depends to a considerable extent on how it contributes directly or indirectly to human betterment. In a society where institutions do not operate *a priori* to serve desirable social ends, there is an incentive to believe that better technology tends to shift the outcome in favor of serving those ends, that new knowledge has intrinsic value. Consequently, many medical researchers pursue answers to problems for which other solutions, such as changing social conditions, are lacking or are at least beyond their control. Some people, for this reason, may have an unduly optimistic outlook on recombinant DNA research. Others in science have careers whose success requires the rapid exploitation of scientific discovery. The advantages include publications, appointments, the realization of creative potential, esteem with family and colleagues, recognition by institutions and officials, and ultimately, entry into business and government circles. It is clear that in situations where advances are imminent, the personal benefits and risks for some scientists—as with investors—can very understandably differ from those of most working people.

Popular Critical Awareness on Technical Issues

Because technical issues cannot be resolved by reference to an "objective," neutral stance, it is especially vital that public policy* in science be determined by a process based on popular awareness, organization, and control. One form this could take would be labor unions with strong member participation and control, with extensive education programs, and with active involvement in defining and enforcing government policy and corporate behavior. Another avenue for popular control of science policy would be community-based organizations. They would oversee, for example, the health care system, medical research, and human experimentation. Even without organization, however, public discussion, debate and criticism can have a major effect on the existing decision-making apparatus, as it is presently. This process has not been encouraged

---

*"public policy"—fundamental policies laid down by Congress or the Executive branch on which government regulation is based.

by most prominent scientists. As Sidney Udenfriend, director of the Roche Institute for Molecular Biology* and member of the NIH advisory committee on recombinant DNA research, explained: "I'm afraid there's going to be some brush fires if we get communities involved in deciding biohazards. If we permit non-scientists to question our work in one area (DNA), we'll open ourselves up to all kinds of things..."[4]

How can good judgment on scientific issues be exercised by the "masses?" This, we propose, is analogous to the question: How do top government leaders and policy experts decide questions of science and technology policy? *They rely on experts whom they believe to be credible.* The people, too, should be able to evaluate the credibility of experts. What are these experts' views on the general role of technology and on specific issues bearing on the people's interests? How have they contributed to dealing with the real problems of working people, and what are their stakes in these matters? Evaluating experts is an important task for any popular organization. Just as the rulers of the country can pick and choose between experts and the opinions that they espouse, so can the people.

Of course, the ability of the people to evaluate technical opinion would be considerably enhanced by their having more widespread technical knowledge and scientific understanding. This is a goal which progressive science workers and technical experts should facilitate, in contrast to what happens normally.

The Developing Controversy

In 1971 a scientist objected to a colleague's proposal to insert the virus SV40, which causes tumors in some animals, into the bacterium *E. coli* K12. It was feared the hybrid might escape from the laboratory, survive, and result in a new form of disease. The experiment was abandoned. The subsequent, self-imposed moratorium on certain gene-splicing research was partly intended to show that scientists could look after the danger of their own research. The first large scale discussion by molecular biologists of hazards took place in February, 1975, at Asilomar, California, where a rough consensus was obtained on how to deal with the safety question. However, the panel subsequently selected by NIH to write guidelines was made up mostly of scientists already using

---

* The Roche Institute is the "pure research" arm of Hoffmann-LaRoche, the most lucrative drug company in history, maker of Valium, Librium and others.

recombinant DNA techniques or planning to, and some advisors to the panel had direct commercial interests in it.[5] It was a foregone conclusion that the techniques would be developed and used extensively.

With minimal public participation, the NIH guidelines committee plunged forward (with occasional backsliding), buffeted on all sides by threatened feudal science chiefs. One early draft, available at the traditional Cold Spring Harbor phage* meeting in August 1975, was sharply attacked by members of Science for the People (SftP) and others as a retreat from earlier, more strict positions. Meanwhile, the debate went public.

The first large scale public confrontation on recombinant DNA took place at the University of Michigan, Ann Arbor, in early spring of 1976. The casual intentions of the university trustees to invest in a campus-based recombinant DNA facility were unexpectedly dragged into the spotlight. The issue was raised by faculty member Susan Wright, with several other faculty and Ann Arbor SftP members joining in. It generated escalating interest on campus and within the surrounding community to such an extent that the university's Research Policies Committee felt compelled to arrange a full-dress forum, inviting a wide spectrum of experts from all over the country. It lasted two days and attracted a continuous attendance of over 600 people.

The outcome was that the two appropriate faculty committees gave near unanimous approval to proceed with the research, subject to the awaited NIH guidelines. However, far more significant was the effect of the debate locally in revealing the full depth of the criticism of the research, and nationally, in providing a stunning precedent for the growth of the controversy into a movement for popular control of science.

The Cambridge Experimentation Review Board

Just as final NIH Guidelines were about to be issued in June 1976, Harvard University's plans to build a P3** facility came to light. Aware of Harvard's intentions, an interested City Councilor, Barbara Ackerman, attended a low-key "public" meeting called by

---

* Phage: a virus that lives in bacterial hosts, studied because of its relative simplicity.

** P3 is the second highest level of laboratory "containment," ranging P1-P4, for keeping experimental organisms isolated and preventing their escape into the world at large, from which they could never be recalled.

Harvard's Committee on Research Policy to discuss the P3 plans. Simultaneously, the facility was announced in the lead article of a local alternative newspaper and immediately hazardous research in Cambridge became a burning issue, fanned by some local politicians running hard to catch up. They included Mayor Al Velluci who gained national attention for his efforts.* Thus recombinant DNA research became the focus of lengthy City Council meetings at which numerous opposing presentations were given and to which hundreds of people came, not all of them academically affiliated. An unprecedented six month moratorium on P3 and P4 recombinant research resulted, an act heard "round the world," and equally startling, a citizen's review committee made up of non-experts was created to advise on the research hazard.

The experience of the Cambridge Experimentation Review Board (CERB) warrants close inspection as an example of public participation in making science policy. CERB, at the City Council's direction, was selected by the City Manager and consisted of people with neither personal interest in recombinant DNA research nor related professional interests, contrary to research scientists. Board members—all Cambridge residents, with an equal number of men and women—included a nurse, a saleswoman, a university faculty member, a homemaker and an engineer. Taking its narrow assignment of dealing only with the immediate public health-safety issues, CERB met in both open and closed sessions biweekly for over 4 months and heard 75 hours of testimony ranging from NIH dignitaries and renowned advocates of the research to lab technicians and members of Science for the People. The board's final

---

* The response of the politicians reflects more than just awareness within their constituencies of recombinant DNA issues. Cambridge has long been dominated by the imperial giants of Harvard and M.I.T., usually with cooperation from most city politicians, with effects which have included the removal of most of Cambridge's industrial employment and the constant encroachment on traditional working class neighborhoods by university expansion and housing for students, faculty, and the technological elite. In the 60s and early 70s, extensive industrial properties were bought up by the M.I.T.-government-aerospace team to be transformed into an electronics, computers, and weapons research center. (Technology Square, for example, is a former site of numerous manufacturing plants.) The details of this process are contained in *Harvard, Urban Imperialist,* 1969, published by the Anti-expansion, Anti-ROTC committee at Harvard. The rent control law, finally passed in the late 60s with little help from most politicians, was a significant victory reflecting the widespread anger of the people against institutions like Harvard and M.I.T. The recombinant DNA issue was for the people of Cambridge but another example of imperial decision-making, and many politicians could not afford to let it pass.

position allowed the research to proceed but with significantly stricter requirements than NIH. These included strengthening institutional biohazards committees, monitoring escape of vectors,* conducting local epidemiological studies, and setting up a city-wide biohazards committee. In addition, CERB recommended that the federal government extend the NIH Guidelines to cover industry, maintain a registry of workers in recombinant DNA labs, and fund health monitoring. CERB rejected assurances from Harvard and NIH scientists that the voluntary NIH Guidelines were a more-than-adequate protection against exceedingly improbable or inconceivable events. The CERB deliberations led to a city ordinance incorporating their recommendations and were in part responsible for the near-passing of another law banning P3 and P4 research indefinitely (defeated 6:5). CERB's most important contribution was to show that non-experts could judge experts and make creditable public policy judgments. The CERB report revealed that public policy issues were not allowed to be obscured by the technical debates.[6] This critical evaluation of the claims being made by experts is in sharp contrast to how the Science Court would function, as it has been proposed.**[7]

There were deficiencies in the CERB conclusions, but first let's examine how CERB was able to do what it did. CERB avoided becoming beholden to Harvard, M.I.T., or the science establishment in part because of the selection process that formed the board, in part because the development of an authority structure or hierarchy was minimized. For example, the original chairperson, who was also Acting Commissioner of Health and Hospitals in Cambridge, removed himself as a voting member on grounds of possible conflict of interest. In addition, all members were encouraged to take part in defining unresolved issues. Finally, at least some members of the committee had a clear perception of political power and the people's interests, as well as an active commitment to working for those interests.

It is evident that the selection procedure which formed CERB cannot be counted on routinely in selecting citizen's boards since the success of this procedure depends on the orientation of the executive officers of, in this case, municipal government. But even

---

* Vectors: organisms containing, in this case, hybrid DNA.
** In the science court concept for resolving disagreements among experts, as originally proposed by A. Kantrowitz, chairman of AVCO Everett Research Laboratory, a panel of scientific experts chosen in the usual manner of elite boards, would cross-examine technical claimants on the "facts," never venturing to examine broader questions.

randomly selected committees of interested working people will not escape the problems of elitism, professionalism, and science mystification that affect all of us in contemporary society, unless some members have had experience in combatting this ideology.

The shortcomings of the CERB report reflect conditions which no citizens' committee could have easily overcome. It is unlikely that any representative committee (feeling the immense weight of world attention on its actions) could have strayed very far from the middle of the road in the absence of a visible migration of popular opinion on the issues. While there is considerable consciousness of the hazards possible in recombinant DNA research, very little organization or examination of the issues in political terms has developed on a mass scale. Thus it would be bizarre indeed if the committee had, at its own initiative, broadened the scope of its enquiry and pursued in depth questions we believe to be central, such as: the likely specific uses of genetic engineering in *class* terms; the ecological or evolutionary dangers (in terms of infectious disease, soil ecology, and other specific areas); and benefits and risks in broad social terms—who really stands to gain, what are the indirect costs, who is at risk, what alternatives are being ignored?

Actually, many Cambridge residents were suspicious and concerned over the proposed research at Harvard, according to two City Councilors. An outright ban on the research was favored by some. Had this awareness been better articulated and publicized, perhaps CERB would have taken a stronger stand. The progressive forces in the Cambridge debate could have been very effective in assisting communication between CERB and Cambridge residents.

A major factor in CERB taking a critical approach, aside from the nature of the committee itself, was pressure from a significant opposition minority within the local science "community" and the radical microcosm within Cambridge, both challenging the N.I.H./Harvard/M.I.T. front. The availability of opposing experts— including technicians—allowed the committee to perceive the political nature of the debate on recombinant DNA research.

There are therefore two main lessons from CERB: (1) With some essential but rarely achievable prerequisites, a citizens' committee can acquire substantial critical expertise free of direct control by nearby institutions and can to some extent reject dominant and respected views. (2) Without a developed progressive movement concretely involved in similar or related issues locally, there are severe limitations to what even a well-selected citizens' committee can do in forging an advanced position. This of course confirms the basic strategy of relying on "mass work"—going to, and being part of, the general populace rather than concentrating on influencing law makers, policy-level scientists, or other persons in high places.

A National Forum

Since the Ann Arbor and Cambridge excitement, there have been many smaller replications of the same debate.[8] In March, the National Academy of Sciences (N.A.S.) sponsored a forum to end all forums on recombinant DNA. The N.A.S., the most select organization of elite science,[9] was probably concerned over the course the debate was taking and wished to present a moderate appraisal, especially for congressional staffers and the press. The panel of speakers was relatively balanced; the workshops were dominated by pro-recombinant forces, but the agenda was improved by the heavy turnout of counter-forces: members of the People's Business Commission (formerly the People's Bicentennial), the Environmental Defense Fund, and the Coalition for Responsible Genetics Research. The only person at the N.A.S forum speaking for organized workers was an official of the Oil, Chemical, and Atomic Workers union, who pointed out that the N.I.H. guidelines were ludicrous as far as protecting workers in industry is concerned.

Several developments were apparent. One was recognition of the extent of commercial inroads into recombinant DNA technology; a number of people argued that this technology, based on publicly funded research, should not be exploitable for profit. Another was the isolation of the most self-righteous and adamant proponents of the research from even mainstream, establishment scientists (who were a little embarrassed by this group). By then, in fact, the tide had already started to turn, and forces were being redeployed to the legislative field.

Legislative Shelter in a Storm

Some academic scientists and drug companies who previously had vigorously opposed legal controls on recombinant DNA research emerged in favor of national legislation at the N.A.S. forum. Their position changed because they sought future protection from actions such as occurred in Ann Arbor and Cambridge. Many other people saw the legislation as necessary to cover industrial applications of recombinant DNA technology since the

N.I.H. guidelines applied only to government-funded research. As a result, California and New York are both considering legislation to cover the work. Two bills pending in Congress would essentially write the N.I.H. guidelines into law with stiff penalties to enforce them.

The right of local communities to enact their own ordinances is an important issue. But the recent interagency report from the federal government emphasizes that national regulations must pre-empt local or state ones, and many scientists and pharmaceutical firms see this as the main value of the legislation.[10] The bill before the U.S. Senate, sponsored by Edward Kennedy (D., Ma.), gives local communities a real option to enact more strict legislation. Even Joseph Califano, Secretary of H.E.W., has felt the need to state publicly that he supports a local option.

While federal legislation will clearly give scientists the protection and sanction they need for recombinant DNA work, many are very resentful of the government's interference in their affairs. Philip Handler, president of the National Academy of Sciences, raises the specter of "constraints that will swathe the research with bureaucratic complexities...and generally frustrate a career in research. If (regulation is) pursued yet further, science could be shattered."[11] A majority of the molecular biologists attending a Gordon Conference in June of this year were greatly aroused by the possibilities of arbitrary government interference in their affairs and stated publicly that earlier warnings by them and others concerning hazards had been exaggerated.[12] Nevertheless, representatives of the Pharmaceutical Manufacturers Association concede that dealing with federal inspectors will be nothing new to them. Politically aware scientists at the N.A.S. forum felt similarly. Donald Kennedy, newly appointed commissioner of the Food and Drug Administration (F.D.A.) and a former Stanford biology professor, went further and said,"Why should there be more regulation? The simple answer, I think, is because it is politically inevitable... How much regulation are we going to have? Answer: As much as people insist on, in light of their own social value calculus." Biologist Clifford Grobstein, prominent in the debate in California, noted there were many at the N.A.S. meeting who felt that "science has become too consequential to be left to the self-regulation of scientists or to be allowed to wear a veil of political chastity."[13]

Still Congress may give power to regulate the research to the same agencies—H.E.W. and N.I.H.—that provide most of the funding for the research. The "Recombinant DNA Research Act of 1977," introduced by Carl Rogers (D., Fl.) of the House Committee on Science and Technology, gives the Secretary of H.E.W. full power to make regulations for the research and to license those who undertake it. Just as the Atomic Energy Commission was unable to both promote and regulate nuclear technology, so too, H.E.W., which runs the N.I.H., will have a conflict of interest.

The proposed federal regulations may frighten scientists, but it is doubtful they will eventually stymie research. Federal inspectors, according to Kennedy's bill, could examine any laboratory materials and could destroy or confiscate dangerous recombinant organisms as well as recommend heavy daily fines, but enforcement would remain difficult. Inspectors would be hard pressed to see through the mass of laboratory paraphernalia in order to use their power meaningfully. As an alternative, Rogers' bill calls for local biohazards committees to be given the prime responsibility for enforcing the regulations, rather than federal inspectors. Such committees would have one third of their members from outside the regulated research institution and might possibly be more responsive to community concerns than a powerful federal bureaucracy.

Will federal legislation make the N.I.H. guidelines more effective? The guidelines ask biologists to understand and follow relatively strict microbiological techniques which few have been trained in. Molecular biologists, especially, are used to treating the bacteria they study as harmless. Thus the guidelines are certain to suffer from much day-to-day negligence, especially from workers who are convinced there is no clear and present danger.[14] In one typical laboratory, the guidelines reportedly are often ignored.[15] Both congressional bills ask that employees who raise questions about safety be protected from loss of their jobs, but such a provision would be hard to maintain without strong local unions and safety committees.

The federal government is also trying to limit the liability of institutions doing the research. One bill, Rep. Ottinger's H.R.3191, no longer under consideration by Congress, made it clear that institutions would be liable for an accident whether or not they had violated regulations. The federal task force on recombinant DNA research, however, concluded that if liability were unlimited, then the work might not proceed due to the costs of insurance. Already one contractor, Litton Industries, has bowed out of a government contract involving a high-containment P4 facility in Fredericksburg, Maryland, claiming it cannot get liability insurance.[16] Limiting liability would require legislation similar to the Price-Anderson Act which placed a ceiling on the liability of a power company for a nuclear power accident. Although the act was ruled unconstitutional recently in a federal court, similar provisions might still be written in the case of recombinant DNA research. At the moment, Kennedy's bill states that federal legislation shall not limit a citizen's right to sue over an accident.

Conclusion

Whether or not strong, meaningful laws are passed, requiring the slow, careful development of recombinant DNA technology—and whether they are enforced—depends on the critical consciousness of the people. The task of progressive science workers is to facilitate this process. Furthermore, this objective makes sense only if it is broadened to include all interrelated areas, e.g., medical research priorities, occupational and environmental health, and genetic engineering uses. So too, the value of citizens' committees depends on informed popular opinion and agitation. Conceivably, legitimate citizens' committees could be arranged by coalitions of organizations in communities, independent of government, to help clarify technical disputes.

Evaluating experts is a political process. However, there is obviously no guarantee that politically progressive and responsible experts will necessarily have more reliable technical opinions and interpretations of fact. Ideally then, experts should be experienced in collectively defining positions and principles—participating with other, non-expert, working people. In this way the technical discipline and political sensitivities of experts will grow in good directions, along with everyone else's. Organizations are therefore needed in which both experts and non-experts can collaborate in non-elitist and anti-sexist practice toward progressive goals.

When working people begin to routinely and systematically evaluate the credibility of experts, the face of technology will change: governments and business will be less free to design our future against our interest.

FOOTNOTES

1. Applications include: industrial microorganisms which may transform chemical and pharmaceutical industry, production of biological materials not now available, plant varieties with unique abilities, e.g., nitrogen fixation. Potential hazards include: disease-causing bacteria never before encountered, ecological disruption, and new diseases of genetic regulation, e.g., cancer. For a more detailed discussion of the hazards, precautions, and alleged benefits, see paper entitled "Social and Political Issues in Genetic Engineering," by the Recombinant DNA Group of SftP, available from the SftP office: 897 Main St., Cambridge, MA 02139.

2. H. M. Cleaver, "The Contradictions of the Green Revolution," *Monthly Review*, June 1972; Nicholas Wade, "Green Revolution (I): A Just Technology, Often Unjust in Use," *Science*, 186, December 20, 1974, pp. 1093-1096.

3. Jon Beckwith, "Recombinant DNA: Does the Fault Lie Within Our Genes?" *Science for the People*, May-June, 1977, p. 14.

4. *Drug Research Reports*, December 3, 1976, p. 6.

5. Francine Robinson Simring, "The Double Helix of Self-Interest," *The Sciences*, May-June, 1977, p. 10.

6. Report of the Cambridge Experimentation Review Board, *Bulletin of the Atomic Scientists*, May, 1977, p.22.

7. Task Force of the Presidential Advisory Group, "The Science Court Experiment: An Interim Report," *Science*, 193, August 20, 1976, pp. 653-656; Arthur Kantrowitz, "Controlling Technology Democratically," *American Scientist*, September-October, 1975, p. 505.

8. Nicholas Wade, "Gene-Splicing: At Grass-Roots Level a Hundred Flowers Bloom," *Science*, 195, February 11, 1975, pp. 558-560.

9. Nicholas Wade, "The Brain Bank of America: Auditing the Academy," *Science*, 188, 1094, 1975.

10. "Interim Report of the Federal Interagency Committee on Recombinant DNA Research: Suggested Elements for Legislation," submitted to the Secretary of Health, Education, and Welfare, March 15, 1977.

11. Editorial excerpted from Philip Handler's annual report to the National Academy of Sciences, *Chemical and Engineering News*, May 9, 1977, p. 3.

12. Walter Gilbert, "Recombinant DNA Research: Government Regulation," letter to *Science*, 197, July 15, 1977, p. 208.

13. Clifford Brobstein, "The Recombinant DNA Debate," *Scientific American*, July 1977.

14. Richard P. Novick, "Present Controls Are Just a Start," *Bulletin of the Atomic Scientists*, May 1977, p. 16.

15. Janet L. Hopson, "Recombinant Lab for DNA and My 95 Days in It," *The Smithsonian*, June, 1977, p. 55.

16. "Scientists-Critics Less Fearful Now of DNA Research," *The Boston Globe* , July 18, 1977.

Scholars for Dollars
—Charles Schwartz

The image of the university as a neutral ivory tower, a center of value-free scientific research that fosters dialogue from all sides is debunked in this article, which details the unambiguous connections between university professors, the military and industry.

Eight hundred years ago, at the University of Bologna in Italy, professors had to obtain permission from their students and post bond in order to leave town on private business.[1] No such requirements impede the travels of entrepreneurial professors in modern United States. While it is widely known that university faculty sometimes hire out their special expertise as private consultants, the full nature and scope of this activity has generally been kept well hidden from public view. While it takes the professor's time and interest away from teaching and other academic pursuits, and even though consulting fees earned by professors for time spent working elsewhere require no surrender of academic salary, college officials do not look upon consulting as "moonlighting." Aside from espousing the vague tenet that outside consulting should not interfere with basic teaching commitments, universities generally take a completely *laissez faire* attitude toward it. In reply to a query about consulting practices, Dr. George Maslach, Provost of the Professional Schools and Colleges of the University of California, Berkeley, said:

I have no knowledge of the extent of outside consulting by faculty and others; I have no knowledge of how many people consult, nor do I know how they have spent their time. There is no indication of how I can obtain this information in any easy way.

The rather startling information presented in this article shows that a large number of academics serve not only as ordinary paid consultants to private industry, but actually sit on the boards of directors of major business corporations. It is proposed that all consulting-like activity by faculty should be treated the same as any other research or academic activity: scrutinized, evaluated in terms of objectives sought, interests served, and publicized and criticized accordingly. It should be another focus of political struggle in academia.

Some Data on Consulting

A survey conducted by the Carnegie Commission on Higher Education in 1969 shows how widespread is the practice of faculty outside consulting. Forty-one percent of the faculty surveyed devote between 1 and 10% of their work time to consulting, with or without pay; 14% devote between 11 and 20% of their time; and 5% devote more than 20% of their time. The recipients of paid consulting services were diverse: federal or foreign government (2%); local business, government, schools (18%); national corporations (17%); nonprofit foundations (11%); research projects (10%). Only 42% of all the faculty had done no paid consulting during a two-year period. Of all sources of supplemental earnings reported by faculty, consulting was the leading type but other types—such as summer teaching and research, private practice and royalties and lecture fees—were also significant.

An earlier survey, covering the academic year 1961-62, gave data on the outside earnings of faculty broken down according to their academic discipline.[2] The overall fraction of faculty having outside earnings was 74%, the highest being in Psychology (85%) and the lowest in Home Economics (44%). The average amount of outside earnings was highest for Law ($5,297) and next highest for Engineering ($3,197); the average for all areas was $2,165.

A reported survey of the Harvard Faculty indicated that nearly half of the senior professors had outside incomes that exceeded one-third of their college salaries; and a leading economist at a major Ivy League school was quoted as saying that he charged about $200 a day and added as much as $12,000 a year to his regular income: "I simply need the money," he explained. "Our nine-month salary is not adequate for the standard of living we like."[3]

Information on individual professors' consulting connections is not publicly available in any systematic form. The standard biographical reference books (*Who's Who* for the very elite, or such professional listings as *American Men & Women of Science*) sometimes list business firms or government agencies for which the individual biographee is a consultant; but these sources, relying as they do on the voluntary contributions of the persons listed, are often incomplete. Numerous cases of academics' consulting relationships, verified through other sources, are not mentioned in these published biographies.

There are, however, two special kinds of consulting relationships for which one can find published listings of the individuals involved. The first kind covers people who serve on advisory committees to the federal government. According to a law passed by Congress in 1973 (PL 92-463) the President must give an annual report of the activities and membership of the more than 1,400 advisory committees that serve the various departments and agencies of the Executive Branch. The first such report was issued in 1973 and it included an index of committee members, arranged by institutional affiliation as well as by name.[4] Quoting from the Senate Subcommittee press release that accompanied the publication of this index,

> Approximately 24,500 individual positions on advisory committees are identified in the index. The Department of Defense had more representatives on advisory committees—713—than any other agency. The university with the most representatives on advisory committees was the University of California (374), followed by Harvard (130) and Columbia (108). Companies with large numbers of representatives on advisory committees include the following: RCA—93; ITT (and affiliates)—92;...

The index includes 78 names of UC Berkeley faculty and staff serving on a wide variey of government committees, from agriculture and military affairs to science and poetry. (One notes that Provost George Maslach, who had "no knowledge" of faculty consulting activities, is himself listed as a member of two advisory committees in the Department of Defense.) Rich as this index is in information, it should be noted that there are other types of government consultantships which are not covered by the public-disclosure requirements of this law. Also, it appears that this index has not been prepared for years later than 1972.

The second kind of consultantship for which one can find thorough tabulations involves a very special relationship to private industry: being on the board of directors of a sizeable business concern. *Dun & Bradstreet's Million Dollar Directory*, published annually and available in many libraries, contains an alphabetized index of directors and top officials in U.S. companies worth over $1 million. The data in this volume is generally one or two years old; and one must take care to verify the identity of persons who are named as directors. This has been done using several published sources: corporation annual reports and stock prospectuses, the biographical books mentioned above, and newspaper items. (*The Wall Street Journal* has a very useful index for this purpose.) This searching can be a very tedious task: however, it has yielded some surprising results.

Table 1 (p. 186) presents some data on the University of California (UC) showing the faculty and adminstrators who sit on the boards of directors of sizeable corporations, including some of the country's largest industries. (This is not an exhaustive list since this search was not carried out for the entire faculty, numbering several thousand persons.)

Similarly, a survey of the boards of directors of the 130 largest corporations in the U.S., as ranked by *Fortune* in 1974, shows that academics serve as directors in fully *one-half* of these giant companies. These findings are presented in Table 2 (pp. 187-190). This listing could readily be extended by further research in this area.

While the job of an ordinary consultant to private business is to help that business solve some particular technical problems, the job of the board of directors is to set and supervise overall company policy, with the express objective of maximizing profit for the company's shareholders. Thus, the data presented in Tables 1 and 2 suggest that the academic world is integrated into the structure of corporate power at all levels.

Not only do some academics consult for private industry and others serve as advisors to government, but some academics do both. These situations present the most obvious possibilities for traditional conflict-of-interest: for example, consulting for industry while advising the government agencies which regulate those industries. Of course the potential conflict between a particular industry or business enterprise and the government in general is a

"We want you to do some pure disinterested fundamental research into something immensely profitable."
(*SftP*, Sept-Oct: 1977)

relatively minor one, usually limited to disagreements about standards, product claims, legal requirements, etc. Nevertheless, such conflicts can be critical for profits. Thus academic consultants, usually promoted in government advisory circles as experts supposedly independent of special interests, are valuable for business to cultivate, especially when they have intimate knowledge of government operations, policy-making, etc.

Recently, a study of the two highest science advisory bodies in the federal government found, not surprisingly, that the great majority of the people appointed to these bodies were academics as opposed to people from industry or government agencies. However, what was surprising was that more than half of these academics have significant personal ties to big business, mostly in the form of directorships in large corporations.[5]

The data given so far whet the appetite and make one eager to find out more about this vast unexplored territory of faculty consulting activities. It is difficult to believe the comment quoted earlier of Provost Maslach (formerly Dean of the School of Engineering) that he has "no knowledge of the extent of outside consulting by faculty." Rather it seems clear that this subject has a certain taboo associated with it. When a Physics Department chairman was asked about looking into this subject of faculty consulting he declined, referring to it as "a whole can of worms." When a faculty member suggested that a faculty Senate committee be given the task of reviewing campus policies and practices regarding outside consulting, the response was as follows:

> The policy committee has thoroughly discussed the arguments in your letter of April 13, 1973, and finds itself unpersuaded that a useful purpose would be served by Senate surveillance of faculty consulting.

> When we considered examples of specific proposals that might emanate from a committee charged with such responsibility, we were unable to imagine situations where positive consequences were plausible. Questions of conflict with University duties are already covered both by Administrative regulations and the Faculty Code of Conduct. The only effective safeguard we can see against the more subtle dangers in consulting is the conscience of the individual faculty member.[6]

Nevertheless, a few people in positions to know what is going on have been willing to discuss the subject of consulting in at least some detail. Prof. Richard H. Holton, Dean of the School of Business Administration, UC Berkeley, in an interview with students looking into consulting activities, expounded as follows:

> First I should tell you that I have only a vague notion of how much consulting is done in our school; no records are kept. Right now we're reevaluating our promotion and consulting policies...We're looking at the whole reward system that the faculty works under. There is an argument now that faculty in the College of Letters and Sciences have an easier time with promotions than faculty in the professional schools. The greatest emphasis for promotion is on research, with teaching closing fast. Neither University and public service nor professional competence is assigned much importance...Not much is done with consulting in the area of professional competence, and faculty don't keep their files up to date on their consulting activities...Our rule of thumb is that 1 day a week of consulting can be carried without problem. The

desirable kind of consulting is the sort that reinforces research and teaching, not competes with it. Consulting can strengthen teaching by providing real case studies and a close look at live management problems...I would guess that perhaps 50%, plus or minus 10%, do some consulting. Most of this would be for business, but many faculty do unpaid work for government and for non-profit organizations.[7]

Examples of Faculty Collaboration in the Corporate World

The following is a sampling of some of the more celebrated cases of dedication to corporate and/or government service on the part of consulting faculty.

1. When the Federal Trade Commission was trying to get ITT-Continental to clean up its fraudulent advertising, the government's position was attacked in a series of learned speeches by Professor Yale Brozen, a University of Chicago economist; the Brozen speeches were printed in full by *Barron's*, the financial weekly, and full page ads containing the speeches appeared in the *New York Times* and other newspapers. Hordes of ITT PR men called on financial editors all across the country to acquaint them with Prof. Brozen's views. It turned out that Brozen was on the payroll of the PR firm handling ITT-Continental's account and that he was paid for making the pro-ITT speeches.[8]

2. In January 1975, 32 eminent U.S. scientists, all of them Nobel Prize winners, issued a public call for a national energy policy which strongly emphasized nuclear power. Their statement, widely reported, appeared as a 3/4 page ad in the *Wall Street Journal* (paid for by Middle South Utilities System), and was displayed in full on the editorial page of the *San Francisco Chronicle* where it listed the signers of the statement and their institutional affiliations. Twenty-six out of the 32 were identified with universities and only 2 with private industry. However, a little research established that 14 out of the 26 academic scientists listed have been shown to have served as consultants. The companies to which these academics had connections included several with large investments in energy.[9]

3. A notorious episode in California concerns the famous oil leaks in the Santa Barbara channel in 1969. The state's chief deputy attorney general publicly complained that university experts on this problem had refused to testify for the state in its multi-million dollar damage suit against the oil companies, and that petroleum

engineers at U.S. campuses indicated fear of losing industry grants and consulting arrangements. One Berkeley professor was quoted as saying, "We train the industry's engineers and they help us."[10]

4. A recent newspaper story revealed that "equipment and personnel from the University of California's Lawrence Berkeley Laboratory are now being used in exploratory tests for geothermal steam on a ranch near Calistoga—providing valuable services at no charge to the private interests involved...An official with another company that specializes in geothermal exploration estimated the work would cost as much as $100,000 if it were undertaken by his or other private firms." The American Metals Climax Co. had made this advantageous arrangement "through a faculty contact." According to the article, the Dean of Engineering on the campus "declined to comment on the propriety of the arrangement."[11]

5. Many academic scientists serving on National Academy of Science committees have ties to industry that are difficult for an outsider to detect. Thus a committee that issued a report in 1971 on the biological effects of airborne fluorides was composed entirely of scientists from universities and research laboratories that were seemingly independent of industry influence. It was later revealed that the four scientists who had written most of the report had close ties to the aluminum industry, which is a major emitter of fluorides. Some had written publications for the Aluminum Association, received research support from the industry, or testified for the industry in hearings on fluoride standards.[12]

6. In 1965, Dr. Robert H. Ebert was appointed Dean of the Harvard Medical School, and in 1969 became a member of the board of directors of Squibb-Beech Nut Corporation, owners of the large drug company E. R. Squibb & Sons. However, some months later following a protest by medical students charging a serious conflict-of-interest between his loyalty to Squibb and his loyalty to the principles of medical practice and teaching, Ebert resigned his directorship. Squibb then gave the vacant seat to Dr. Lewis Thomas, Dean of Yale's Medical School.

Three years later, Dean Ebert and Dean Thomas appeared together as expert witnesses in a hearing before the Food and Drug Administration, arguing against the banning of one of Squibb's lucrative drug products. When questioned by the press, a Squibb official stated that neither dean had been paid a special fee for his appearance, since both of them had been retained on the company's payroll for a number of years. This revelation raised another brief flurry on the Harvard campus; however, when one student was bold enough to propose a university-wide "audit" of faculty consulting, this idea was branded as "McCarthyite" by a prominent administration official.[13]

7. The Jason group is an elite gathering of mostly academic physicists who provide consulting services for the Defense Department. Little is known about Jason; however, the publication of the Pentagon Papers revealed their role in the creation and promotion of the "electronic-battlefield" strategy in Vietnam.[14] Several Jason members ran for elective office in the American Physical Society; along with the ballots came long lists of their professional achievements and honors. It was later pointed out that none of them had acknowledged their connection with Jason—although several did list their consultantships with the more dovish Arms Control and Disarmament Agency.[15]

Conflict of Interest in the University

The purpose of the modern university is usually proclaimed in high and noble terms: to search for truth, to transmit knowledge and critical skills to students, and to do all this for the betterment of society as a whole. Research and teaching, the twin primary jobs of the professor, are expected to advance civilized society with both short-term and long-term benefits. We now want to ask how outside consulting is said to fit into this imaginary scheme. In the basic UC policy statement on outside services, it is spelled out that this activity by faculty may be justified provided that: (1) it gives the individual experience and knowledge of value to his or her teaching or research; (2) it is suitable research through which the individual may make worthy contributions to knowledge; or (3) it is appropriate public service.[16]

Thus as Dean Holton indicated in his interview, outside consulting is supposed to be an adjunct to the professor's primary tasks of teaching, research and public service. Certainly there are examples of faculty consulting work that meet this literal standard (using traditional definitions of "value," "worthy" and "appropriate"), and just as surely there are cases that would fail this test. It is equally clear that university administrators have no particular desire to meddle in these matters. But what of the more basic conflict of interest for which there are not only no rules but also no admission of its existence? What are the many ways that consulting for private corporations or government agencies influences the content of teaching, directions of research, and allocation of university resources? Ways which have nothing to do with "knowledge," let alone serving progressive social purposes, but instead serve the pressing needs of special interests. The real conflicts-of-interest involved in the practice of outside consulting can thus be identified in the following areas:

1. Teaching: Some professors may be distracted from conscientious teaching because of their frequent involvement in outside enterprise; a commitment to some outside business or government agency might distort the presentation of course material. For example, in the words of one student: "I remember taking a forestry course which repeatedly emphasized how the public should leave the big forestry companies alone and trust them to harvest safely; afterwards I found out that the professor was consulting regularly for the big timber companies."

2. Research—"the unfettered search for truth": The outside connections a professor has may readily influence the choice of research topics, especially if the availability of research funding is scarce; and may also have the effect of slanting the research analysis or limiting the types of solutions that may be considered for acknowledged problems. For example, the well-documented history of scientific studies on the health hazards faced by asbestos workers shows how an industry can essentially purchase the kind of research it needs for its own uses.[17]

3. The Role of Universities in General: Students, parents, taxpayers, and legislators are paying professors' salaries for the same time for which outside income is being earned in the service of private clients. As an illustration, the University of California pays its vice president, Dr. Chester O. McCorkle, Jr., an annual salary of $53,000 (more than the state pays its governor); but at the same time Dr. McCorkle is working for two large agribusinesses—Del Monte Corp. and Universal Foods Corp.—as a member of the board of directors of each ($10,000 a year in director's fees is typical for a corporation the size of Del Monte). Thus any citizen can see the contradiction that emerges when the university presents itself as an institution dedicated to the broad public interest while at the same time its faculty is being subsidized to do consulting for the private interests of outside employers, or worse, to help direct entire enterprises. At a time when cutbacks are forcing large scale reductions in hiring, academic programs, etc., free handouts to corporations and government continue. It is in this last category that the issue of conflict of interest is most fundamental, because it allows us to look beyond the activities of individual faculty

members to focus on the roles of social institutions.

As we might expect, there are common arguments in defense of current consulting practices. First, there is the elitist-pragmatic view, held by many academics, that the ideal of service to the whole society is merely propaganda (which it is), designed to placate the masses but never taken seriously in practice. The consulting privileges of faculty were obtained during past years when money was plentiful and top rank experts were rare; the universities had no choice but to allow these "big time operators" a free reign, and they had no objection in principle. This position deserves no comment.

Apologists, on the other hand, will claim that the participation by academics in the powerful institutions of our society will provide an enlightening influence and is thus to be praised. (This parallels the arguments advanced in favor of ROTC programs on campus and the justification given by many liberal professors involved in reactionary research programs such as counterinsurgency work for the Pentagon.) The problem is that academics in this position can only work to assist powerful institutions to achieve their goals whatever those goals are; they can try to modify the means (as by suggesting the electronic battlefield alternative to massive bombing in Vietnam) but they must support the ends (military victory) as given.

Another example of this view is the case of Dr. Clifton R. Wharton, Jr., president of Michigan State University, who recently accepted positions on the board of directors of Ford Motor Co. and Burroughs Corp. He announced that he would consider himself to be a "public director" and turned over all his directorship fees to the university. Interviewed about this in *Business Week*, he said, "I view my role as a person who can exercise the responsibility of the directorship to make a profit, and bring to it a broad social and public concern."[18] Left unsaid is what he will (or can) do when these two stated objectives, corporate profits and social good, come into collision with each other—as they surely do.

This criticism of consulting is not to say that all consulting by faculty would be abolished in a different social order where private corporate and illegitimate governmental institutions no longer dominate. We should not imagine the university as an ivory tower; it should be interactive, it should serve society. We should struggle for a society in which education, research and production are much more integrated than at present. Private-property restrictions on knowledge, production-technology and future-research planning would be replaced by public access, discussion and critical evaluation. Consulting would not be an activity of elite, highly privileged individuals who happen to monopolize specific technical knowledge, but rather a communication process involving large numbers of people in all institutions.

A Proposal for Action

While much of the data on faculty consulting presented in this report is new, the broad issues raised are embedded in a rich history of criticism.[19] During the 1960s the U.S. campuses were hotbeds of protest against racism, against imperialism, often against the universities themselves, which were seen as instruments serving those evils. Radicals analyzed the relationship of the university to the larger powers in the society and saw the flow of government dollars into campus research for weapons of war and subtler means of social control, saw the predominance of big business leaders on the boards of trustees or regents that ruled the campuses, saw the calling in of police power to repress student movements that seemed to present any palpable threat to the existing order of things, and saw students being educated not for the "glory of knowledge" but to meet the call for highly trained workers that the corporate system required.

This present study, concentrating on the area of faculty consulting, is intended to illuminate one more aspect of the integrated relationship that exists between the university and the mainstream of U.S. power, showing the outflow of the special expertise of professors into the service of the large corporations and their allied institutions.

The next question we consider concerns action. Extensive public discussion on consulting may itself bring about formal public disclosure of consulting activities. University administrators might decide that it is best for them to institute such a procedure themselves rather than risk too much attention from students, legislators, etc., looking into the cozy arrangements. Under a public disclosure scheme, every faculty or other staff member who engages in outside consulting should make an annual report of this activity for public inspection. It should include the name and location of each person or organization served, the amount of time spent and the compensation received for each consulting job, a brief description of the work done along with copies of any written reports. There is a precedent for this kind of public disclosure in the

Freedom of Information Act, which itself resulted from increased public interest and probing into the activities of government bureaucracies.

Examination of consulting activities in detail would have the effect of stigmatizing the most odious kinds of service and forcing much wider accountability than now exists for others. As has happened to some extent with defense contracting, consulting in some areas would become much less frequent. Of course there is always the recourse of seeking faithful consultants at universities not normally inconvenienced by critical debate—as has also happened in many areas of sensitive research work—but this merely reinforces the need to encourage that debate everywhere.

There will naturally be indignant outrage with this sort of program. Claims will be made about the "invasion of privacy" of faculty members, about "where you draw the line" on these kinds of issues, the bureaucratic burden, and "harrassment and embarrassment." Prof. Luis W. Alvarez, then president of the American Physical Society, explained why a proposal for disclosure of consulting activities was rejected by the Society as follows:

> I do not see how one can find a proper cutoff point for information if one does not restrict it to information concerning one's ability to serve the Physical Society. I think that if I happen to be a member of the Board of Deacons of the local Presbyterian Church, it would be none of the Physical Society's business. I feel the same way about my directorship on the board of the Hewlett-Packard Company, which is known to most of my friends and associates.[20]

It doesn't take much insight to see the difference between a local church and a 500-million dollar electronics-manufacturing corporation, as far as significant consulting involvement is concerned.

It is not difficult to predict which groups would oppose the disclosure of consulting activities. The faculty establishment, university administrations, business and government organizations all have benefited from these traditional arrangements. However, on the other side are students, working people, most consumers and taxpayers in whose interest it would be to see this program aimed at consulting actively pursued. These people are not usually able to hire university "experts" to advance their causes (however, they would usually *benefit* from publicity if they did so), and in fact are often the victims of big business and government agencies that do make use of the professor's special talents.

Generally it would seem that a political program that addresses current consulting practices should attempt to reveal what really goes on to more people, and restrict the freedom that private interests have to utilize these resources unencumbered by public discussion. We should also try to show the way toward a different social order where "consulting" (among other things) would serve the people. In exposing and publicizing the consulting situation, it is especially important to reveal the activities of the most elite faculty members, some of whom as participants in the rule of major institutions have graduated from being servants of the ruling class to being members. We should incorporate critical examination of consulting into broader debates concerning teaching and research goals at every university.

This program can come about only if there is strong and determined effort from students, with some support from faculty, in educating, organizing and agitating. This can go forward in open campus debate and inside faculty committees where students have a voice. Groups can form to work in individual departments to investigate, generate discussions, pool findings and build pressure for changes. Many students have first-hand experience with the professor-away-consulting syndrome and have extensive knowledge of such activities.

Lastly, it should be recognized that this is a systematic problem that can't be solved with a patch-up job of treating symptoms instead of the real problem—ruling class control of society. Only by making this clear and attacking the full spectrum of problems can we bring about a society in which consulting can be done in people's interest.

FOOTNOTES

1. Robert Reinhold, "Academic Jet Set, Schedule is Hectic, Reward is High," New York Times, June 18, 1969, pp. 49, 95.

2. See Seymour E. Harris, "A Statistical Portrait of Higher Education," The Carnegie Commission on Higher Education, New York, McGraw-Hill, p. 532.

3. Reinhold, op. cit.

4. Federal Advisory Committees, First Annual Report of the President to the Congress, including Data on Individual Committees, March, 1973,

Printed for the Committee on Government Operations, Subcommittee on Budgeting, Management, and Expenditures, United States Senate, 93rd Congress, 1st Session, Parts 1-4, May 2, 1973; Part 5, Index January 7, 1974.

5. Charles Schwartz, "The Corporate Connection," *Bulletin of the Atomic Scientists,* October, 1975, pp. 15-18.

6. Letter from Prof. Jeoffrey F. Chew to the author, 4/24/73.

7. From students' notes, 1974.

8. Jack Anderson, *The Anderson Papers,* New York, Random House, 1973, p. 20.

9. For details, see Charles Schwartz, "Corporate Connections of Notable Scientists," *Science for the People,* May, 1975, pp. 30-31.

10. John Walsh, "University Links Raise Conflicts of Interest Issue," *Science,* 164, 25 April 1969, p. 411.

11. William Moore, *San Francisco Chronicle,* 1 June 1974, p. 5.

12. Phillip M. Boffey, *The Brain Bank of America,* New York, McGraw-Hill, 1975, p. 77.

13. David Ignatius, *The Washington Monthly,* October, 1973, p. 21; *San Francisco Chronicle,* November 23, 1972.

14. See "Science Against the People: The Story of Jason," Berkeley SESPA, 1972.

15. Charles Schwartz, letter entitled "Conflicts of Interest," *Physics Today,* January, 1975, p. 13.

16. "Principles Underlying Regulation No. 4," University of California, Office of the President, June 23, 1958.

17. David Kotelchuck, "Asbestos: Science for Sale," *Science for the People,* September, 1975, p. 8.

18. "A Director Everybody Wants," *Business Week,* February 17, 1973, p. 69.

19. Upton Sinclair's book, *The Goose Step: A Study of American Education,* (published by the author, Pasadena, 1923) is remarkably fresh and relevant despite its age. From recent years there is James Ridgeway's *The Closed Corporation: American Universities in Crisis,* New York, Random House, 1968, which contains much information pertinent to this present study. There have also been a number of local campus pamphlets: "Who Rules Columbia?" etc.

20. Schwartz, *Physics Today, loc. cit.*

Table 1

SOME U/C ADMINISTRATORS AND FACULTY ON THE BOARDS OF DIRECTORS OF SIZEABLE CORPORATIONS

Vice Presidents

Chester O. McCorke, Jr.	Del Monte Corp.**
	Universal Foods Corp.*
James B. Kendrick, Jr.	Tejon Agricultural Corp.

Chancellors

Daniel G. Aldrich, Jr. (Irvine)	Buffums, Inc.
	Stanford Research Inst.
William D. McElroy (San Diego)	Southern California First Natl. Bank*
Charles E. Young (Los Angeles)	Intel Corp.*

Faculty, Berkeley

Luis W. Alvarez (Physics)	Hewlett-Packard Co.*
Melvin Calvin (Chemistry)	Dow Chemical Co.**
Richard H. Holton (Bus. Admin.)	Rucker Co.*
	Dymo Industries, Inc.*
	Northwestern Mutual Life Insurance Co.**
Kenneth S. Pitzer (Chemistry)	Owen-Illinois, Inc.**
Glenn T. Seaborg (Chemistry)	Dryfus Third Century Fund
Edward Teller (Physics)	Thermo Electron Corp.
Charles H. Townes (Physics)	Perkin-Elmer Corp.*
	General Motors Corp.**
Theodore Vermeulen (Chem. Eng.)	Memorex Corp.*
John R. Whinnery (Elec. Eng.)	Granger Associates

Faculty, Los Angeles

Neil H. Jacoby (Management)	Occidental Petroleum Corp.**
Willard F. Libby (Chemistry)	Nuclear Systems, Inc.
	Research-Cottrell, Inc.*
Harold M. Williams (Management)	Signal Companies, Inc.**
	Norton Simon, Inc.**
	ARA Services, Inc.**
	CNA Financial Corp.**

*Corporations having over $100,000,000 in annual sales or total assets.
**Corporations having over $1,000,000,000 in annual sales or total assets.

TABLE 2

ACADEMICS ON THE BOARDS OF DIRECTORS OF THE 130 LARGEST U.S. CORPORATIONS

From a search of the annual reports of the companies listed by *Fortune* in 1974: the 100 top industrials and the 5 top companies from each of the other 6 categories. Many of the individuals listed below also sit on the boards of other, lesser, corporations.

University of California
Melvin Calvin (Prof. Chemistry, Berkeley)—Dow Chemical
Neil H. Jacoby (Prof. Management, Los Angeles)—Occidental Petroleum
Kenneth S. Pitzer (Prof. Chemistry, Berkeley)—Owens-Illinois
Charles H. Townes (Prof. Physics, Berkeley)—General Motors
Harold M. Williams (Dean, Management, Los Angeles)—Signal
  Companies

University of Michigan
W. J. Cohen (Dean, Education)—Bendix
Morgan Collins (Prof. Emer. Business)—S.S. Kresge
Robben W. Fleming (Pres.)—Chrysler
Paul W. McCracken (Prof. Business)—S.S. Kresge
William E. Stirton (Vice Pres. Emer.)—American Motors

Harvard University
Donald K. David (Prof. Business)—Xerox; Great A&P Tea Co.
Lawrence E. Fouraker (Dean, Business)—RCA; First National City
  Bank, NY
Jean Mayer (Prof. Nutrition)—Monsanto Chemical
Frederick J. Stare (Prof. Nutrition)—Continental Can

Massachusetts Institute of Technology
Howard W. Johnson (Chm.)—Du Pont; J.P. Morgan; John Hancock Life;
  Champion Int'l.
James R. Killian, Jr. (Hon. Chm.)—General Motors; AT&T
William F. Pounds (Dean, Management)— Sun Oil
Jerome B. Wiesner (Pres.)—Celanese

Columbia University
Courtney C. Brown (Dean, Emer. Business)—Borden; American Electric
  Power
Grayson Kirk (Pres. Emer.)—IBM; Consolidated Edison
William J. McGill (Pres.)—Texaco; AT&T

California Institute of Technology
Robert F. Bacher (Prof. Physics)—TRW
Harold Brown (Pres.)—IBM

Cornell University
John E. Deitrick (Dean, Emer. Medicine)—Prudential Life
Franklin A. Long (Prof. Chemistry)—Exxon

Duke University
Juanita M. Kreps (Vice Pres.)—J.C. Penney
Terry Sanford (Pres.)—Cities Services

Northwestern University
John A. Barr (Dean, Management)—Esmark
Donald PL Jacobs (Prof. Finance)—Union Oil

Princeton University
Burton G. Malkiel (Prof. Economics)—Prudential Life
Courtland D. Perkins (Prof. Engineering)—American Airlines

Purdue University
Frederick L. Hovde (Pres. Emer.)—General Electric, Inland Steel
Mary Ella Robertson (Prof.)—John Hancock Life

University of Rochester
Robert L. Sproull (Pres.)—United Aircraft
W. Allen Wallis (Chancellor)—Eastman Kodak; Esmark; Metropolitan Life

Stanford University
Arjay Miller (Dean, Business)—Ford
J.E. Wallace Sterling (Chancellor)—Shell Oil

Barnard College
Martha E. Peterson (Pres.)—Metropolitan Life

Brown University
Donald F. Hornig (Pres.)—Westinghouse

Bryn Mawr College
Katherine E. McBride (Pres.)—New York Life

California State University
Brage Golding (Pres., San Diego)—Armco Steel

Carnegie Institution of Washington
Caryl P. Haskins (Pres.)—DuPont

Case Institute of Technology
T. Keith Glennan (Pres. Emer.)—Republic Steel

Emory University
E. Garland Herndon, Jr. (Vice Pres.)—Coca-Cola

Hunter College
Robert C. Weaver (Prof. Urban Affairs)—Metropolitan Life

University of Illinois
John Bardeen (Prof. Physics)—Xerox

Illinois Institute of Technology
John T. Rettaliata (Pres. Emer.)—Western Electric; International
    Harvester

University of Leiden
Ernst. H. van der Beugel (Prof. Int'l Relations)—Xerox

Marquette University
Charles W. Miller (Prof. Business)—W.R. Grace

Meharry Medical College
Lloyd C. Elam (Pres.)—Kraftco

Michigan State University
Clifton R. Wharton, Jr. (Pres.)—Ford; Equitable Life

University of Nebraska
Durward B. Varner (Pres.)—Beatrice Foods

New York University
James M. Hester (Pres.)—Union Carbide

Notre Dame University
Theodore H. Hesburgh (Pres.)—Chase Manhattan Bank

Pepperdine University
M. Norvel Young (Chancellor)—Lockheed

University of Pittsburgh
Marina vN. Whitman (Prof. Economics)—Westinghouse; Manufacturers
  Hanover Trust

Pomona College
David Alexander (Pres.)—Great Western Financial

Rensselaer Polytechnic University
Richard G. Folsom (Pres. Emer.)—American Electric Power; Bendix

Rutgers University
Margery Somers Foster (Dean, Douglass College)—Prudential Life

University of Southern California
Norman H. Topping (Chancellor)—Litton

Syracuse University
William P. Tolley (Chancellor, Emer.)—Colgate Palmolive

Tulane University
Herbert E. Longenecker (Pres.)—CPC International; Equitable Life

Tuskegee Institution
Luther H. Foster (Pres.)—Sears & Roebuck

Virginia Polytechnic Institution
T. Marshall Hahn (Pres.)—Georgia-Pacific

Washington University
William H. Danforth (Chancellor)—Ralston Purina

Wayne State University
Edward L. Cushman (Vice Pres.)—American Motors

Wesleyan University
Edwin D. Etherington (Pres. Emer.)—American Express

Yale University
John Perry Miller (Prof. Economics)—Aetna Life & Casualty

Statistical Summary

68 academic people, from   44 universities, holding 85 directorships, on the boards of 66 corporations. When these findings are compared with the tabulation given by Ridgeway (14) we find that the presence of academics on the boards of these largest corporations has increased by 65% in the seven year interval between these two studies.

Computers in South Africa:
A Survey of U.S. Companies
—Richard Leonard

Yes, computers are only tools, but tools with tremendous power. This article describes some of their myriad uses in controlling the lives of the black population to the advantage of the 16% white minority of South Africa. Here is the prototype of big brother and the details of U.S. corporate and government complicity.

Computers are a vital part of the apparatus the South African government uses to enforce apartheid. They are also critical to the operation of the modern South African economy. U.S. corporations are the major source of computers for South Africa. The white-controlled government is the largest computer user in South Africa.[1] Computers are used by the South African military for nuclear development, in internal security, for the administration of apartheid, and by private and state-owned corporations.

In 1971 the managing director of the South African subsidiary of Burroughs Corporation, a giant U.S. computer manufacturer, told two American researchers:

> We're entirely dependent on the U.S. The economy would grind to a halt without access to the computer technology of the West. No bank could function; the government couldn't collect its money and couldn't account for it; business couldn't operate; payrolls could not be paid. Retail and wholesale marketing and related services would be disrupted.[2]

The vulnerability of South African computer users and the entire South African economy to sanctions or boycotts affecting computers was recently stressed by the head of computer operations of South Africa's Anglo-American Corporation:

> No other sector of the economy is as utterly dependent as the computer industry is on the multinationals, and it is a sector through which a stranglehold can be applied on the whole economy.[3]

This study will examine the role of U.S. computer companies in South Africa, the ways that computers strengthen apartheid, and it will critically analyze the claims made by the companies in justifying their South African operations.

The Corporations Involved

U.S. computer companies do not manufacture computers in South Africa; their operations consist of sales and service.

Five U.S. companies which produce large "mainframe" as well as smaller computers have operations in South Africa.[4] They are: International Business Machines, Burroughs, Control Data, NCR, Sperry Rand.

Several U.S. companies manufacturing small computers also have operations there: Hewlett-Packard, Data General, Datapoint, Computer Automation.

Some U.S. companies sell computers to South Africa but have no offices in the country.[5] These include Digital Equipment and Foxboro Corporation, which sold at least two of its FOX 1 computers for use at the highly strategic Valindaba uranium enrichment plant.[6] Honeywell has withdrawn from the computer market and Singer Business Machines has sold its South African subsidiary to the British computer company International Computors Ltd. (ICL).

History of Usage

South Africa was late among the industrialized nations to enter the computer era. The first computer, a British one, was installed in 1959. During the 1960s the computers industry in South Africa expanded at a rate of more than 30% annually,[7] and by 1970 there were an estimated 400 computers in the country with a value of some $100 million.[8]

During this period U.S. computer companies came to dominate

the South African market, with IBM leading the field. By 1974 the total number of computers was estimated at more than 1000, with a value of $365 million.[9] While the South African economy slipped into a recession in the mid-70s, computers were described as "selling like hot-cakes," with market growth estimated at between 20% and 30% a year.[10] In September, 1976 South African Interior Minister Mulder said that there were 1500 computers in the country representing an investment of more than $500 million.[11] *Management* magazine of South Africa noted in its December, 1977 computer survey that only the U.S. and Britain spend more than South Africa on computers as a percentage of GNP.

A recent article in the *Sunday Times* of Johannesburg estimates the growth rate of the computer industry in South Africa at 20-25 percent for 1978, with the market for minicomputers expanding the most rapidly.[12] The annual value of computer sales is estimated to be approximately $100 million. The three largest companies (IBM, ICL, and Burroughs) have combined annual sales of some $200 million, but this includes software, service, and other business equipment.

IBM is ranked as having the highest revenue from computer sales in South Africa. ICL of Britain is ranked second in sales while claiming to have sold the largest number of computers. Burroughs is ranked third in computer sales.

The Apartheid System

In South Africa the white minority, 16 percent of the population, controls all aspects of the political, economic and social life of the country. South Africa is a country based on racial domination, maintained by violence and terror.

Apartheid is enforced by repression on a massive scale. Amnesty International in its recent report *Political Imprisonment in South Africa* says that repression is implemented by "...a system built upon detention without trial, banning and banishment, the widespread and systematic use of torture, and frequent judicial and extra-judicial killings by the government."[13] For the first time in its history, Amnesty International refused to include specific recommendations to the government, saying "no reforms in the present structure will be sufficiently far reaching to remove the causes of political imprisonment unless the whole system of apartheid is dismantled."

This society, in which U.S. computers play an integral part, is in the midst of revolt. The black majority refuses to accept apartheid. The brutality of the white regime's efforts to crush the black movement for liberation has been shown by the police killings of an estimated 1000 young black men and women involved in protests against apartheid that began in Soweto in 1976.

The underlying postulate of the Nationalists' apartheid policies is that the African majority are not South African at all, but foreigners, residents of ten fragmented, impoverished, rural "bantustans" created by the white regime and comprising only 13 percent of the country's land area. Africans are being denied citizenship and political rights in the country they were born in and helped to build. Millions of Africans who have lived in urban areas for decades, serving in the modern economy and demanding equal political rights, have been forced back to these bantustans. They are allowed in white areas only as transient migrant workers to serve the labor needs of whites.

Apartheid has demanded an extraordinary degree of repression and administrative control by the regime. The white government, industries, mines, and farms are all dependent on black labor. Virtually all white homes are served by black domestic workers. At the same time, black South Africans are kept powerless and under white control. Some 200 laws govern all aspects of black people's lives: where and how they may work, eat, live, sleep, drink, be born, die and be buried.[14]

It is not surprising that U.S. computer companies have found a booming market for their products in South Africa. Computers have become a valuable tool in numerous government agencies administering apartheid. Further, while apartheid has built white prosperity at the expense of the black majority it is also placing a tremendous strain on white resources. Faced with the emergence of black revolution at home and growing international isolation, the white regime has been pursuing a policy of "strategic investment" to promote development and self-sufficiency in defense, atomic energy, oil, and electricity; telecommunications; transportation; mining; and the production of steel and aluminum.[15] Computers are used by the government and private corporations to promote the greater sophistication of operations in all these strategic fields.

Thus the significance of the role of U.S. computers in South Africa is not restricted to their use by repressive agencies of the government. The web of apartheid laws extends throughout the government administration and state-controlled corporations into the operations of business and all spheres of social life. The modern corporate economy served by U.S. computers meets the strategic needs of the white regime in many areas.

How Computers Work

Computers are electronic machines for doing numerical calculations. On the basis of computer programs (called "software") they can store, manipulate, and retrieve information ("data") as called for by their users. What all computers can do, with variations depending on size, are prodigious feats of "number crunching," making monumental calculations in a fraction of a second. Computers can be linked on a world-wide basis to exchange information, and data may be provided to and retrieved from a central computer office by decentralized local computer terminals.

It is true that computers are just another tool and must be programmed and maintained by people. But they are tools with tremendous power. Technological advances of the '70s have produced the microprocessor (computer "chip"), a single piece of silicon 1/6" x 1/8" which can contain 2,250 transistors and has the power of the first room-sized computer or a 1960s computer the size of a desk. "In theory the same chip could do everything, from guiding a missile to switching on a roast," says a recent article in *Time* mazagine.[16]

Military and Police Use of Computers

The military applications of computers are vast and include areas as complex as missile guidance and as mundane as menu planning. Computers are used in the production of nuclear weapons. Computers are used in early warning systems and to track and control satellites and aircraft. Computers are used for navigation, for military codes and communications, for logistics and planning, for surveillance and intelligence, and for procurement and recruitment.

Some idea of the military importance of computers can be gained by the listing below of the rank in the top 100 U.S. Department of Defense contractors for 1976 of those U.S. computor companies which have operations in South Africa.[17]

Computer	DOD Contractor Rank, 1976	U.S. Contracts ($ million)
Sperry-Rand	14	505.5
IBM	30	255.9
Control Data	47	121.7
Burroughs	88	55.5
Hewlett-Packard	97	44.5
Computer Sciences	98	44.3

Computers are the basis for the "electronic battlefield" developed by the U.S. in Vietnam. In 1969 General William Westmoreland described the new military technology.

> On the battlefield of the future enemy forces will be located, tracked, and targeted almost instantaneously through the use of data links, computer-assisted intelligence evaluation, and automatic fire control.[18]

A recent *New York Times* report notes that the major part of the $44.4 billion Pentagon budget request for military research and development for 1979 is earmarked for this type of program.[19]

Computers have also come to play important roles in police work. As in the military, computers can add efficiency in all phases of administrative and logistical work. Computers allow the maintenance of centralized data banks of arrest records, criminal convictions, and other information which can be linked with local terminals.

A recent article entitled "Why Police States Love the Computer" by journalist Hesh Weiner underscored the repressive potential of computer usage. "Pioneered by the wealthy and technologically advanced democracies, the use of computer systems for police, political, health, and economic administration is now a high priority for every dictatorship."[20]

U.S. COMPUTERS IN SOUTH AFRICA*

South African Military

In 1971 a South African newspaper report described some of the areas in which computers were being used by the South African military forces:

---

*The basic source of information for this section, unless specified otherwise, is the December, 1977 survey of computer users in South Africa compiled by the South African magazine *Management*. This listing is not totally comprehensive, though the magazine notes that it was easier to account for large computers than small ones.

In its previous November, 1974 survey *Management* had specified the computers being used by military, arms production and procurement, and atomic energy agencies. In the 1977 survey these agencies are cited only as using "various" unspecified computers.

In December, 1977 the Control Data Corporation provided the Interfaith Center on Corporate Responsiblity with a listing of its major computer installations in South Africa. However, this listing did not specify Control Data computers being used by the South African Atomic Energy Board and South African Airways that are listed by *Management*.

Computers have been built into the South African Air Force early-warning system to make it far more sophisticated and effective...

The computers have been incorporated in the underground nerve-center of the Northern Air Defence Sector at Devon, in the satellite radar station at Ellisras, near the Botswana border, as well as at Mariepkop on the edge of the Transvaal Drakensberg escarpment commanding the Lowveld and the [Mozambican] border.

A computer also functions in the latest equipment of the Mobile Radar Unit—a branch of the Strike Command.[21]

The origins of the computers described in this article have never been established. However, the 1974 *Management* survey lists the South African Defence Department as using *all* IBM equipment (an IBM 360/40, and IBM 370/145, and an IBM 370/145, and an IBM 370/158) for "personnel, financial, and stock control" purposes.

In Congressional testimony IBM vice president Gilbert Jones stated that no IBM computers sold to the South African government have been used for security functions or military use, but only for accounting, payroll, or administrative purposes.[22]

Defending this claim, IBM chairman Frank Cary stated at the corporation's 1977 annual meeting:

We have investigated each instance brought to our attention [*note*: including the question of use by the South African military], where there has been any reason to believe that [IBM computers] might be used to abridge human rights, and...we have found no instances in which they have.[23]

Yet only a few moments before Cary had said:

We would not bid any business where we believe that our products are going to be used to abridge human rights. However, we do not see how IBM or any other computer manufacturer can guarantee that they will not be. The facts of the matter are that we do not and cannot control the actions of our customers...

...there are thousands of experienced customers throughout the world who do all their own work, and there are customers who for reasons of business or national security *do not allow anyone to know what they are doing.* [italics added]

Thus, IBM spokesmen have themselves admitted their inability to control the uses to which their computers are put.

IBM has also argued that in compliance with U.S. arms embargo restrictions they have not suppled "military" computers to South Africa, such as those developed by their Federal Systems Division for the U.S. military.[24] However, ordinary commercial computers can be put to military use. An article in *Armed Forces* magazine of South Africa on the use of computers in military air and marine simulators (for training) notes that the requirements for military simulators have been built into computers which are commercially available in South Africa.[25] [IBM has taken out at least one full-page advertisement in *Armed Forces*.[26]]

Further, any computer used by the South African armed forces, even for purely administrative purposes, adds to their efficiency and capabilities. As a specific example of this, a deserter from the South African armed forces has recently given information reaching the Interfaith Center on Corporate Responsibility that a computer is being used in the South African military draft program.[27]

The 1977 *Management* survey states that the South African Defense Force is implementing a $13 million computer-based army logistics project (code-named Convor), which is expected to save the South African military $55 million annually when installed. The kind of computer being used is not revealed.

The South African Defence Research Institute was listed in the 1977 survey as using "various" computers. This agency was not listed in the 1974 survey.

The National Institute of Telecommunications Research does both civilian and military (in areas such as radar and communications) research. In the 1977 survey this institute was listed as using an IBM 1130 computer (linked to the IBM 370/115 at the Council for Scientific and Industrial Research), and six Hewlett-Packard computers.

Arms Manufacturing and Procurement

ARMSCOR (the South African Arms Development and Production Corporation) is the state-owned corporation which has developed South Africa's domestic arms production capabilities. ARMSCOR was listed as using "various" computers in the 1977 survey. With the aid of foreign technology and licensing agreements South Africa can now manufacture a wide range of sophisticated modern weapons, such as Mirage jet fighter planes and Panhard armored cars, as well as smaller arms. Any computer used by ARMSCOR must inevitably strengthen South African military capabilities by direct involvement in armaments production.

The agency handling military procurement is the South African Armaments Board. In 1974 the Armaments Board was listed as using an NCR C100 computer and a Hewlett-Packard 2116 computer, and in 1977 as using "various" computers. The most significant aspects of South Africa's arms procurement are conducted on the international scene and are shrouded in secrecy, but a report last year by Sean Gervasi[28] indicates that South Africa has been able to build a powerful modern military machine based on weapons provided by Western powers in violation of the voluntary international arms embargo imposed by the UN since 1963 and made mandatory by the Security Council in 1977. U.S. computers used by the procurement agency must thus serve to streamline procurement procedures and add to South African military strength.

Nuclear Power

South Africa's nuclear arms potential caused an international incident in 1977 when the USSR and then France reported a nuclear weapons test site in preparation in South Africa. The U.S., which has long aided South Africa's nuclear development, confirmed these assertions and President Carter warned South Africa not to proceed with any tests. South Africa, which claims that it is able to make nuclear weapons, was apparently dissuaded by U.S. pressure from conducting tests. But the nuclear capacity of the white regime threatens both neighboring African countries supporting black liberation in South Africa and the fabric of world peace, and South Africa has consistently refused to sign the Nuclear Non-proliferation Treaty.

The importance of computers for South Africa's nuclear programs was underscored in a 1977 article by journalists Tami Hultman and Reed Kramer of *Africa News*:

> In 1973, two Foxboro Corp. engineers left their Massachusetts factory for South Africa, where they supervised the installation of two FOX I computers purchased by the South African government. Negotiations for the deal, code named Project Houston, had been conducted with extraordinary secrecy. Not until the two engineers were

in South Africa did they learn that the Foxboro equipment was the key to an experimental uranium enrichment
plant, a highly clandestine facility outside the network of
international nuclear safeguards.

Two years later, South Africa successfully brought
the Valindaba enrichment plant near Pretoria into operation, thereby propelling itself into that elite club of nations
that had mastered the secret of transforming raw uranium into a form usable in nuclear reactors. South Africa
has passed the major hurdles towards making nuclear fuel
and weapons—an event of enormous importance in a
world both energy hungry and insecure.[29]

The two Foxboro computers had been sold to the South
African government-owned Uranium Enrichment Corporation
(UCOR) for $1.8 million for the stated purpose of the "operation of
experimental facilities and pilot plants for nuclear research and
development." The deal was approved without question by the
State Department and the Commerce Department. If U.S. officials
were untroubled by the prospect of aiding South Africa in its drive
to become a nuclear power, they were curious enough to have the
CIA question the Foxboro engineers about this secret facility on
their return. They said they had been restricted to the plant's
computer area and closely watched.

In addition to the obvious military threat posed by South
Africa's nuclear potential there are other strategic considerations
as well. South Africa has no known oil reserves. Nuclear power is a
valuable alternative energy source, a buffer against oil sanctions.
Also in a world facing future energy shortages South Africa's
ability to market enriched uranium, essential for the production of
nuclear energy, is a strong economic asset.

"We now have the bargaining power of any Arab country with
a lot of oil," claimed South African Atomic Energy Board vice
president Lou Alberts in 1974. South Africa claims to have scaled
down its immediate plans for the construction of a commercial
enrichment plant at Valindaba because of excessive costs, but its
potential capacity to produce enriched uranium gives it important
economic and military leverage with the U.S., Japan and countries
of western Europe.[30]

South Africa's atomic research program has drawn on the
resources of several U.S. computer companies. The Pelindaba
atomic research facility, for instance, was at one time equipped
with an IBM 360/40 computer. In a letter of March 14, 1978 IBM

informed the American Committee on Africa that this computer was no longer operating the research facility. The letter confirmed, however, that the South African Atomic Energy Board owned an IBM 370/155 computer, which, according to IBM, was installed to assist with "reactor development" and is also used for "recording and controlling the industrial use of radioactive materials."

The 1974 *Management* survey had noted the IBM 370/155 in use by the Atomic Energy Board, as well as a Control Data 1700, a Hewlett-Packard 2115 and three 2114's, and a Computer Sciences Varian 620L computer. There was no 1974 listing for the Uranium Enrichment Corporation. In 1977 both agencies were listed as using "various" computers.

In 1977, the director of Control Data's South African subsidiary informed the Interfaith Center on Corporate Responsibility that the Atomic Energy Board, the Uranium Enrichment Corporation and the Department of Defense each have "a wealth of computers (all IBM)," but this does not gibe with the information noted above.[31]

Apartheid and Internal Security

The *Department of Justice* administers the legal structure of apartheid, the laws which mandate white domination and which make South Africa a police state, by proscribing groups opposing apartheid and allowing detention and banning (similar to house arrest) without trial. Department of Justice policies have implicitly sanctioned police violence and torture against opponents of apartheid—both black and white. This agency was listed in 1977 as using six Data General Nova computers, reportedly for financial and statistical purposes.

The *Department of Prisons* is a pervasive part of the picture of apartheid. South African prisons hold hundreds of political prisoners and detainees held without trial. Many of the leaders of the South African liberation movement are imprisoned on the notorious Robben Island. The majority of South African prisoners are those charged with violations of the apartheid pass and "influx control" laws—there were 250,000 blacks arrested under these laws alone in 1976.[32] The Department of Prisons uses an IBM 360/20 computer for "financial" purposes.

The *Department of Interior* has been preparing the "book of life" identity document to be used for South African Whites, Coloureds, and Asians. This agency is using two IBM 370/158 computers and

IBM has admitted that its computers are being used for the "book of life" program. IBM also admits that it had bid for the program to computerize the "passbook," the central instrument of imposing apartheid on Africans. It lost out to ICL, the British company.

According to IBM, its involvement with the "book of life" does not constitute support for apartheid or the abridgement of human rights, but this claim cannot be sustained in light of the importance of identity documents in a police state based on race.[33]

The Department of Labor controls the terms and conditions of work under apartheid. Under this agency's policies no black worker may supervise a white worker. Black trade unions are not recognized and most strikes are illegal. The Department of Labor uses two Data General Nova computers.

The East Rand Bantu Administration Board is using a Burroughs 3700 computer. These boards administer apartheid on the local level by decree. They administer the pass laws, and are instrumental in destroying Africa family life by preventing a man from living with his wife and children when he goes to work in a town. They have the power to herd people into compounds, destroy unauthorized housing areas, and arbitrarily raise rents. They are a hated symbol of apartheid for blacks and were a prime target during the student rebellion in 1976. A number were wrecked, their files and records burned in an effort to destroy their control of black lives. By computerizing its systems, however, the white regime can keep local apartheid records securely stored in central computers, ready to be put into use when the need arises.

The Bantustans represent the ultimate goal of apartheid: the dispossession of the African majority to be accomplished by forcing them to become citizens of "independent" bantustans comprising only 13% of the country. The administrations of the Ciskei and the Transkei are using British ICL 2903 computers.

A growing number of *local municipal administrations* are also using computers to solve problems of control. NCR has been a principal supplier for these purposes and NCR computers are being used by the following municipalities: Bloemfontein, King Williams Town, Parow, Pietersburg, Stellenbosch, and Worcester. The town councils of Alberton and Witbank also use NCR computers. The Randburg municipality is using a Control Data Cyber 18 computer.

The Transvaal Provincial Administration is using a Burroughs Dual 6700 computer for administrative functions and "law enforcement."

Business and Industry

Computers are an essential part of modern business and industry and are widely used in the private sector in South Africa, both by U.S. and other foreign companies, and by local South African businesses. Defenders of business have claimed that economic expansion would bring about the breakdown of apartheid, but the economic boom of the 1960s and '70s has brought prosperity, "for whites only," while the Nationalist regime has intensified repression to maintain white control. Business prosperity is seen among white South Africans as vindicating the apartheid policies of the Nationalist Party and strengthening their capacity to rule.

U.S. Government Imposes Restrictions

The export of computers to South Africa first began to be subject to restrictions in December, 1976 (aside from the terms of the United Nations arms embargo adhered to by the U.S. since 1963) when general licenses for the export of large computers to "free world" countries (including South Africa) were revoked. To export large computers to South Africa, companies had to apply for individual licenses. Approval of the export of computers for sale to South African police, military, or atomic energy agencies was subject to review by the State Department.

In February, 1977 President Carter announced that in cases of applications to export computers to foreign police agencies, the Department of State would recommend denial of the export license where it was believed that the computer would be used in the suppression of human rights.

In November, 1977, following the passage of a mandatory arms embargo resolution by the UN Security Council, the Carter administration announced new curbs affecting computer sales to South Africa, which were implemented in Department of Commerce regulations published on February 16, 1978. The new regulations prohibit, in furtherance of the administration's policies "supporting human rights," the sale of *any* U.S. commodities or technical data to military or police entities in South Africa and Namibia. The restrictions apply not only to the direct sale of commodities, but any form of indirect sales or other arrangements through subsidiaries or other companies as well.

However, the impact of these restrictions is limited in several ways. First, while computers are vital for the South African police and military, these agencies account for only a small number of the total number of computers in use throughout the country, many of which also play strategic roles in assisting apartheid. Second, it is not clear how tightly the new restrictions can or will be enforced.

An example of the enforcement problem arose soon after the restrictions were announced, when the manager of IBM's South African subsidiary was reported by the British magazine *Computer Weekly* to have said that IBM in South Africa would continue to supply spare parts and service to any affected military or police computers as long as parts supplies lasted.[34]

The West German subsidiary of IBM and the Japanese company Hitachi were both reported to have offered to supply parts to service embargoed IBM equipment.[35] Such arrangements could violate the new restrictions, however, since the parts would be made under IBM patents registered in the U.S. and such products appear to be included in the embargo. It is clear that IBM's concern is to maintain its servicing contracts with affected South African agencies, rather than try to operate within the human rights spirit of the Carter regulations.

The IBM computers which have been listed as being used by the South African Department of Defence, for example, are commercial models similar to ones being used by other government agencies and corporate customers. It could therefore be difficult for the U.S. government to monitor spare parts that might go to the restricted agencies, as opposed to those going to other customers.

There are further problems as well. The South African government has already established a cooperative network among its computer sections in eight separate government offices.[36] Therefore, work for military or police agencies could be farmed out to other agencies with computers.

A range of other suggestions about ways to blunt U.S. curbs was made by the head of Anglo-American Corporation's computer company, interviewed in an article entitled "How to Beat the Computer Seige" in the *Sunday Times* of Johannesburg, March 26, 1978. These include setting up local computer leasing and maintenance companies, as well as government funding for the local South African manufacturers of small and medium-sized computers. He noted that there is a good deal of spare computer capacity in existing government installations and "there is no way the U.S. authorities will be able to prevent this capacity from being shared by the defence and police departments."

The ability of the South African government to exploit

existing computer installations extends into the private sector as well. Under the terms of the 1970 National Supplies Procurement Act, the government may order any company operating in South Africa to deliver products to the government that it determines are necessary for national security. Thus any computer or computer part in the country could be seized and used by the white regime.

Another South African law could make it difficult for the U.S. to openly monitor the actions of the South African government as they relate to U.S. computer companies in South Africa. The Second General Law Amendment Act of 1974 makes it a criminal offense for anyone in South Africa to provide information in reply to any request from outside South Africa concerning any aspect of business, without government permission.

A striking example of the ineffectuality of U.S. legal restrictions relating to corporate operations in South Africa is the well documented charge that Mobil Oil's South African subsidiary was a continuing supplier of petroleum products to Rhodesia in violation of U.S. criminal law enforcing economic sanctions.[37]

In the face of investigations by the Treasury Department and the Senate Foreign Relations Committee, Mobil claimed that the South African Official Secrets Act and other laws prevented it from obtaining information from its own subsidiary. Under South African law, it seems, Mobil's subsidiary is *required* both to hold all the information about these transactions confidential.

This case shows that subsidiaries of U.S. corporations in South Africa are free to flout U.S. legal restrictions, under the protection of South African law. Thus the current restrictions imposed by the Carter administration, while posing a minor hindrance to the white regime, cannot be seen as a serious effort to end the involvement of U.S. computer companies with apartheid, and make the need for corporate withdrawal all the more urgent.

Computer Manufacturing in South Africa

Another development which weakens the effects of U.S. restrictions on computer sales to South Africa is the emergence of a local South African computer industry. While South Africa remains totally dependent on foreign companies, mainly U.S., for large computers, several minicomputers are now being produced and sold in South Africa. These include the Commander made by Messina, the 800 produced by Anker Data Systems, Hamac produced by the Mercedes group, and the Syfa produced in South Africa by the Commercial Systems Division of Computer Automation of the U.S., which may go on the South African market in 1978.[38]

The military and strategic implications of these developments can be seen by the fact that Standard Telephone and Cables of South Africa (in which ITT of the U.S. holds a large interest) is working on a military-specifications minicomputer using Israeli components, which they expect to have in production by the end of 1978.[39]

Yet the South African computer industry still remains dependent on foreign sources. The Messina Commander, for example, has a foreign content of 40% materials and labor.[40] A South African expert has noted, "the U.S. does have considerable muscle in this field as the bulk of computer patents in the world are still held in the United States."[41] This indicates that to be most effective in the computer field, sanctions by the U.S. would have to not only mandate withdrawal by U.S. companies, but also block licensing and patent technology agreements with South Africa.

*****

Computer Companies Defend Their Role in South Africa

The growing international pressure for corporations to stop collaborating with apartheid and withdraw from South Africa began in the early 1960s after the Sharpeville massacre. Computer companies in particular have faced public scrutiny and protest for their role in strengthening apartheid. Since 1971 when the American Committee on Africa published a fact sheet condemning IBM's operations in South Africa and national church agencies called for IBM to disclose information on its operations in South Africa, IBM has continued to face strong protest. Other companies such as Control Data and Burroughs also have been cited and criticized for collaboration with racism and repression in South Africa.

The computer companies have justified their decisions to remain in South Africa by claiming that their operations do not constitute material and moral support for the apartheid system. Ultimately the companies are caught in a contradiction. They contend that they would not sell computers for repressive use, but also admit that they cannot control the actions of those who use their equipment. All computer companies exporting their products to South Africa must certify to the U.S. Commerce Department how that equipment is to be used; if there is a deviation from uses specified in the export license, the company or distributor could face legal penalties.

So in order to defend and facilitate continued sales the companies imply that they do know about and can control the end-uses of their equipment. In fact this almost certainly is not true, for, once delivered, computers can be adapted to very different purposes from those originally specified.

Summarized below are the arguments most commonly put forward by the companies, followed by a brief critical analysis of each position.

Claim 1: Computers supplied by U.S. companies are not used for repressive purposes.

Control Data announced in October 1977 that it would not increase its South African investment, citing government repression as a factor, but it has not ended all sales. Frank Cary, Chairman of IBM, told the 1978 IBM annual meeting that "on more than one occasion" IBM had not bid for a computer contract because the company thought the computer might be used for repressive purposes. But these are very general statements, and all guidelines have been left very vague.

Chairman Cary had told the 1977 annual meeting:

> You know also that I have said time and again that we have investigated each instance brought to our attention where there was any reason to believe IBM computers might be used for repressive purposes, and we have found no such use.

Asked about IBM's criteria used in such investigations, Mr. Cary said at the 1978 annual meeting that the company could not use general critieria, and that investigations are made on a case-by-case basis, but "we know what repression is when we see it."

Analysis

This study has shown that IBM computers, and those of other U.S. and foreign companies, are being used pervasively by the South African government and in South African industry in ways that contribute to enforcing apartheid and building the strategic power of the regime.

IBM has adopted the convenient tactic of first defining the crime, then conducting the trial, and finally declaring itself innocent. Further, Mr. Cary admitted at the 1977 IMB annual meeting that during the previous year the question of the use or misuse of IBM computers in totalitarian societies such as South Africa or Chile had never been on the agenda of the meetings of the Board of Directors.[42]

IBM finds itself not guilty by arguing that although certain South African government agencies use IBM computers, they do so only for "administration." Since "administration" in IBM's view would not seem to constitute repression, then the computeres are not put to repressive use. This approach ignores the nature both of the government and of the administrative tasks being performed.

Claim 2: We care about human rights, and will do nothing to abridge them, but it is not possible to control all final uses of our computers.

Thus, IBM Chairman Frank Cary told the April 1977 annual meeting that:

> We would not bid any business where we believe that our products are going to be used to abridge human rights. However, we do not see how IBM or any other computer manufacturer can guarantee that they will not be. The facts of the matter are that we do not and can not control the actions of our customers, and it would be grossly misleading to espouse a policy that we cannot enforce.

Similarly, Control Data said in October, 1977 that "no U.S. company would want its activities to lend support to the abridgement of human rights anywhere." But Chairman William C. Norris qualified that statement in a recent dialogue with church leaders by saying:

> You can't place restrictions on a computer that you put in someone else's country...(or) you wouldn't be selling computers in this world. You do the best you can, and that's exactly what we're doing.[43]

Analysis

IBM's statement marked a new departure in the company's rhetoric: previously it had emphasized its willingness to sell computers anywhere not prohibited by U.S. law. But despite the language voicing a concern for human rights, current statements by IBM and Control Data do not extend beyond the human rights considerations recently incorporated into licensing procedures for the export of computers by the Carter administration. The companies are still willing to make available to South Africa just as much as the U.S. government will allow.

And despite their expressed concern for human rights, the computer companies are still reluctant to go along with one aspect of the adminstration curbs—the regulation placing an embargo on the supply of spare parts for computers already in place. The Computer and Business Equipment Manufacturers Association (of which both IBM and Control Data are members) wrote the Department of Commerce complaining that the new regulations, which will force them to default on their existing maintenance contracts, "will be seriously injurious to their reputation and their ability to sell throughout the world."[44]

In qualifying their claims about human rights, both IBM and Control Data argue the impossibility of controlling the use of their computers. Somewhat contradictorily, Control Data has said that it can monitor the major ways in which its computers are used, while IBM has claimed that it cannot. Both agree that they cannot control the ultimate uses to which their computers are put. As indicated earlier, this casts serious doubts on the reliability of the information companies provide to the Commerce Department in order to obtain export licenses.

The principal control a company has in this situation is control at the source. It can refuse to do business; it can cut off new supplies, maintenance, and spare parts; and it can thus insure that it is not supplying equipment in conflict with human rights considerations.

Claim 3: Ending Computer sales would not affect the South African government's philosophy and therefore could not induce any change.

In a letter of May 6, 1975 written to Dr. Sterling Cary, then-president of the National Council of Churches, IBM Chairman Frank Cary stated:

> Apartheid is abhorrent. However, there is no reason to believe that halting the sale of one company's line of computers would affect the government's philosophy.

Analysis

The call for computer companies and other foreign companies to withdraw from South Africa is not based simply on the hope that this would prompt a change in "philosophy" or policy on the part of the regime. It is first of all based on the premise that foreign corporations strengthen apartheid by their business activities in South Africa. The problem of creating a new society in South Africa, of changes of philosophy and policy, is a problem for South Africans themselves to solve. The call for withdrawal is a call to stop collaboration which strengthens apartheid and thereby to make it easier for the people of South Africa to eliminate apartheid.

Mr. Cary is ready to deprecate the role of "one company's line of computers" in South Africa, but this has not been borne out in this study of the impact of IBM and the other companies, which perform vital functions for the government and business.

Claim 4: Corporations do not and should not engage in politics.

IBM has stated:

Critics have called upon IBM and other corporations to take a variety of actions to change the policy of the South African government: from refusal to sell anything to the government, to refusal to sell any South African organization any product which could conceivably strengthen the economy. For a corporation to undertake any of these actions for political rather than economic motives is to inject itself into the conduct of foreign policy...corporations should be free to do business in any country acceptable to the U.S. government where they can operate profitably and treat their employees fairly.

Analysis

This argument is faulty in several respects. A private U.S. company may conduct its business in any manner it chooses so long as it does not violate U.S. laws or laws of countries in which the company operates. No U.S. laws compel companies to do business in South Africa, or would be violated by a withdrawal from South Africa.

It is a myth that corporations never "inject" themselves into the conduct of foreign policy. Corporations are free to express their views on U.S. foreign policy and frequently do. For example, to quote two minor examples, the Computer and Business Manufacturers Equipment Association protested the Carter administration's imposition of curbs on spare parts for computers now in use by the South African police and military, and William Norris, Control Data Chairman, went to the banking committee in Congress to oppose the revocation of Export-Import Bank credits and loan guarantees for business with South Africa. Control Data, he said, would not even consider withdrawing from South Africa.

Business interests and business concerns about foreign policy issues are in fact an influential element in the development of U.S. foreign policy affecting many areas.

Claim 5: Companies promote progressive change by staying in South Africa.

Both IBM and Control Data have propounded versions of this standard corporate argument in defense of their doing business in South Africa.

> IBM believes it should continue to do business in South Africa and to play a role in the development of black employees there. That role is to assist in their education, their employment in dignified and meaningful jobs, their opportunity to advance and to improve their leadership skills. Withdrawal would eliminate our chance as a company to set examples in job opportunity, equal pay for equal work, paid benefits, personal development and support for education.[45]

> Control Data's objective in South Africa has been to assist, through its business and employment practices, progress in the living conditions of the deprived populations there. To that end, for example, we have allocated funds and are seeking cooperation both in South Africa and the United States for the establishment of a large scale computer-based education system for black people in South Africa. We see this development as the most practical means of bringing about a significant closure in the education gap and hence the opportunity gap of millions of under-privileged people.[46]

Analysis

This justifications are a distortion of the significance of the companies' efforts to employ and train black workers and to promote special education and assistance programs.

In all these fields, the actions of the computer companies must fall within the parameters set by apartheid. Many of the companies have endorsed the "six principles" formulated by Reverend Leon Sullivan, who is on the board of General Motors. These call for desegregated work facilities, equal pay and employment practices, professional training programs for blacks, and business-sponsored programs to improve black housing and transportation. But the Sullivan principles and other benefits established for black employees pose no challenge to apartheid and have in fact been endorsed by government Interior Minister Mulder.

The cause of black poverty, lack of education and training and problems in housing and transportation lies in the system of apartheid, which denies to the black majority all political and social rights as South Africans. The Sullivan principles and the computer companies do not confront this central issue.

The records of IBM and Control Data reveal the limitations of the programs they propose. By December 1977, after more than five years of criticism by social activists and company assurances that change was being initiated, IBM's total black work force (African, Coloured and Asian) amounted to 131, some 16% of a total work force of 1,427.[47] Most IBM workers, and almost all workers doing more skilled jobs, were white.

Control Data reported in December 1977 that it had 19 black workers out of a total of 190 employees.[48]

It is hard to take seriously corporate claims of progress in light of these figures.

Control Data has proposed a computer-run education program to solve problems in black education. But such a program would have to be administered under the rigid control of the white regime which is responsible for the systematic denial of education to blacks and the corruption of educational institutions by the poison of the apartheid ideology of black inferiority and white dominance. In these circumstances it is naive to see Control Data's plan as a program that would change apartheid.

CONCLUSION

The computer companies have continued to contend that withdrawal from South Africa would mean the abandonment of the hope for progress of their (few) black employees and of the influence of their social programs. In fact, however, the only way that there will be any true hope for the advancement of black workers and solutions to South Africa's social ills will be through the elimination of apartheid. The computer companies, by allowing their products to be used under apartheid, are acting to strengthen apartheid and so intensifying the violence and suffering that lie ahead on the road to freedom in South Africa.

N.B. For more information on computer companies in South Africa, write The Africa Fund, 198 Broadway, New York, NY 10038.

1. *Financial Mail* (South Africa), 25 February 1977.

2. C. Cotton, managing director of Burroughs South Africa, interviewed by Tami Hultman and Reed Kramer in Johannesburg, 3 March 1971: in *IBM in South Africa* (New York: National Council of Churches, 1972), p. 3.

3. Morris Cowley interviewed by Tony Koenderman ("How to Beat the Computer Siege," *Sunday Times* (Johannesburg), 26 March 1978.

4. *Management* (South Africa), December 1977, p. 57.

5. *Financial Mail* (South Africa), 25 February 1977.

6. Tami Hultman and Reed Kramer, "South Africa's Rising Nuclear Prowess," *Los Angeles Times*, 28 August 1977.

7. *Sunday Times* (Johannesburg), 6 December 1970.

8. *Financial Mail* (South Africa), 23 January 1970, and *Sunday Times* (Johannesburg), 6 December 1970.

9. *Management* (South Africa), November 1974, p. 33.

10. *Financial Gazette* (South Africa), 29 March 1975.

11. *South Africa Digest* (South Africa), 1 October 1976, p. 12.

12. *Sunday Times* (Johannesburg), 29 January 1978.

13. *Political Imprisonment in South Africa* (London: Amnesty Internatl., 1977).

14. See for documentation publication series *Notes and Documents*, Centre Against Apartheid, United Nations, NY.

15. William N. Raiford, "International Credit and South Africa," in *U.S. Corporate Interests in Africa*, U.S. Senate Subcommittee on Africa, January 1978, p. 52.

16. "The Age of Miracle Chips," *Time*, 20 February 1978.

17. "The Defense Department's Top 100," (New York: Council on Economic Priorities, 1 August 1977).

18. General William Westmoreland, speech to the Association of the U.S. Army, 1969, quoted in Michael Klare, *War Without End* (New York: Vintage, 1972), pp. 203-4.

19. *New York Times*, 26 February 1978, p. 4.

20. Hesh Weiner, "Why Police States Love the Computer," *Business and Society Review*, #22, Summer 1977.

21. *Sunday Times* (Johannesburg), 26 September, 1971.

22. Testimony of Gilbert H. Jones, vice chairman of IBM, 29 September 1976, summarized in *U.S. Corporate Interests in Africa, ibid.*, p. 224.

23. Transcript of IBM annual meeting, 25 April 1977, pp. 55-56.

24. W.E. Burdick, IBM vice president, reply to questions at National Council of Churches hearing, 20 November 1974 in *The Role of IBM in South Africa* (NY: Interfaith Center on Corporate Responsibility, 1974), p. 16.

25. G. Cassidy, "Military Applications of Digital Simulators," *Armed Forces* (South Africa), vol. 2, no. 7, July 1977, p. 34.

26. *Armed Forces* (South Africa), vol. 2, no. 9, October 1977.

27. Information provided to Timothy Smith, Interfaith Center on Corporate Responsibility, March 1977.

28. Sean Gervasi, "Arms for Apartheid," (NY: The Africa Fund, 1977), p. 1.

29. Tami Hultman and Reed Kramer, "South Africa's Rising Nuclear Potential," *op. cit.*

30. "Unenriched, but Rich," *The Economist* (U.K.), 25 February 1978.

31. Letter from Gary H. Lohn, Control Data, to Frank White, United Presbyterian Church, 13 December 1977.

32. "South Africa Fact Sheet," (NY: The Africa Fund, 1977), p. 4.

33. Congressional testimony of W.H. Jones, *op. cit.*, and letter from Frank Cary of IBM to Timothy Smith, Interfaith Center on Corporate Responsibility, 7 December 1977.

34. *Computer Weekly* (U.K.), 31 March 1978.

35. "Firms May Defy U.S. Ban" (Agence France Press, 7 March 1978) reported in *Facts and Reports* (Holland Committee on Southern Africa) item 575, 1978.

36. *Financial Mail* (South Africa), 25 February 1977.

37. Fred Armentrout, "Mobil's Oily 'No Comment' on Rhodesia," *Business and Society Review*, no. 21, Spring 1977, pp. 52-55. "Mobil's Unctuous Silence; Rhodesia's Endless Ordeal." *Business and Society Review*, no. 24, Winter, 1977-78, p. 36.

38. *Management*, (South Africa), December 1977.

39. Tony Koenderman, "U.S. Embargo Not Likely to Hurt S.A." *Sunday Times* (Johannesburg), 26 February 1978.

40. "R.S.A. Designed and Manufactured Micro-Computer." *Armed Forces* (South Africa), August 1977, p. 25.

41. Tony Koenderman, "How to Beat the Computer Siege." *Sunday Times* (Johannesburg), 26 March 1978.

42. Transcript of IBM annual meeting, 25 April 1977, p. 29.

43. "Revolution Sayth the Churchman. One Soul at a Time, Says a Businessman," *Forbes*, 6 February 1978, pp. 31-3.

44. *Financial Mail* (South Africa), 10 February 1978.

45. IBM report operations in South Africa, November 1977.

46. Control Data Corporation statement, 24 October 1977.

47. Information from J. O'Connell, IBM, May 1978.

48. Letter From Gary H. Lohn, Control Data Corp., *op. cit.*

III
WORKING IN SCIENCE

Introduction

The Yale Professor   *—Mary Mackey*
I met a man who had been to China
he was a communist, he said,
I talked with Chairman Mao, he said,
I went, he said, to serve the people
Science for the People
Science for Everyone
(he was a professor
at Yale
his research involved implanting electrodes
in the human brain
to control
human emotions
he was a Yale professor
tenured
divorced
with four mistresses
all graduate students
he was a Yale professor
funded by the CIA.)

sleep with me, he suggested
putting his feet up on his desk
I've been to China
sleep with me
serve the man who serves the people.

While advertisements in scientific journals present images of young bright scientists (usually white males with glasses) engaged in solving "advanced" research issues, the ads do not tell much about who is doing the actual work. The truth is not very glamorous: the everyday work in science is done by graduate students, laboratory assistants or technicians hired to work on a particular project. The scientific world is a male hierarchical world where work relations are clearly delineated and competitive and aggressive behavior is rewarded for the principal investigator, the "he" scientist. Women and racial minorities perform much of the boring, repetitive technical work that allows the male scientist to build his theories and present his results. Sexism is rampant in the scientific milieu. Mary Mackey's poem, "The Yale Professor," powerfully depicts a situation not uncommon for women who work in science. Even scientists who would be considered "progressive" because of their ideas and commitment to political issues will exhibit sexist behavior.

In general, scientists are trained not to see their work in the context of the society they live in, and they are supposed not to get involved in the issues of the day. In fact, such involvement is regarded with suspicion and is considered unprofessional. Most scientists live and work isolated from the rest of society and are unable to communicate their thoughts in lay language. For scientists involved in the movement for social change the pressure to conform can be overwhelming. Many questions arise: Is it really possible to contribute to the movement for a new science while working in the scientific establishment? Where does one draw the line? What are the areas that deserve priority and should be attended to first?

The articles in this section address the issue of work in the scientific world from widely different perspectives. They all contribute to the demystification of that world and help us see it in a clearer way.

SftP January 1978

The Science Establishment
—Robert Yaes

In this article an unemployed high energy physicist discusses the long training in a field with low priorities on creativity and independence and high ones on hard, fast driving success. He discusses the relationship between these priorities and the financial and social structures of the field.

Technological progress is a mixed blessing, as ecologists, consumer advocates, and social critics have abundantly demonstrated. The physics community has been afflicted with its own kind of progress for more than a generation.

Since one cannot shed one's own preconceptions and prejudices, it is best to bring them out into the open. I therefore admit that my view of the situation is colored. I have not been able to find a regular faculty position in physics in the United States, so I consider myself to be a victim of recent developments. But then the quite different views of the members of the National Academy of Sciences who prepared the Bromley Report, *Physics in Perspective*, were influenced by the fact that they have benefited from these same developments. This does not make my perception less accurate or my judgment less valid than theirs.

The scientific revolution was made possible by the invention of moveable type in the middle of the fifteenth century; thousands of copies of scientific reports and treatises could henceforth be printed and distributed throughout the world. For several centuries, the great scientists—Copernicus, Kepler, Galileo, Newton,

Darwin, Maxwell—published their life's work in books or monographs. However, by 1900, scientific journals began to supplant books as the principal means for publishing new knowledge. Journals disseminated information faster and more broadly; scientists no longer were forced to wait until they had enough material for a book in order to publish.

Today, even journals are not fast enough for the publication of "exciting" discoveries. We now have "letters" journals: *Physical Review Letters, Physics Letters,* etc. Because the letters journals impose limitations on length, one does not have to wait until one has enough material for a regular article in order to publish. Because the rejection rate of the letters journals is high, many scientists deem it a significant accomplishment just to get published in them. Whether a particular paper needs fast publication is a highly subjective judgment. The referee's decisions are often arbitrary and capricious. Since publication in a letters journal yields extra Brownie points when one is up for tenure, promotion, or an increase in salary, letters authors argue with the referees of *Physical Review Letters* long after their papers could have appeared in *Physical Review.*

For high-energy physicists, even letters journals are not fast enough. We now have "preprints." As soon as papers are typed, hundreds of preprint copies are made and sent to colleagues all over the world, a practice that has been given some formality by the distribution of preprint lists by Stanford Linear Accelerator Center (SLAC) and European Organization for Nuclear Research (CERN). By the time a paper is actually published in a journal, it is likely to be obsolete.

Acceleration of publication has been more than matched by the increase in numbers of publications. A typical SLAC list (April 5, 1974) includes 101 preprints received in one week in just one subfield, high-energy physics. This amounts to more than five thousand papers a year. No human being can read and digest all this information. One must limit oneself to a very narrow sub-subfield, and even then carefully select what one has the time to read.

The journals, too, have been growing at an exponential rate. *Physical Review* is now divided into six parts, each of which is larger than the whole journal was a dozen years ago. Many individuals have stopped subscribing because they have run out of shelf space. Paradoxically, as the number of papers increases, the pressure on academics to publish still more papers also increases, since the worth of each paper, like inflated currency, becomes less and less as more and more are printed and put on the market.

The most significant development in the "growth" of physics, and the one from which all others seem to have flowed, has been the massive, unprecedented input of federal money into this science since World War II, due largely to the contributions physicists made to the war effort, the most important, of course, being the atomic bomb. Government money has been accompanied by the hope—seldom explicitly stated, but always implied—that additional basic research will lead to more important contributions to national defense.

Few physicists have tried to dispel this misconception. On the contrary, physicists have done everything to foster the relationship between research and the Department of Defense. Prominent physicists accepted funding for their "pure research" directly from the Air Force Office of Scientific Research and from other Department of Defense agencies before the Mansfield Amendment put a stop to the practice. Some have even been able to pick up extra pocket money by working directly on military problems in their spare time.

An example of the latter phenomemon is the JASON division of the Institute for Defense Analyses, which received worldwide attention after the exposure of its counterinsurgency studies during the Vietnam war. Since JASON had a disproportionately large percentage of high-energy theorists, someone in the Department of Defense must have believed that prominence in high-energy physics guarantees expertise in counterinsurgency and guerilla warfare.

Federal funding is a new phenomenon. Before World War II, direct government support for scientific research was almost nonexistent. Researchers scraped by on whatever support they could get from their own institutions or from private individuals and foundations. Historically, some of the most important scientific achievements came from well-to-do eccentrics—men like Gibbs, Lavoisier, and Cavendish—who supported themselves out of their private funds and did scientific research as a hobby.

The amounts of money involved in science are astronomical. Construction costs for the large high-energy accelerators run into hundreds of millions of dollars: $250 million for the National Accelerator Laboratory (NAL); $113,600,000 for SLAC; $51,400,000 for the Argonne Zero Gradient Synchrotron (ZGS); $30,600,000 for the Brookhaven Alternating Gradient Synchrotron (AGS). The annual operating expenses are also astronomical: for fiscal 1975, they are $36 million for NAL; $24,700,000 for SLAC; $24,600,000 for the AGS; $13,900,000 for the ZGS. The figure for annual operating expenses of these installations—commitments that must be met year after year—are the most important ones to watch. However, whenever such facilities are proposed, it is usually only the construction costs that are given to the public. As a result, Congress appropriates funds to build a particular facility but not for its year-to-year operation. This seems to have happened in the case of the NAL with disastrous effects on the rest of the high-energy physics community. Not only have the Cambridge Electron Accelerator and the Princeton-Penn Accelerator been shut down, but many university-based research groups have had their contracts cut or eliminated since the NAL commitment was made.

The operating expenses for the National Accelerator Laboratory are rising. In fiscal year 1974, they were $28,400,000; they are expected to reach sixty million dollars eventually. If one allows for only a fifteen-year lifetime at its full budget, the NAL will represent a total commitment by the federal government of considerably more than a billion dollars, a figure that does not include the support of the NAL's "users" groups.

What have the taxpayers received for their money? For simplicity, I will concentrate on the NAL, not because I think that it is more of a waste of money than the Argonne ZGS, the Cornell Electron Accelerator, or various cyclotrons, synchro-cyclotrons, and Van de Graafs scattered around the country, but because the NAL seems to be costing more than all of these put together.

Here is the way *Science* described the NAL in December 1973:

> The facility is still far from complete. The accelerator still does not yet operate at its full intended power; none of the three experimental areas is complete; and a logjam of approved but not yet done experiments is causing problems for university physicists and other users of NAL.... Experimental physicists who hurried to prepare experiments for NAL have found themselves facing delays of two tô three years....

> Magnets are still breaking down at the rate of about one a week.... Nearly half of the one thousand magnets in the accelerator's main ring have been replaced, some more than once. There have been difficulties with accelerating cavities, power supplies, and the equipment for extracting the beam of protons from the accelerator. In the first ten

months of 1973, unscheduled repairs closed down the accelerator for a total of 2,514 hours, a period slightly longer than the operating time for research. The accelerator, in short, has simply not worked well in its initial year of operation.

The problems at NAL appear to have been aggravated by what in retrospect were unrealistic promises about the performance of the accelerator.... Operation at four hundred Gev is by no means routine yet; the utility that supplies electricity to NAL is concerned about voltage fluctuations on its power lines at that energy.... In the absence of any striking new discoveries, however, more quantitative measurements are planned.

It is the "absence of any striking new discoveries" in the NAL operation, rather than the breakdown of the magnets or the trouble with the power lines, that is the heart of the matter. Now CERN has decided to build a machine basically identical to the NAL, presumably to make sure that Europeans will be able to experience in time the same "absence of any striking new discoveries."

In the United States, new harebrained schemes for wasting the taxpayer's money are popping up everywhere. The NAL wants both an energy doubler and colliding beam facilities. Brookhaven is pushing its Isabelle colliding beam proposal. And SLAC is proposing a large "e+e– colliding beam project." Judging by statements like R.R. Wilson's that "the ultimate capability of the National Acceleratory Laboratory is limited only by the size of the site," the people proposing these facilities seem not to be thinking responsibly about costs at all.

One of these schemes is the "very large array" (VLA) radio telescope system being built in the New Mexico desert at a cost of seventy-five million dollars. Unfortunately, this facility is being supported by the National Science Foundation which, until now, has been the mainstay of "little science." (The Atomic Energy Commission has tended to support "big science.") It is too early to tell how many other projects will be cut or terminated in order to fund the VLA, but if I had an NSF grant in astronomy, astrophysics, general relativity, or cosmology, I would be more than a little worried.

The Bromley Report refers to these major facilities as "high leverage situations" because "in the case of major facilities, such a large fraction of the total funding is required to keep them in operation that even small fractional changes in funding are reflected as very large changes in the research component to which scientific productivity is much more directly coupled." The Bromley Report therefore concludes that these major facilities should continue to be funded at a high level, even at the expense of other programs.

I would draw the opposite conclusion; we should avoid getting trapped in "high leverage situations" that destroy the flexibility needed to deal with the changes caused by new scientific discoveries and shifts in over-all federal funding policies. Instead of thinking about what kind of new "high leverage" major facilities we can construct, we should be thinking about which of those we already have we can proceed to shut down.

Why, in the face of declining research funding and increasing unemployment, are influential people calling for even more and larger major facilities? Some may really believe that these facilities are necessary for scientific progress. However, while these facilities may be detrimental to the physics community as a whole, some individuals stand to profit greatly from them. Some experimentalists may be hoping for a repetition of what may be called the "Chamberlain-Segre effect," i.e., the opening of a new energy range by a new accelerator which led to a Nobel Prize. Owen Chamberlain and Emilio Segre won the prize for the discovery of the antiproton. Even discounting what Oreste Piccioni had to say on the matter, the more cynical experimentalists may feel that this discovery resulted not so much from the experimenter's ingenuity as from the fact that the Bevatron for the first time provided enough energy to produce an antiproton. They would probably not go so far as to suggest that perhaps the prize should have gone to the Bevatron itself, or, even more to the point, to the taxpayers who paid for it.

These major facilities also provide power bases and lucrative positions for their directors, associate directors, and assistant directors. This will not make much of a dent in the unemployment lines, since it is not likely that any of these administrative positions will go to new PhD's who would otherwise be out of work. If, as I believe, the cost of these major facilities does result in an over-all increase in unemployment, it will not be the members of the National Academy of Sciences—i.e., those who recommended their construction—who will lose their livelihood.

As these major facilities come up for a vote in Congress, they tend to get the pork-barrel treatment. When the Senate voted on the NAL in July, 1967, its chief proponent was Senator Everett Dirksen of Illinois, who was concerned mainly with obtaining another federal government installation for his state. The NAL's principal opponent was Senator John O. Pastore of Rhode Island; his main concern, one surely shared by all civil libertarians, was the fact that the open-housing laws in Illinois would cause hardships for the NAL's black employees.

Only the editors of the *New York Times* seemed to be concerned primarily with what the NAL would actually be used for and whether or not its cost was justified. In an editorial, after agreeing with Senator Pastore's stand on the open-housing issue, the *Times* said:

> But there is an even more basic objection to any commitments or expenditures for this expensive research tool at this time. That objection is simply the irrelevance of a two-hundred-billion-electron-volt accelerator to any real, present national problem...Millions of Americans lack proper housing, adequate medical care, and essential educational opportunity. The budget cutters are now in full cry demanding reductions in already inadequate expenditures for human needs. It is a distortion of the national priorities to commit many millions now to this interesting but unnecessary scientific luxury.

Congressmen do not, of course, have the technical expertise to make decisions on scientific matters. To a large extent, they must rely on experts from government agencies, who sometimes are more interested in touting their own agency's projects than in giving disinterested advice.

The amounts of money given to selected university-based research groups are also substantial. For fiscal year 1972, among the Atomic Energy Commision contracts in high-energy physics, there were eight for more than one million dollars each. In fiscal 1971, there were nine such million-dollar-plus contracts. I do not know how the AEC determines that the work of one group is worth $1,333,679, that of another group only $25,000, while presumably the work of other groups whose applications were rejected is worth nothing at all. All one can do is look for regularities in the data and try to construct hypotheses that might explain them.

One regularity is that in both fiscal year 1971 and fiscal year 1972 there were more AEC contracts of over one million dollars than there were under fifty thousand dollars in high-energy physics; and in fiscal year 1972 the smallest grant made was twenty-five thousand dollars. If you come to the AEC asking for only a few thousand dollars for publication costs, computer time, and maybe one postdoctorate assistant, they think you do not regard yourself very highly and that your work probably isn't very good anyway. If you ask for an exorbitant amount, say $1,333,679, they think that your work must really be good, and you will probably get what you ask for.

Another regularity is that the 1971 and 1972 lists generally consist of the same groups receiving just about the same amounts of money. This illustrates a rule common to all bureaucratic systems, the law of inertia; i.e., the best reason for supporting a given group at a given level in a given year is that the same group was supported at the same level the preceding year.

In a high-energy experimental group, the bulk of the contract funds pays for equipment, technicians, and other expenses. But some of the money helps to improve the standard of living of the funded physicists. More important than the free trips to conferences—often held in sunny places like Hawaii or Florida, or exotic places like Kiev—is the phenomenon of "summer salaries." A principal investigator, taking advantage of the fiction that a faculty member is paid by his university for working only nine months of the year, pays himself—and perhaps a few of his friends and colleagues—a salary from his research grant or contract, during all or part of the remaining three months, for what he would have been doing anyway. The conflict of interest when the principal investigator is both the employer and the employee is conveniently overlooked.

Also overlooked is that if a university is paying a person thirty thousand dollars or more a year, and if this person is teaching only one course, the university is really paying only for research work, and thus there is no reason why the government, too, should pay for the same research. Nevertheless, if a principal investigator must decide which will be more useful to his research, hiring himself for the summer or hiring another postdoctorate assistant, it is clear what he will do.

Some people claim that universities take summer salaries into account when drawing up faculty pay schedules, and thus summer salaries tend to redress imbalances that would otherwise exist. However, summer salaries are directly proportional to winter salaries, so they tend to aggravate existing imbalances rather than to mitigate them. The practice of paying oneself summer salaries—

just a bit shady in the best of times—is certainly unjustifiable when hundreds of physicists are out of work. It is difficult to muster sympathy for a principal investigator who, without his summer salary, would have to scrape by on a mere twenty-five or thirty thousand dollars a year, and have trouble keeping up the payments on his motorboat, second car, or summer home.

Another interesting item on research budgets is "overhead," money destined to reimburse universities for indirect costs incurred by their faculty members in doing research. However, nobody seems able to determine what these costs are or how they arise; so overhead is usually figured as a percentage of the salaries paid by the contract.

What overhead seems in fact to be is an indirect way of giving universities government subsidies. In itself, this would be laudable. However, most of these subsidies seem to be going to institutions like Harvard, Yale, and Princeton, which do not need them, or which need them less than the smaller colleges and universities. If the government wants to subsidize institutions of higher education, it would be more efficient to do so directly rather than by the stratagem of "overhead expenses."

Overhead payments also affect whether professors get hired, fired, tenured, or promoted. University advertisements of positions in *Physics Today* often ask for professors with a proven record of sponsored research, individuals who can bring in money. Many administrators also believe that a professor's ability to obtain research support is *ipso facto* evidence of his research excellence.

It is instructive to compare the funding policy of the United States Atomic Energy Commission with that of the National Research Council of Canada (NRC). The NRC awards no million-dollar contracts, pays no overhead, and gives no summer salaries. In fact, no one who holds an NRC grant can receive any compensation from his or her own or anyone else's NRC grant other than dollar-for-dollar reimbursement for out-of-pocket expenses incurred in the performance of research. Very large grants are almost nonexistent. But any faculty member at a Canadian university who is engaged in a viable research program can count on a grant of at least several thousand dollars to support the work. In the United States, it is not uncommon to find *theory* groups, in whose research budget the two largest items of expense are overhead and summer salaries. In Canada, research funds are used for research.

I cannot blame the recipients of summer salaries for accepting them. If somebody threw a bucket of money at me, I would grab it too. The fault lies with the AEC and the NSF for paying summer salaries. The blame also lies with the National Science Board, the High-Energy Physics Advisory Panel to the AEC and other committees that are supposed to advise these agencies. The members of these committees are usually among the chief recipients of summer salaries and other forms of AEC and NSF largesse. The Bromley Report, for all its concern with what should be funded and should not, makes no recommendation regarding summer salaries; it does not even acknowledge their existence.

At a time of declining research support, available funds must be used as effectively as possible. In such a situation, everybody must be asked to make sacrifices, not just the recent PhD's. One must judge the sincerity of the members of the physics establishment by their concern for the welfare of physics as a discipline, a concern best displayed not by the sacrifice they demand of others, but by the sacrifices they are willing to make themselves. At a time when hundreds of physicists are being forced out of the field, it would not be unfair to ask tenured faculty members, who have absolute job security, to give up their summer salaries.

The availability of big money has changed the way in which physics is done. Physics used to be a profession of individualists. Today, experimental papers are sometimes written by fifty or more authors from six or seven institutions in three or four countries. Apparatus used to be whatever one could put together with string and sealing wax. Now one piece of "apparatus" can cost $250 million to build and $60 million dollars a year to run. At one time, the important international physics gathering was the Solvay Conference, attended by thirty or forty people. Recent conferences on high-energy physics alone, at Vienna, Kiev, and Batavia, have drawn hundreds of physicists.

As physics has become both richer and larger, calls for the application of business management methods have become more insistent, as if scientific discoveries could be stamped out the way Detroit stamps out car fenders. F.T. Cole, in *New Scientist*, says: "In many ways, high-energy physics has come to be the epitome of Big Science. In the large laboratories, the management of people and resources is as important as in a large enterprise." D. Wolfle, in *Science*, says: "Scientific research and its applications are becoming more management-intensive, and this trend will probably continue.... As a signal of this trend, the RANN (Research Applied to National Needs) program of the National Science Foundation has twice the manager-to-dollar ratio that the research division has." If this trend continues, we may reach the point where all the resources of science will go into research management and none into research itself.

The "management-of-people" effect on physicists seems certain. Bright young physicists will be able to exercise little of their ingenuity, originality, and creativity when they are junior members of enormous research groups and are told exactly what to do. Many physicists are now being subjected to a routine of boring, meaningless, alienating work. The purpose of science is to gain an understanding of the nature of the universe; the purpose of business is to make a profit. A scientist thinks of a grant as a means to carry out research; a person in business tends to regard research as a means to obtain a grant.

Along with business management methods comes business management morality, according to which, as Lord Keynes said, "We must pretend that fair is foul and foul is fair, for foul is useful and fair is not." As the stakes become higher and the competition keener, physicists are forced to be less scrupulous in their methods. As ideas become "property" and colleagues become competitors, physicists become more secretive and less cooperative. We seem to be forgetting that we are working toward a common goal.

Two Nobel laureates in molecular biology have described the situation. First, J.D. Watson:

> Generally the first reaction to the prospect of being scooped is a combination of despair and hope that your opponent, X, will fall dead.... So it is hard not to think about retooling your effort to try the same approach as your competition. Even though you are behind, by being a little more clever, you may overtake him. He, of course, might then become hellishly mad...so you can be almost certain of making a long-term enemy.

And S.E. Luria:

> The availability of ample research funds from government and foundations creates an entrepreneur system with either the university or an individual researcher as entrepreneur. A new style enters the picture: that of the Mr. X who is an expert at finding out not who will pay for what X and X's institution want to do, but who needs and will pay for something that X knows how to do (or more often how to direct a team of technicians to do).... But the entrepreneurial system does lend itself to opportunism. Insofar

as it resembles a competitive production system, graduate students become employees and project directors become fund-raisers.

A subtle change in ethical standards follows: not necessarily a loss of integrity, but a shift in responsibility from scholar to the entrepreneur.... For example, if someone published some good work, other scientists used to allow him to develop it alone, at least for a few years. Now, eager researchers rush back from professional meetings to perform the obvious experiments that a speaker had not yet had time to do. Nothing strictly unethical, of course— not according to the ethics of competitive enterprise.

There is at least one person, not a Nobel prize winner, but one who works in applied physics, who believes that this competitiveness is a good thing and who is actually afraid that it will disappear. According to Harvey Brooks (in *Science* 174, 21, 1971):

The achievement of scientific excellence is highly dependent on the Protestant ethic of work and individual achievement. Although the scientific community is one of the most open of all social systems in terms of all criteria other than its own internal standards of performance, its insistence on individual excellence and on rigorous interpersonal valuations runs strongly counter to contemporary egalitarian trends and rejection of all competition and comparisons between people, especially among youth. In current jargon, science is inherently an elitist activity and its success as a social institution is highly dependent on a rigorous selection and ranking of its practitioners by their colleagues and their seniors.

As science has become professionalized in the last generation, its competitiveness has, if anything, increased. Some very able and talented people seem to be rejecting the "rat race." Although some of the more extreme forms of competitiveness caricatured in Watson's book (*The Double Helix*) are certainly not necessary to a healthy scientific system, the advance of science does depend on a process of natural selection of ideas and people, not unlike biological evolution; and without this selective pressure, truth cannot avoid being swamped by error in the long run. Just as biological evolution runs against the average trend of the Second Law of Thermodynamics, so does science run strongly against the social second law of the least common denominator.

Since Harvey Brooks was chair of the NAS Committee on Science and Public Policy, under whose auspices the Bromley Report was prepared, and therefore an ex officio member of the committee that prepared it, the above quotation is an important indication of the preconceptions that these committee members brought to their task.

The forerunner of the large, modern, efficient, well-funded high-energy or nuclear physics laboratory of today was, of course, E.O. Lawrence's Radiation Laboratory at Berkeley in the 1930s. The single-minded drive to larger and larger machines, producing higher and higher energies, more for their own sake than for the research that could be done with them, was begun by Lawrence. It is still being continued by many of his former students and associates who hold positions of the greatest power and authority in the United States' physics community.

The central issue in large, massively funded laboratories is how well they perform in making significant new scientific discoveries. On that score, and without detracting from Lawrence's genius in inventing the cyclotron, or from the importance of his cyclotrons as research tools in providing beams of unprecedented energy and intensity as well as isotopes for medical research and treatment, the Radiation Laboratory's success was not as great as might have been expected.

When Lawrence and Livingston had achieved 1 Mev with the eleven-inch cyclotron, they were more concerned with building a larger machine than with doing experiments with the one they had. As a result, the first nuclear disintegration, of lithium and beryllium, was achieved by Cockcroft and Walton at the Cavendish Laboratory, using their electrostatic generator that could produce protons of only several hundred thousand electron volts. Livingston is quoted in Nuel Pharr Davis' book, *Lawrence and Oppenheimer*: "We had the energy for atomic disintegration, but we lacked the instruments for observing it."

The same sort of thing happened with the discovery of artificial radioactivity. Davis writes: "The Geiger counter was wired to the same switch as the cyclotron. By turning them both on and off at the same time, one could get on with operations faster.... On February 20, 1934...Lawrence came running through a door waving a French journal with an article by Frederic Joliot.... 'It is now possible for the first time to create radioactivity in certain elements,' Lawrence translated haltingly for the staff. Joliot had used a petty apparatus powered by a minute quantity of radium.... The staff changed the wiring on the Geiger counter and swung a carbon target into the beam. Five minutes later, they shut off the cyclotron. Click, click, click went the Geiger counter."

The Berkeley group also missed out on being the first to make the most significant discovery in nuclear physics, namely, fission, even though they had the world's most intense neutron beam, which they were using for cancer therapy and in trying to produce transuranium elements. The discovery of fission was made because of the work of Hahn, Strassman, Meitner, and Frisch in Europe. The first that the Berkeley people knew of it was when Luis Alvarez read a newspaper article about a lecture on the subject given by Neils Bohr in Washington.

In short, even though the Berkeley group had beams of energy and intensity that could not be duplicated anywhere else in the world, the most significant discoveries were still being made by the "string and sealing wax" people in Europe.

In the case of artificial radioactivity, the Berkeley group missed out on making the discoveries, not so much in spite of the fact that their laboratory was large, efficient, well-run, and well-managed, as because of it. This criticism is not aimed only at Lawrence and his co-workers. Any high-school student can cite other examples. Alexander Fleming, for example, left a bacterial culture out where it became contaminated by a mold spore. Being curious, he waited to see what would happen and found that the mold was emitting a substance that killed all the bacteria in its immediate area, leading Fleming to the discovery of penicillin. Today's researcher with a neat, orderly, managerial mind would take one look at the contaminated culture and toss it in the trash can, cursing the fungus that "ruined my experiment."

It is inherent in the nature of science that we never know exactly what it is that we are looking for, or even where to look for it. So it is absurd to maintain that we can determine, a priori, the most efficient way to find it. Roentgen was not looking for X-rays; Becquerel was not looking for radioactivity; and Kamerlingh Onnes was not looking for superconductivity, but they discovered them. Geophysicists were not seeking to verify the theory of continental drift when they set out to study magnetic anomalies of the ocean floor; nevertheless the data they obtained gave them the key to the whole question. Hahn and Strassman were not looking for nuclear fission; they were trying to create transuranium elements.

This point, which is so obvious, seems to have been lost on many senior physicists today. For example, the Bromley Report tries to determine the intrinsic merit of the various subfields by asking such questions as, "To what extent is the field ripe for exploration?" That is like asking, "Which horse will win the fifth race tomorrow at Santa Anita?" At the end of the nineteenth century, many physicists were saying that the only thing left for them to do was to measure some physical constants to a few more decimal places. For them the whole field of physics was definitely not "ripe for exploration," a judgment soundly disproved by events in the next twenty or thirty years.

If business management techniques are applied to science, they will rationalize and homogenize everything, make everything predictable, eliminate the accidental, and, in the process, eliminate much of scientific progress as well. Harvey Brooks has said, "the first question is how to organize staff and direct the search for knowledge so as to obtain the greatest rate of scientific progress for a given investment of human and material resources." Unfortunately this question can only be answered in retrospect.

Mr. Brooks' "first question" implies that somehow the rate of scientific progress will be proportional to the resources invested. Any research scientist can come up with dozens of counter-examples. I cite just one. The most significant single scientific paper of this century was written by a clerk in the Berne Patent Office in his spare time. Einstein's paper cost absolutely nothing.

His theory of relativity also refutes the thesis that important advances in science are always the result of new data. (Physicists today are concerned primarily with obtaining more and more data at higher and higher energies.) Most textbooks on relativity begin with the Michelson-Morley experiment, but that is mainly for the benefit of the students, since Einstein does not even mention the experiment in his original paper. In fact, special relativity was implicit in Maxwell's equations, and anyone since Maxwell could have obtained it—if he were clever enough. Certainly the Lorentz transformations, which form the mathematical basis of special relativity, were, as their name implies, known to Lorentz—and to Poincare too—but they did not know how to interpret them. The only additional experimental fact one needed in order to proceed from special to general relativity is the equality of inertial and gravitational mass. This fact was well-known to Newton; and it becomes one of the cornerstones of Newtonian mechanics. Einstein, however, was the first to appreciate its full implication.

The same can be said about the Copernican revolution. It was not the sudden appearance of "exciting new data" that led Copernicus to favor the heliocentric system. Data indicating that a heliocentric system would be simpler than a geocentric system were known in the time of the Greeks. Of course, once the initial breakthrough was made, additional data from Tycho's laboratory were necessary for Kepler to conclude that the orbit of Mars is an elipse. But, before Copernicus, such additional data would have been useless. The fact that the positions of the planets could be fitted with geocentric epicycles and heliocentric epicycles, as well as heliocentric ellipses, proves another point: you can always fit the data with the wrong theory if you have enough parameters; the right theory, however, will fit the data in a particularly simple and elegant way.

Where breakthroughs do arise from new data, these data are likely to come from an unexpected direction. I have already mentioned how the study of magnetic anomalies of the ocean floor led to the theory of plate tectonics that has revolutionized geophysics. Another example is the development of quantum mechanics. Before Niels Bohr, people were trying to understand atomic spectra in terms of models with springs and weights that would vibrate at the right frequencies. They could have called for larger and larger spectroscopes to produce more and more spectroscopic data. However, the essential clues came, in a way that nobody could have predicted: they came from black body radiation, the photoelectric effect, and Rutherford scattering. Of course, after the first breakthrough, additional data were quite useful in developing a perturbation theory and determining that the electron had spin. But, before Bohr, any additional data, over and above that necessary to obtain the Rydberg law, would have been useless, if not confusing. It is also clear that measuring everything that is not nailed down is not very useful, since it is necessary to isolate the simplest system—in this case the hydrogen atom—in order to understand what is going on. Spectroscopic data for helium, copper, molybdenum, or whatever would not have been very useful to Bohr.

An excursion into the past is only useful to the extent that it helps one to understand the present. In high-energy physics alone, there has probably been expended in the last thirty years more money and "man-hours" than had been spent on all of physics from the dawn of civilization. Despite that, the only honest answer to the question: "What do we know about the basic interactions between elementary particles that we did not know thirty years ago?" is "Not very much." It is true that a major advance was made about 1950 in renormalized quantum electrodynamics. But as far as the strong and the weak interactions are concerned, we seem to have obtained mainly a number of significant, semi-empirical facts, such as that baryons tend to fall into SU(3) multiplets, and that weak interactions violate parity, whose implications are still not entirely clear.

In the absence of sustained progress in any direction, the phenomenon of researchers rushing back from scientific meetings to develop someone else's ideas has become a way of life in high-energy theory. The question now is not so much whether this practice is slightly unethical as whether it is counterproductive.

It has produced a succession of fads that have become as popular as hula hoops and maxi-coats and lasted about as long. Axiomatic field theory, dispersion relations, double dispersion relations, N/D calculations, SU(6)W, $\bar{U}(12)$, super-convergence relations, finite energy sum rules, current algebra, dilitation invariance, and even dual resonance models are not as popular as they once were. They have been abandoned by many, if not most, of their former practitioners, not so much because they were proved to be "wrong," but because, as time went on, the mathematics became more complicated, results were harder to obtain, and the connection with the real world became increasingly tenuous.

Faddism is encouraged by some of the practices already mentioned—hasty publication in letters journals and the even hastier distribution of preprints. Most younger physicists realize it is to their advantage to get in on one of these fads as early as possible, whether or not it turns out to be truthful, and whether or not they really believe in it. The model of a modern high-energy theorist is not the person who conceives and develops his own original ideas. Today's system rarely rewards one for sticking one's neck out. It is small satisfaction to be proved right, five or ten years from now, when you have to find a job for the year after next.

Some physicists may even begin to think that the increased "selective pressure" of Harvey Brooks' neo-Darwinian academic world tends to reward the high-energy theorist who has little imagination but a sharp enough mind and the technical mathematical skills that enable him/her to obtain cute, but not necessarily significant, results from whatever fad happens to come along; that it tends to favor those with the social skills necessary to butter up members of the Establishment; in short, that it tends to "select" those physicists who are hard to distinguish from young business executives on the make. If high-energy physicists begin to think this way, they will act accordingly. And that may be a clue to the "absence of any striking new discoveries" not only in the NAL, but also in high-energy physics in general.

My knowledge of the subfields of physics other than my own is limited. It is also difficult to make comparisons because high-energy physics is oriented almost exclusively toward producing "new knowledge," whereas fields like nuclear physics, and condensed matter physics, not to mention optics and acoustics, are concerned with immediate practical applications. To that extent, they are reaping the fruit of the "new knowledge" obtained in the first part of this century. Some of the recent spectacular results of condensed matter physics, lasers, masers, transistors, tunnel diodes, and even recent advances in the theory of superconductivity, are of interest as much for their practical results as for their intrinsic value.

Nevertheless, there is evidence that the same diseases that affect high-energy physics are beginning to appear elsewhere. The fact that the Los Alamos Meson Physics Facility (LAMPF) has been built and that the very large array telescope has been funded indicated that the misplaced emphasis on "major facilities" has not been limited to the subfield of high-energy physics.

Astrophysics, concerned primarily with obtaining "new knowledge," is experiencing conceptual difficulties. Astrophysical models for exotic objects—pulsars, quasars, neutron stars, black holes, and even the universe itself—involve assumptions, such as those about the behavior of ultradense matter, which cannot be tested directly. It now appears that even models for conventional stars, such as the sun, which were thought to be firmly based, may be in serious trouble. The prediction made by astrophysicists for the solar neutrino flux, on the basis of nuclear reactions that were thought to take place in the sun, has proved to be wrong. Somebody will have to go back to the drawing board. And the very large array radio telescope system will not be of much help in solving that problem.

It should be clear by now that exponential growth in people-power, publications, and funding could not continue indefinitely. Still, the recent falling off of funding and the increase in unemployment came as a shock to many. My own job-hunting experience gives no cause for optimism. Last year, I received an offer of a five-thousand-dollar-a-year postdoctoral position in England, a one-year extension of my present visiting appointment, and 245 negative replies. I received no offers from any institution in the United States. My experience is not unique.

Some of the negative replies said there would be no hiring at all for the next few years; some said that a number of present faculty must be dismissed; and one said that applications from "over three hundred qualified persons" had been received for the one open position. The brain drain of scientists from other countries coming to the United States has not only stopped, it has reversed. Many U.S. physicists, like me, now must go to other countries to find even temporary employment.

So far, the reaction of the members of the physics establishment, if the Bromley Report can be taken as a fair reflection of policy, has been one of trying to sweep these problems under the rug. It is well-known, for example, that, though hundreds of millions of dollars are spent by the federal government on high-energy physics, the highest unemployment rate is to be found among high-energy theorists. The Bromley Report washes its hands of the problem with this statement: "There is no reason to expect that the total operating expenses for high-energy physics are closely correlated with the number of theorists." What the report does not say is why. As previously noted, most of the "operating expenses for high-energy physics" go to run the large accelerators; most of what is left goes to support "users groups" of the same machines; and even most of the money meant for the support of theory goes to summer salaries and overhead. Naturally there is almost nothing left to hire theorists; and the Bromley Report's statement follows from that. The Bromley Report obviously does not recommend any change in these policies.

In fact, theorists are in a double bind, since they are not even able to get jobs at small colleges where they might be able to continue their work. Such colleges always seem to be looking for someone who can set up the junior laboratories, and they do not think a theorist can do that. Because virtually no high-energy theorists can obtain new tenured positions, the typical high-energy theory group twenty years from now will be demographically similar to a home for the aged.

There is, as one might expect by now, a physics establishment, a small group of individuals in whom most of the decision-making power in physics is concentrated, whom the government agencies listen to most of the time, and whose identity of interest and outlook is quite complete. One gets a glimpse of this concentration of power when an establishment figure runs for office in the APS and lists the committees on which he serves. D. Allen Bromley, for example, is: professor of physics and chairman of the Physics Department at Yale, director of Yale's Wright Nuclear Structure Laboratory; consultant to Brookhaven, Los Alamos, Oak Ridge, IBM, and High Voltage Engineering Co.; consulting editor for McGraw-Hill; associate editor of the *Physical Review, Annals of Physics*, and *American Scientist*; a member of the boards of directors of United Nuclear Corporation and Labcore, Inc.; a delegate to the International Union of Pure and Applied Physics; a member of the National Research Council and of the executive committee of its Division of Physical Sciences; chairman of the National Academy of Sciences Committee on Nuclear Science; and, of course, chairman of the National Academy of Sciences Physics Survey Committee that prepared the Bromley Report.

What can be done about these problems? I have no magic wand to produce solutions. I can speculate, however, about the forms that the solutions might take.

First, business management methods should be left to businessmen. Scientists have little use for, say, the "salesmanship" and "styling" that have given us the vinyl landau roof and the opera window instead of a practical, economical, safe, nonpolluting automobile.

We should leave cutthroat competition to the businessmen also. The purpose of physicists is to work together to unravel the laws of nature, not to cut each other's throats.

We should de-emphasize the letters journals. After all, the laws of nature have been around for billions of years waiting for us to come along and discover them; a delay of a couple of months in publication will not matter that much.

We should end the headlong rush to build every possible major facility. Facilities should be built only when old facilities become obsolete and useless, or when technological advances make it possible to build them at reasonable cost.

I am not saying that the NAL is not useful to high-energy physics. But in a time of exceedingly tight budgets, the point is not usefulness in an absolute sense, but the ratio of usefulness to cost. The NAL is the most expensive single scientific instrument ever constructed. Sixty million dollars a year could support three or four thousand theorists, instead of the three or four permanent theoretical positions there seem to be at the NAL. The cost of just the thirty-two thousand liters of liquid deuterium for the NAL's fifteen-foot bubble chamber would be enough to endow a chair or two.

The best policy would seem to be a moratorium on the construction of any new "major facilities" until economic conditions improve.

Closely related to this is the question of the relation of theory to experiment. Theory is as necessary to scientific progress as is experiment. Theorists should not be treated as appendages to experimental groups or as poor relations who should be content with whatever crumbs the experimentalists may let fall from their table. I would be amazed if as much as five percent of the government support for high-energy physics goes to theory (it is impossible to come up with the exact figures, because theorists and experimentalists are often supported by the same contract). Theorists tend to do their most imaginative work at a relatively young age. It is therefore at least as necessary to maintain a steady stream of young people entering the field as it is to build new major facilities, or even to maintain the old ones. Funding agencies should institute policies directly creating more jobs for young theorists. Failing that, both the quality and the quantity of young people entering the field will drop drastically.

We must reverse the trend toward the concentration of bureaucratic and decision-making power in the physics community. We must have a broader representation on the advisory committees to the government, on the American Physical Society council, and among the associate editors and referees of the various physics journals. Among other things, this would allow the establishment figures who are now carrying most of the load to devote more of their time to research, and it will restore the confidence of many younger, non-tenured physicists as they take part in the decision-making process.

My most important recommendation concerns the best way to increase the probability of making new, exciting scientific discoveries. It is to allow the "centers of excellence" to become lean and hungry again, and to concentrate support instead on the "centers of mediocrity." Support will then be more broadly distributed, not only because "centers of mediocrity" vastly outnumber "centers of excellence," but also because they are much cheaper to run. It is only at a "center of mediocrity" that a bacteria culture would be left out where it could be contaminated by a mold spore, or where people are inefficient enough to wire the cyclotron and the Geiger counters to separate switches, or where a patent clerk would have enough spare time to write scientific papers.

Concerning tenure, I am of two minds. On the one hand, it does not promote the progress of physics. Tenure allows many positions to be held by people who are no longer productive. Still, I do not believe the solution is to throw some people out of work to provide jobs for others. The job security provided by tenure is a good thing; it is a shame it applies only to a few older, privileged academics. Why not give tenure to post-doctoral people and assistant professors as well, or, for that matter, to plumbers, truck drivers, secretaries, and janitors?

The ostensible reason for tenure is that it protects politically outspoken people. However, these are a negligible minority of all those who hold tenure. If this is really the purpose, then tenure is given to the wrong people, since most dissidents are found among the younger, nontenured staff. If a person has become skilled at keeping his or her mouth shut for six or seven years in order to obtain tenure, it is not likely that s/he will change this habit upon receiving it.

I do not have tenure and I am sticking my neck out by saying what I have said here. I have done so because I am convinced that nobody with tenure will come out and say these things. What is at stake is no longer the construction of a five-hundred-Gev this or a very-large-array that, or even the livelihood of hundreds of young physicists like me. Rather it is the very survival of physics itself as a viable, vigorous, productive discipline.

What is a People's E-Tech?
—Bob Broedel

This article describes the everyday difficulties of being an electronics technician at a southern university, of being committed to social change and working in an atmosphere permeated by sexism and racism.

I work as an electronics technician for a science department of a southern state university. It is one of the larger departments on campus and has a fairly good reputation around the country. The department has several shops that support teaching and research efforts. There is a wood shop, a machine shop, an illustration/photography shop, a glass shop, a precision maintenance shop, and an electronics shop. I work in the "E-Shop" where we repair all kinds of electronic and electro-mechanical equipment from simple heat controllers to computer controlled data collecting instrumentation. We also do a bit of interfacing and basic design, referred to by some people as engineering.

There are three people in the shop, all of them white and male. I have been an electronics technician for ten years but have been in this particular shop for only four years. Both of my co-workers have been here for over eleven years. Both have working class roots and conservative politics. My boss is an okay guy considering the circumstances and is, in fact, the best boss I have ever had. He dabbles a bit in the stock market, and as his small part of the "American Dream" owns shares in South African gold. Though the faculty on this campus is unionized, the career service employees are not and both of my co-workers adamantly prefer it this way.

Electronics is an exciting field, and the work I do is very interesting. All things considered, this is the best job I have ever had. Since many of my friends do not like their work, I consider myself quite lucky. But basically it's a job and not a soap box for doing political work. Since most of the people I work with are very conservative, it would be very easy to be singled out as the "department's radical." The department does not presently have one. Since the forces of reaction are such that people identified as radical are vulnerable (as far as future employment goes), my politics are not "up front." The "good old boy" syndrome is quite strong here so racist jokes, sexist attitudes, anti-union raps (on state time of course), anti-Asian war stories, general red-baiting, etc., all have to be dealt with on a daily basis. Some things one struggles with, others one lets go for another day. I prefer long-term persistence as a style of work rather than short-term "burn-out"-prone therapeutic triumphs. The struggle in this workplace will definitely be a long one.

I consider myself to be a committed political activist and have been so for nearly ten years. At first, I was involved with anti-war organizing, but after having discovered *Science for the People* (August 1970), I have been shamelessly committed to the progressive science movement in one way or another ever since. Almost every evening I do political work, mostly as a science activist, but none of it is done in the workplace. In this respect, I lead two lives. Some of my more militant movement friends criticize me for taking this stance, but to me it is a realistic style of work.

For three years I worked as an electronics technician at a university media center doing repairs on audio-visual equipment. In the local movement community I am an obvious person to contact about setting up sound systems, running the projector for films, etc. One of my more moving encounters with feminism came as a result of my being asked by a group of feminists to give a workshop on the operation and basic preventive maintenance of the various 16 mm movie projectors available locally. A few years previous (April 1971) I had produced an extensive bibliography/resources guide to the women's movement that had a press run of 50,000 and that was distributed both nationally and internationally. As a result of it, I received positive feedback from various women. It was assumed that I would be advanced enough to be able to give the workshop in a non-sexist way. This proved not to be the case, so what resulted was a lengthy (and rather tedious) discussion about technical training in our society and about the sexism incorporated into it. They wanted to know how it happened that a guy like myself (i.e., a "plastic feminist") ended up with the privilege of having such extensive technical training. Since we all learned a lot from that discussion, I think it would be worthwhile to review it briefly.

I grew up with two brothers on a small family farm in southeastern Ohio. My major area of study in high school was agriculture. I was a member of the Future Farmers of America (FFA) and the One-Hundred Bushel Corn Club (members had to have grown over 100 bushels of corn per acre). Though it was assumed I would be a farmer, my parents did encourage me to study chemistry, mathematics, etc. (although the only math I had was algebra, no trigonometry, geometry, or calculus). My uncle was a chemist, and one Christmas, my brothers and I were given a chemistry set. It was first set up in the basement of our house, but perceptive parents soon encouraged that it be set up in an abandoned house trailer, used at other times as a chicken house. Natural gas, electricity, and later even a telephone extension were set up in the trailer. It was a perfect set-up that served as a chemistry lab (we made a lot of model rockets and explosives) and an electronics shop. It was our pride and joy, and it was a rallying point for many of the neighborhood kids—except, of course, for "girls," who were not allowed in. This was the climate of the times (late 1950s to the early 1960s). One day while we were at school, it caught on fire and exploded.

College was not a serious option for myself, and it was not really clear to me at the time why I left the farm. But it was a rather small farm and it was rapidly being surrounded by suburbia, and we did in fact have an imperialist war to fight. So I signed up for a four-year hitch in the Air Force. The first year consisted of six hours per day of technical training in electronics. There were no women in the school. In a class of twenty or so people, there was one black student who had dropped out of an engineering school because of a lack of money. The last three years consisted of more training via correspondence courses, in-service training sessions, evening courses in a college level bootstrap program, etc. More important, however, was that fact that we had three years of hands-on experience with the most sophisticated electronics in the world. Most of the electronics technicians I know got their training in the military. It is usually considered a prerequisite for the kind of work I do.

The training was sexist through and through, and I will give several examples to illustrate this point. Many electronic components are color-coded. The way one is usually taught to remember the color code is by the following mnemonic device: **Bad Boys Rape Our Young Girls But Violet Gives Willingly** (B=black=0, B=brown=1, R=red=2, O=orange=3, Y=yellow=4, G=green=5, B=blue=6, V=violet=7, G=gray=8, W=white=9, colors going from dark to light). A "bleeder" resistor is a resistor that insures that high voltages are "bled" off to discharge filter capacitors when equipment is turned off. This avoids accidental shock. As an in-house joke (totally unrelated to the function of the resistor) many instructors are sure to hold up a large resistor that has "Kotex" written on it. One of the most widely distributed periodicals that electronics technicians learn from is *Radio-Electronics*. Until just recently it was subtitled, "For *Men* With Ideas in Electronics" (emphasis mine). Periodicals dedicated to introducing new electronic components (such periodicals are free to qualified people) are often oriented to macho types in that they often find it difficult to sell electronic components unless they are being held by a woman in provocative clothing. This type of advertizing is not as prevalent nor as blatant as a few years ago. Advertisements for correspondence courses often show a white man doing the lessons while he is being observed admiringly by a woman. Many of the cartoons in the magazines that E-techs must read are degrading to women. This is particularly true of

This is an example of the kind of sexist cartoon that appears in electronic products' magazines.
November 1978
Electronic Products

periodicals oriented to television repair and industrial electronics. Though things are getting better in this respect, not all that long ago it was almost impossible to enter a television repair shop without seeing a sexist poster on the wall. Such posters were made available free by distributors and manufacturers of components and equipment.

The lesson learned...young white men are encouraged to enter technical fields, and young women are discouraged. If they choose to enter it anyway, they will continually run into sexist barriers. The only physical requirement I have heard about is that one cannot be color blind, but this occurs less with women that men anyway. Seldom are heavy items handled, so upper body strength is not a general requirement. It's pretty easy work if you can get it. As I remember, the only aptitude requirement is a minimal background in mathematics and normal intelligence (the "whiz kid" requirement is a myth). It should also be mentioned that a society that gets so many of its technical people as a by-product of an anti-people military apparatus has some very serious problems with educational priorities.

Though it is true that few women are working in the kind of electronics I do, it should be pointed out that the electronics industry in general is very dependent on women to do the dirty work. How ironic that this space age industry is still so firmly rooted in laborious hand work. Electronics is so labor intensive that, measured by assets per employee, it ranks next to the most labor-intensive industries—textiles and apparel. Over half of all electronics workers are production workers. And in some sectors, such as component production, almost two-thirds are production workers. The overwhelming majority of these production workers are Third World women. It is the fifty-year-old woman in Massachusetts, the young Latin and black women workers in California, and still other women in Indonesia, Singapore, the Philippines, Thailand, Malaysia, Mexico and Brazil who perform the tedious and intricate steps that turn ideas into reality. Women workers are extensively used as a way to keep the overall wage rate low. Employers who argue that women are innately better at the intricate, monotonous, eye-straining work typical of electronics production know they will be able to hire women at a lower wage rate than men, since so many other jobs are closed to women. In 1975, 41% of all electronics workers were women—the overwhelming majority of them are in low-skilled, low-paying jobs. In the branches of electronic components and radio and television, over half of the workers are women, whereas only 29% of the higher skilled and better paid computer workers are women. In a Western Massachusetts plant visited by NACLA,* for example, 80% of the employees are women who virtually all work on the assembly line. In the same plant 90% of the technicians (testers) are men.

But there is more to electronics than just sexism. Racism also is an integral component. Though it sometimes takes a different form than sexism, it is just as deeply ingrained.

Working for a state university is much like having a regular state job, and most state jobs require that some kind of proficiency exam be taken before one gets an interview. But for electronics technicians there is none except the Federal Communications Commission (FCC) test for those who plan to work on transmitters. When I applied for a job, my proficiency was based on my word, an "honest face," and the fact that I had military training. In fact, I thought that was the normal way of doing things (really!) until the shop I worked in a while back had a vacancy and a black technician came to apply for the job. He had military training in electronics as well as several years experience as a TV repairperson. Later I was to find out that the way he got the TV experience was to start his own shop since none of the established shops in town would hire him because "whites would have problems with a Negro coming into their homes." He knew all the "buzz words" that we E-techs seem to be so fond of—the guy was obviously into electronics as a life-time venture. I had never seen it happen with a white applicant, but before he could work in the shop with the rest of us, he was required to go to the local technical school and pass their final exam for advanced electronics, and he had to pass a "hands-on exam" by fixing something in front of the supervisor. And this was for a job that was not all that advanced. In fact, many whites have bluffed their way into such jobs with hopes of learning the skills on the job. Supervisors did not get all that uptight about the bluff as long as the applicant was a friend of theirs and "a really fine guy."

---

* NACLA—North American Congress on Latin America (151 West 19th Street, New York, NY 10011). I have been wanting to put in a good word for this incredible group for years. They have consistently done some of the very best anti-imperialist research on multinational corporations that has been done. As one would expect from such a group they did a special issue on the electronics industry. "Electronics: The Global Industry," *NACLA's Latin America & Empire Report* (vol. XI, no. 4, April 1977), $1.25. All information about women production workers came from this issue.

In the part of the country where I grew up (rural Ohio), there were very few blacks. In the high school that I attended, in a graduating class of approximately 200, there was only one black student. It was not until I got to the university (at the age of 23) that I encountered black activists on an on-going basis and in a positive political environment.

Living in a university setting has its distinct benefits because it is relatively easy to obtain institutional funds to bring progressive speakers to the campus. As a result, I have had the privilege of meeting people like Sam Anderson, who came to speak about "Black Studies and the Natural Sciences," and with Tapson Mawere, chief representative to the U.S. of the Zimbabwe African National Union (ZANU), and others. These black leaders all stressed the importance of the contributions that can and are being made by progressive technicians and scientists. I have gained much from talking with such people about the roles electronics technicians can play to aid the black liberation struggle here in the U.S. and in other countries. A long list of productive tasks could easily be included in this essay, but rather than do that I will relay a message that kept coming out of the discussions. Electronics technicians and all other technical workers should place a major emphasis on political education in order to do a political critique of their own fields and be more informed about *all* of the various people's movements in the U.S. and around the world. Once they are politically aware, they themselves will be able to produce long lists of tasks that technical people can do to advance the struggles of the working class.

Progressive technicians should take note of this important message and we should help to set the pace of progressive social change. "What we need now is an enthusiastic but calm state of mind and intense but orderly work" (Mao Tse-tung, 1936). If political education is not emphasized by technical people we could easily fall into a state where we are described as Albert Speier of the Third Reich described himself: "Basically, I exploited the phenomenon of the technician's often blind devotion to his task. Because of what seems to be the moral neutrality of technology, these people were without any scruples about their activities. The more technical the world imposed on us by the war, the more dangerous was this indifference of the technician to the direct consequences of his anonymous activities."*

---

* *Memoirs* by Albert Speer, MacMillan Co., 1970. He was in charge of armaments and war production under Hitler. Master technocrat, for a time the second most important man in the Reich, at Nuremberg he was the

We have much to do, and it is time to get started on the tasks at hand. But one lesson that I have learned from the last few years is that it is also important that the tasks be done in a non-technocratic, non-mechanistic, and non-dogmatic way. *Style of work is very important.* Many of the splits in the progressive movement and much of the failure to reach a larger constituency can be attributed to character conflicts, personality problems, etc. of the people sparking the organizational models. Progressives should ask themselves, Are we "real people?" Are we really trying to reach people with an alternative to the present situation? If more progressive technicians think about these kinds of things and act on them, then we can make a significant contribution to what is happening and to what is being written. We will all feel better about ourselves, our people, our work, and our world.

---

only defendant to admit his share of the guilt in the crimes of the Third Reich; he was sentenced to 20 years in prison. At the end of the book when he was asked how all of the crimes of the Nazi regime were possible, his answer was that for the first time in history "unlimited personal power was combined with new devices provided by modern technology."

"Ladies" in the Lab
—Angela Corigliano Murphy

The article tells a typical story of a young female research assistant who finds her initial bright-eyed excitement at working in science quickly evaporating as the realities of the social relations of the laboratory becomes clear. Although written in the early '70s, the article provides insight into the work life in the lower echelons of the scientific hierarchy.

To many people, a career in the life sciences appears to be a glamorous one—full of exciting discoveries and high salaries. In the next few pages, I hope to show how great a misconception this is. I also hope to show how, behind the "men in white," stands a veritable army of underpaid assistants, most of whom are women, who do not know, or will not admit, the magnitude and roots of their oppression.

I have worked in two kinds of laboratories: a medical, or clinical laboratory in a hospital, and a research laboratory in two different universities. I have not worked in an industrial laboratory, but I have friends who have, and I gather that except for a generally higher pay scale and less variety in the work, roughly the same conditions prevail. Some technical and clerical workers in hospitals are unionized; most in universities and in the chemical and pharmaceutical industries are not. Why is all this so? Let me detail some actual conditions and use these to reveal the roots of the problem.

My first job after college and a six-month stab at graduate school in chemistry was in the clinical biochemistry laboratory of Memorial Hospital for Cancer and Allied Diseases in New York

City. Our laboratory, which handled blood and urine chemistries, had a very large workload—as many as three hundred samples needing multiple analyses in one day. The year was 1960. There were twenty-one technicians, eighteen women and three men. Two or three secretaries and a dishwasher ("laboratory aide" is the euphemism) completed the staff. Starting salary for technicians was $4,050 a year. Overtime was paid for, but frowned upon.

A newly hired technician was handed a large black spring-back book. In this book were the "recipes" for the tests we did. For three months we learned to do the tests with control serum before we were allowed to do them with the blood of the patients. As we learned to do each test satisfactorily, we did it "for real" and then went on to another test. Speed was as important as accuracy, but honesty was even more important, for people's lives were at stake.

We were expected to do good work, but rarely was the theory behind it ever explained. If something went wrong, we had to wait on others for an explanation. Some seminars were given in the fall, but those who came at other times had to be content with working mechanically, like robots. What was worse was not knowing what the doctors had in mind when they ordered the tests. Such knowledge would have made for better patient care, by allowing us to work in concert with the doctors. It would also have kept us from getting bored, as we all eventually did. After learning a certain number of basic tests, we were put into rotation with the other technicians. We did a new test or series of tests every week. To alleviate boredom we were allowed to learn a new, more difficult test as we became more proficient, or possibly to learn how to use a new instrument. In this way also we were not always repeating tests.

But what of those who had been there for years, who had learned all there was to learn? How did they stand it? For now they were just members of an assembly line, more educated than most, but assembly line workers just the same. Why didn't they rebel? One answer is that most of them were women—single women waiting for marriage or graduate school and married women waiting for pregnancy (one friend concealed hers until the fifth month, for she knew it meant the end of learning new tests). Then there was the rationalization that care of the patients came first, that this was a non-profit institution, etc. Another was the implicit message that we were "professionals"—and that precluded certain ways of acting for us. I will say more of this later, because it really becomes an explicit attitude in university surroundings.

As for me, after eight months at Memorial, I had had enough. I had saved as much as I could out of my bi-weekly pay check, and when I thought I had enough money, I applied to Columbia University's Graduate School of Chemistry. I spent a year there trying to get a Master's degree, but my background really didn't qualify me for Columbia, and I didn't know whether I wanted to be a woman or a chemist, which kept me from fully concentrating on my studies. At the end of the year I was told to leave without the Master's and not to return. I needed money, so I went job-hunting. I had heard some good things about the Rockefeller University (then the Rockefeller Institute), so I tried there. I was lucky. There was an opening and I was sent to see the professor involved. His work seemed interesting, and he seemed nice, so I thought that if he was satisfied with my credentials, we were all set. He was. He warned me, however, that I would not be doing independent research, for he needed me to work on his project, and he hoped I would become as interested in it as he was. I stayed seven years—until I left to get married.

To solve my identity crisis (which a women's liberation movement would have helped), I had some psychotherapy. Gradually I began to find myself as a person and a woman. I found I could better cope with my job and with the world in general. I became quite interested in my work, found out I really liked it, and became more and more expert at it. Because my boss was willing to teach me all the biochemistry and biochemical techniques that I was willing and able to learn, I did finally get to do some quasi-independent research. I even became co-author of a paper in *Biochemical Preparations*. Working conditions in my department were pretty good. However, conditions at Rockefeller in general were pretty poor. Many of the laboratories were not as friendly as ours. They were on a "Miss" and "Doctor" basis, rather than on a first name basis as we were. There was not as much opportunity to learn. Although Rockefeller had become a graduate university in 1953, little of its academic program was open to technicians, dishwashers, mechanics, electricians, animal house employees, nurses, pharmacists, etc. Rockefeller had no system of tuition rebates, as have some other institutions. They claimed that they preferred to put the money into salary increases, but they didn't pay that much better than other institutions in the area. In some laboratories, if a technician went to a lecture during working hours, even if it was directly connected with her work, she had to make up the time afterward. There was no overtime pay. One was supposed to receive time off equal to overtime, but this was more honored in the breach than in the observance.

When I first came to Rockefeller, most of the dishwashers and waitresses I knew were older women, Irish, Italian, German, Puerto Rican, etc., working their way toward retirement. The pension plan at Rockefeller was a good one, but they didn't make enough money to retire early. In 1968, a dishwasher's salary at Rockefeller averaged out to between $75.00 and $80.00 a week. In later years, the membership of both these groups changed drastically. More and more it was young black women, often with small children at home, who filled the dishwasher and waitress jobs.

Health care at Rockefeller was fairly good, but not as good as at Memorial which maintained a free, well-staffed Health Service and provided free diagnostic x-rays and laboratory services. Rockefeller was simply not equipped to offer all that. Therefore, Rockefeller maintained what was called "Social Service," staffed by two nurses and a secretary. One could go to them to have one's throat looked at, or for an antacid, or for other minor first aid. In cases of accidents covered by Workmen's Compensation, one was sent immediately to New York Hospital, across the street. For illnesses, one was sent to a variety of specialists. Rockefeller would pay for a first diagnostic visit for any illness and for any lab work or x-ray connected with that visit.

Both Memorial and Rockefeller had the Blue Cross-Blue Shield and Major Medical plans. Both paid a certain percentage of the Blue Cross-Blue Shield fees. As time passed, an expanded plan became available at Rockefeller which covered more expenses.

The above health care situation sounds pretty good, but the Rockefeller University students had a much better one. Cornell Medical College at New York Hospital considered them the equivalent of its own medical students, and therefore its student health service was completely open to them, free of charge. These students were not paupers. They were paid a $3,500 a year stipend (more if they had children), which under existing federal laws was tax free. Those who held an M.D. degree were given a $5,000 a year stipend, plus dependent allowances, if applicable.

Social stratification and sex discrimination were evident in other ways. Students and faculty members ate together in a large dining room. Technicians, dishwashers, electricians, etc., did not eat there, except by invitation, although if pressed, the administration allowed that they were free to do so. Coats and ties were required for the men and street attire for the women. For *female* technicians and dishwashers, nurses, secretaries, receptionists, pharmacists, etc., and their guests *only*, there was a smaller dining room where the same meals that were served in the faculty room were served, at identical prices. For all those who could not or would not eat in either dining room (and they were in the majority), there was a small cafeteria in the basement of one of the laboratory buildings. It held a maximum of about one hundred fifty people and was not air-conditioned. However, there was a greater variety of foods at inexpensive prices.

Why did we take this? If the older women who were dishwashers, waitresses, etc., could not afford to lose their jobs because of economic necessity, still less could the younger women with children. Nor was it advisable or safe to unionize. Although it is against the law to fire someone for union organizing, one can fire a union organizer without ever alluding to his/her job, but in most cases Personnel had a free hand, and in a few even had the cooperation and support of the supervisor involved. Blacklists are not supposed to exist, but the word gets around to personnel offices about who the "troublemakers" are. If one has a few skills and really needs a steady job, one will put up with quite a lot before rebelling. Even the more skilled workers, secretaries, and technicians would think twice before engaging in activities that might, even temporarily, endanger their livelihood. Furthermore, there was, on the part of secretaries, an identification with the interests of their bosses and their work which tended to vitiate organized action. This was also true on the part of technicians, but more so, there was the myth of "professionalism," of "I am a professional and unions are for dock workers, so I don't need a union. If I have a problem I can work it out with my boss." This myth, this propaganda was so pervasive that even a reading of the legal definition of "professional" under New York State law ("one who has spent four or more years in graduate study after the baccalaureate degree, and who holds a graduate degree, i.e., a PhD or M.D.") did not dispel it. The only hope all of us had to effect change in our salaries and working conditions was to organize and to act together to confront the university administration, but our upbringing and education had all worked to make this seem difficult, if not impossible.

I suppose the fact that we were mostly women had a great deal to do with it. We women, more than men, are brought up to put the interests of others above our own. We are even denied full identity: we are always seen in relationship to someone else, as some man's daughter, or a man's wife, or someone's mother. It is not so very surprising, therefore, that we seem to internalize the prestige of the institution we work for because this makes us feel some of the sense of worthiness we should feel in ourselves as persons, but do not.

When the impetus toward organization and change finally came, it came from some of the more radical graduate students. They felt that they could help us because they were in a more favorable position vis-a-vis the university administration. They could be more active and visible and run less risk of reprisal. Also, they had access to facilities and information that could be helpful to us. They arranged a meeting for us with Local 1199, Drug and Hospital Employees Union, RWDSU, AFL-CIO, and a group of us, mostly secretaries and technicians, attended. That meeting was something of a revelation, for I found that the "professionalism" myth was much more widespread than I had believed. Local 1199 had to have a separate unit within the larger union called the "Guild of Professional, Technical, Office and Clerical Hospital Employees" because these people found it difficult to join a union which included nurse's aides, orderlies, kitchen staff workers, maids, guards, etc. The unit even had to be called a "guild" rather than a "union" because the word "union" had unpleasant, low status connotations for them.

Such people seem to have the idea that some types of work are more "dignified" than others because they involve "brains" rather than "brawn." Intellectual labor seems to have more value than manual labor and to offer more reward ("spiritually" if not materially). It has taken me, at least, until now to see that this is propaganda. All socially necessary work has dignity. This is only "lost" when workers are alienated from the products of their labor and from the full value of these products. It is not the repetitiveness of work that makes it deadening, but the alienation. It is not in their labor that workers lose dignity, but in the exploitation of that labor by the bosses. Unfortunately, the bosses have made us believe otherwise. They have set us laborers against each other so that we will not turn against them. Organizers have a huge task ahead of them when they try to make workers see that they are all in the same leaky boat.

Our attempt to organize technicians and secretaries at Rockefeller failed. Our student friends gave us as much help as they could, but to no avail. We just couldn't rouse enough people to get a viable organization going. Also, as 1199 had warned us, the university moved to co-opt the drive. Salaries were raised, movement was made to fill more of the workers' needs, just enough to impart a sense of well-being. We decided to redirect our efforts toward the more oppressed workers, lower down on the scale than we were. That effort, carried out mainly by the students, finally bore fruit. The electricians of the Power House and animal tenders of the Animal House now have some sort of union representation. They went through a lot to get it.

By this time, I had already left Rockefeller. In June 1968, I married a student at Rockefeller who had just graduated. He was going to Yale on a post-doctoral fellowship, so we moved to New Haven. It was quite a wrench for us, and especially for me. I had reached a position of some responsibility. I liked my job, and the organizational effort with the animal tenders and electricians was just getting started. I had a job waiting in New Haven with a fancy new title—"associate for research"—and at the same salary as Rockefeller (generally Yale's pay scale was lower), but I was frightened by it and sad to leave all the friends at Rockefeller. However, some of our friends at Rockefeller had told us of a political organization in New Haven which had grown out of the anti-war movement there. We contacted it soon after we got to New Haven and found it to be doing important, relevant work that we could share.

I started working for Yale Medical School in early September, 1968. I worked in a building called the Laboratory of Clinical Investigation, an architect's dream and a scientist's and technician's nightmare. It lacked some of the conveniences and even some of the safety features of the older buildings at Rockefeller, and it was brand new. For example, in the labs there was no distilled water tap at the sink, so the dishwasher had to fill up a large bottle from somewhere else, rather than rinse the glassware directly. The washing procedure was that much more time-consuming.

When I read the printed guide Yale gave its employees, I learned some interesting facts. One was that I was supposed to work from 8:30 to 5. For me, this was tantamount to taking a cut in salary, which I had no intention of doing, especially since my work often involved staying until 6 or even later, with no overtime pay. Moreover, we were only paid once a month at Yale. (The more prestigious the institution, the less frequently one is paid, I've discovered.)

Yale had Blue Cross-Blue Shield and Major Medical Insurance, but since I was covered by my husband's policy, I didn't learn much about it. I got the impression that employees at Yale paid the full premiums and that the plans were pretty poor in services compared to Rockefeller's. Yale had an employee health service, but it presented a problem. Yale's campus is in two parts, across town from each other. There is a shuttle bus which runs at fairly frequent intervals between the two parts. The main office of the health service is not at the medical school, but across town from it. There is a small branch office for medical school employees open a few hours each day. During student vacations it is closed, even though medical school employees have to work. If an accident or sudden illness happens at such a time, one has to wait for the shuttle bus or take a cab to the main office. The ironic part of all this is that the health service of Yale-New Haven Hospital is right across the hall from the medical school health service, but it cannot care for medical school employees, even when the medical school health service is closed. I wrote the Yale administration about this situation, but as far as I know it has not been corrected, even though there has been some talk of doing so.

These were general conditions at Yale, but what were the conditions where I worked? After the lab I'd worked in at Rockefeller, I found them oppressive. There were four other technicians besides me, and two graduate students. We were all on a first name basis, but all called the department head and our boss "Doctor." There were also two secretaries.

Other factors were the tensions between my boss and his graduate students—due to the extremely competitive atmosphere at Yale—and the animosities, overt or covert, between my fellow workers. The friendliness of our laboratory at Rockefeller had served as something of a shield from the oppressive administration. Among the technicians, secretary and dishwasher, in addition, there was a sense of solidarity, almost of sisterhood. In short, we were friends. This was not true at Yale. I have never seen so much backbiting, sarcasm, competition for the boss's attention, and snobbery as I saw there. I couldn't even avoid it by hard work, for three of us worked in one room, and I could see and hear the quarrels.

In late October I found out for sure that I was pregnant. My first reaction was pure joy, since this meant deliverance from the lab at Yale. Later I began to feel a certain ambivalence. I had given a half-promise that I would work at least a year; now I'd only be working nine months. Also, before starting work at Yale, I'd played at being a housewife and found I didn't like it that much. It seemed to be a terribly isolated existence. But, I rationalized, I would have the baby and then do political organizing. Still, I felt a vague but gnawing uneasiness. I wasn't a member of a woman's liberation group. There was such a group, but their meetings were on Sunday afternoons and evenings, an inconvenient time for me. Even though Tom, my husband, encouraged me to go, I did not. It was, I think, an unfortunate decision. Had I some women's liberation consciousness, it might have made things easier for me, both in the lab and at home.

My husband and I did not share household tasks equally, and I was smouldering with unspoken anger over this until I talked to other married women at work. The problem was almost a universal one and became more complicated when children were involved, what with illnesses, babysitting difficulties, etc. Whatever anger or rebellion was felt against this tended to be directed toward an individual, i.e., one's husband, rather than toward the economic and social system that conditions people for their roles in life. Yale, of course, like any other employer, did not even acknowledge the existence of such problems for its women employees, and more than one woman with an unreliable babysitter, sick child, etc., was fired for being "late" or "out" too often.

During this time, I heard of a group that met once a week during lunch hours in another laboratory building. I went to one meeting and began a happy association with YNFAC, the Yale Non-Faculty Action Committee, that lasted until shortly before we left New Haven after our daughter was born. YNFAC was a group of workers from all parts of Yale, and they were very clear-headed about what they were up against. They knew that most of Yale's workers were women, that they were frightened, that they were wary of unions, both because of the professionalism and prestige myths, and because of some bad experiences a few had had in a local (New Haven) group. The very name of our group was chosen to show that it was us, workers at Yale—and not "outsiders"—who were conducting this organizing drive. We wrote and distributed leaflets and posters, talked to people, held meetings right after working hours, and we confronted the Yale administration on a variety of problems. Because so many of us were women, we began to make demands that were directly connected with the problems women faced, such as the need for day-care centers at the workplace. We did research that showed that Yale could well afford to meet our demands. We said to Yale: "We want more money. We want job security. We want to make the decisions affecting us at our place of work." We said to our fellow workers: "We are powerless at Yale until we fight together as an organization rather than as individuals." We began to increase in numbers and strength, so much so that in the Spring, Yale forbade us to use its auditoriums and meeting rooms without payment, saying that this use would violate a law designed to protect workers against the establishment of company unions. Such laws, like the laws designed to "protect" women are wonderful things; they always seem to be used *against* the very people they are supposed to protect.

However, despite Yale's harrassment, YNFAC is today a growing, thriving organization which has recently voted to affiliate with a national union, and which may soon win the right, by election, to be designated the bargaining agent for most of the workers at Yale.

Women in the lower echelons of the life sciences have a long and difficult road ahead before they can make significant changes in their situation. One thing they must do is realize that they are not merely an adjunct to science, but a necessary part of it, and as such, deserving of equitable remuneration. They must also rid themselves of the myths foisted upon them by the bosses to keep them docile. Finally, they must realize that individual solutions are not possible, and it is only by getting together and organizing that they will be able to bring about an economic and social system that will enable them to fully realize their capabilities as women and as workers.

As for myself, I will not go back to working in a laboratory, unless driven to it by economic necessity or by the prospect of becoming involved in an organizing drive. I really love biochemistry, but I am tired of being exploited on all sides. I don't want to pay exorbitant babysitting or day-care fees. I consider what I do as a housewife and mother to be socially necessary *work*, a *job* in fact. It is another myth, foisted upon us by the bosses that denies this. Why should I work at *two* full time jobs if I don't have to? However, this is an individual solution. For most women there is no choice, because for them the matter is one of financial necessity. Their husbands can't help, either, because *they* are usually working two full-time jobs. Only the big bosses profit. For things to change, we must get rid of the big bosses.

Women in Chemistry
—Ana Berta Chepelinsky et al.

This article points out the obstacles which interfere with women's achievement in science. It points out that even for the few women who manage to overcome early negative socialization and discrimination, the work situation in science is rough and demanding.

In a country where over half the population are women, why are only 9% of all chemists women? Why do women constitute only 4.2% of all physicists and 0.8% of all engineers? Are we dealing here with those mythical natural interests and capabilities of women? Is the reason irrational discrimination, or is it perhaps a more pervasive social force? Certainly the practice and pattern of discrimination can be well documented in the sciences, but the real problem is indicated by the fact that there really are not that many women who enter science to be discriminated against. Women are excluded in general from the higher paying, "high status" jobs in this culture. The socialization of women tells them that their place is in the home, that their purpose in life is to marry and to raise children. On the other hand, economic reality forces almost all women to work at some point in their lives and at the present time 40% of the women in this country work. Women in our capitalist society form a cheap, flexible labor pool and are responsible for all of the unpaid labor in the home as well. The socialization process assures that women are basically untrained for anything but menial, mind dulling, uncreative jobs. They are not considered to be true members of the workforce and are easily drawn in and out of the job market as demand changes. It is in the interest of the economic structure to maintain the present attitudes toward women. The availability of such a group helps keep business costs down and profits up. We wish to focus here on the mechanisms by which women are kept "in their place"—using chemistry as an example.

Channeling or What's a Nice Girl Like You Doing in a Place Like This?

One can isolate a number of sociological and psychological forces which channel women away from higher education and away from careers. Each kind of pressure, whether subtle or overt, arises from this society's stereotype of character and roles for women and is reinforced by the need to keep women as a cheap labor force. The chronology is probably familiar to most professional women. The American girl grows up in an environment, both at home and at school, that molds her into a sweet, submissive, agreeable form. It confirms the belief that a woman's proper and natural spheres are domestic and maternal. On the other hand, boys are encouraged toward aggressive, independent, and ambitious behavior and are surrounded by images and actualities of the active adult male. In their early school books, on television shows and ads, and in the home, young girls are bombarded by images of what they are expected to be. For example, in a survey of ads for chemistry sets, investigators found that most pictures showed young boys experimenting, with invisible ink and magic solutions. There was only one picture of a girl...making lipstick. By the nature of these social models and prejudices, young girls are not encouraged to develop the traits of the scientist, such as self reliance, inquisitiveness, creativity, or analytical ability. As Rossi notes, "If we want more women to enter science...some quite basic changes must take place in the ways girls are reared. A childhood model of the quiet, good, sweet girl will not produce many women scientists or scholars, doctors, or engineers."[1]

This channeling process intensifies during secondary and college level education. The ambivalence grows, as the possible conflict between personal interests and social expectations becomes more marked. This pressure is reflected in statistics for high school and college graduates and for advanced degrees. For example, Epstein points out that "although there are more girls than boys in high school graduation classes, more boys than girls graduate from college."[2] And although drop-out rates are similar, "boys are more likely to leave because of academic difficulties or personal adjustments," while girls leave to marry. But then, this is what is expected and thus almost imposed upon women. As Weisstein notes, American women are defined as "inconsistent, emotionally unstable, lacking in a strong conscience or superego, weak, 'nurturant' rather than intelligent, and if they are at all 'normal,' suited to the home and the family...(However) In a review of the intellectual difference between little boys and little girls, Eleanor Macoby has shown that there are no intellectual differences until about high school, or, if there are, girls are slightly ahead."[3]

A young woman is discouraged from being intelligent or "over-educated" since this economic investment would only make her unmarriageable. After all, "what a man does defines his status but whom she marries defines a woman's."[4] As many women scientists will acknowledge, they were especially discouraged from entering science or engineering because these fields are viewed as being beyond women's mental capacities, are incompatible with desirable, womanly attributes, and in the long run, are unproductive for them. Distribution in undergraduate degrees reflects the success of this channeling: men receive almost 9/10 of the degrees in the physical sciences and about 3/4 in biological sciences and mathematics, while women generally choose education, English, and foreign languages. Weisstein concludes that "in light of social expectations about women, what is surprising is not that women end up where society expects they will; what is surprising is that little girls don't get the message that they are supposed to be stupid until high school; and what is even more remarkable is that some women resist this message even after high school, college, and graduate school."[5] Interestingly, many of the women who have "made-it-through" are either foreign born or are first-generation American. This can be accounted for by the differences between the channeling of middle-class European women and American women.

If a woman successfully challenges these myths and traditions and enters a graduate school, she encounters another set of barriers. As Epstein suggests, a profession is a microscopic society that depends on "the mutual understanding among its practitioners" and on conformity to "shared norms and attitudes." First, Epstein writes, "the sponsor-protege or master-apprenticeship may inhibit feminine advancement in the professions. The sponsor is most likely a man and will tend to have mixed feelings, among them a nagging sense of impending trouble, about accepting a woman as a protege. Although the professional man might not object to a female assistant—he might even prefer her—he cannot identify her as someone who will eventually be his successor. He will usually prefer a male candidate in the belief that a woman has less commitment and will easily be deflected from her career by marriage and children...Even if she serves an apprenticeship, the woman faces serious problems in the next step in her career if she does not get the sponsor's support for entree to the inner circles of the profession...He may feel less responsible for her career because he assumes she is not as dependent on a career as a man might be."[6]

Epstein also notes that women are often excluded from the informational interactions and means of recognition which are essential to advancement. And unfortunately, "the only possible antidote for the familiarity and lineage which oil the wheels in professional environments is power through rank, seniority, money... women do not often have any of these defenses." The elitism of the professions is evident from Epstein's analysis, and she also shows the overwhelming male domination of our present society.

If all these social pressures fail to discourage a particular obstinate woman, there is always outright, old-fashioned discrimination to fall back on. Incredible as it seems, loud cries of denial still go up from academics and industry people when this question is raised. However, a rather superficial look at the situation is enough to make the point.

Women in Universities or Is Anybody There?

Hiring policies and practices in universities are discriminatory. Most U.S. universities which grant PhD's do not have any women chemistry faculty. From a study of 172 schools with a total of 3,925 PhD faculty members, only 90 were women, or 2.3%. Of these, 39 (or 43%) were of subprofessorial rank (instructor, research associate, etc.).[7] If there were no discrimination, about 6.3% of all faculty members in chemistry would be women (since 91% of women with PhD's were working in the last decade and 6.9% of PhD's in high-ranked schools in chemistry were women).[8] As a striking example one might consider the top five departments of chemistry, which grant 6.9% of their PhD's to women. Top universities are training PhD women, but they are not hiring them:

1. Harvard   0.00% women faculty
2. Cal Tech   0.00% women faculty

3. Berkeley   0.00% women faculty
4. Stanford   0.00% women faculty
5. M.I.T.   0.00% women faculty

Women are in less "prestigious" universities, colleges, and junior colleges. Of those privileged enough to be employed in top schools, 43% are placed in the lower echelons of university ranking.

If any doubt remains as to whether basic attitudes and hiring decisions are discriminatory, the results of a study involving the chairmen of graduate departments of a physical science discipline in colleges and universities will offer conclusive evidence. If a man and woman of equal qualifications apply for the same position on a faculty, the man is more likely to be hired. If a woman with superior qualifications is among the applicants (with men of average qualifications) she is seriously considered for the position, however with reservation. Important factors to be taken into account are her marital status, number of children, husband's occupation, and of course, her compatibility with the male faculty.[9] That discrimination does exist on the hiring decision level is evident.

If a woman surpasses the barriers of hiring attitudes, is she then treated with equality with regard to salary? On the professorial level in all colleges and universities in the United States in 1965-66, the median annual salary for men was $12,768, and for women $11,649. The gap narrows as one goes down the ladder of prestige, but is there at all levels.[10] The same inequality appears in promotions. The study by the Committee on Education and Labor in the House of Representatives concluded that promotion possibilities for women in universities are worse than for men. The proportion promoted is lower at *all* ranks studied and for all time periods studied (1920-40, 1950-69).

Since the proportion of women professionals married to men in the same field is very high, the question of nepotism policies in universities becomes important. Many women in this category whose husbands hold university positions are employed as lecturers (sometimes without pay), research associates or forced to find positions in other departments or different (often "inferior") institutions. In a study of the University of California at Berkeley, most women in such a situation felt that their talents were not fully utilized and that they were qualified for regular positions on the Berkeley faculty. Some of their husbands' comments are rather instructive: "I presume that the University nepotism rules bar her employment here, and so she is consigned to a job vastly inferior in all ways, though her qualifications are equal or superior to my own...and better than many of the people the department does hire." Or "She is employed here, at a lower level than in her previous position and in a temporary position...She has no facilities for research or support for research here and is forced to use my lab, where she has an established reputation as an independent investigator." Wives with B.A.'s and M.S.'s or M.A.'s are affected by nepotism in many ways: in the Berkeley study, some could not be appointed as lecturers in their husband's fields, though uniquely qualified; several could not be hired as secretaries or researchers even with excellent training and qualifications. More frequently, however, wives were found working as unpaid research or editorial assistants for their husbands.

Women in Industry or The Same Old Story

Although less specific information is available on employment of women chemists outside universities, the same trends are apparent. Women are found in lower ranking positions in both industrial and government laboratories. For example, 75% of the women employed by the National Institutes of Health are in ranks of GS9 or below, and no women are in the two highest ranks, GS16 and 17.[11] Recruiters for industrial positions tend to look for men for the positions offering permanence and good opportunities for advancement. Even when women are hired, promotion is slower. Women are less likely to "advance" into management. Only 6% of the women in natural sciences are found in management while 24% of the men in science are working in managerial positions.[12] Starting salaries for women in the chemical industry are slightly lower than those of men, and the gap appears to increase with length of employment. *Chemical and Engineering News* did a study of chemists' salaries in the fall of 1968, which showed that, with seniority held constant, women with PhD's made less than men with only B.A.'s Fringe benefits are often less inclusive for women than for men. In addition, married women often hold a second full-time (unpaid) job as housekeeper. Women with children are particularly handicapped by the lack of adequate child care facilities.

Grants or From Each According to his Ability— to Each According to his Sex

Grants and awards are of crucial importance in the professional development of scientists under the present system. What is surprising is that women scientists are as productive as their male colleagues in spite of the fact that they have more difficulty in obtaining support for their research. The problems encountered by women in this area are shown by a recent study of grants awarded by the National Science Foundation, "In the Senior Postdoctoral Fellowship competition recently held by the NSF, 14 of the 395 applicants were women. Fifty-four grants were awarded—none went to women. In a pilot study of grants and awards given by the NSF to university researchers in a physical science discipline for the years 1964-68, women were only awarded less than 0.03% of the grants although they comprise 5 to 8% of the scientists in the discipline. Furthermore, the mean dollar value of the awards received by women was smaller than that of those received by men."[13] Similar discrimination is shown in the awarding of NIH grants. "Of Postdoctoral and Special Fellowship applications made to NIH in 1970, the rate of disapproval was higher for women than for men"—or 38.8% for men compared with 55.7% for women. The discrepancy was even greater in the disapproval rates for applicants for the Research Career Development Awards for one NIH institute over the years 1949-1965. Only 21.9% of the men were turned down while 70.8% of the women applicants were rejected. In 1970, women were only 2.5% of the NIH Study Panels which review grant applications.[14] In addition, the representation of women on advisory panels of the NIH and other agencies is not anywhere near the actual proportions of women in the scientific fields. As pointed out by Lewin and Duchin, "similarly, out of the 827 Sloan Research Fellowships that have been awarded by the Alfred P. Sloan Research Foundation over the past 16 years, only one or two women were among the recipients."[15]

Myths, or, If You're So Smart, Why Aren't You Married?

Though faced with the facts, many people would still deny that the above is real proof of discrimination against women, but only a result of the various characteristics of women in the job market which make them impossible to accept as serious, dedicated and qualified workers. One of the most durable myths is that it is useless to hire or train women since they only marry and leave. Statistics show that women can be expected to remain on the job about the same number of years as men, regardless of educational level. As an example, 91% of the women who received PhD's in 1957-58 were employed in 1964. Of these, 80% had not interrupted their career during that period of time.[16]

Another stereotype is that women professionals are less productive than men. According to a study, there are no differences in the productivity of men and women scientists.[17] This is despite the fact that women are discouraged in their professional life and placed in positions which indicate their so-called incapability of competing with men. Furthermore, our culture defines women as incapable of abstract thought. In all doctoral fields, a study has shown that women receiving the doctorate are brighter than their male counterparts.[18] However satisfying this may seem to those who have PhD's, it is a discouraging fact. For a woman to succeed she must be brighter than her male colleagues, must work harder and still face the fate of seeing her efforts and talents oftentimes remain underutilized or unrewarded.

Then there is the "superwoman" stereotype. Women who do succeed are seen as "different" from other women. The woman who manages to get by the socialization barriers and to find a job in spite of discrimination may do so at a great psychological price. She may fall victim to the "superwoman" syndrome, attributing her success to her own superior capabilities and perseverance. She implicitly accepts that a woman must "prove herself" by demonstrating far greater capability than a man for the same recognition. Having joined the elite, she does not see women's problems as societal ones, but as personal ones. She accepts the society's rhetoric that every woman who is truly competent and determined can succeed.

One can write, then, convincing critiques of the social and psychological pressures that discourage women from entering science and other professions. One can also make endless surveys of facts and cases that show the extensive discrimination that women face if they somehow "make it through" to a career. However, the treatment of women is fundamentally tied to the structure of our society. The mere recognition of problems will not resolve them. Nor can one depend on the goodwill of societal institutions or on the promises of government to produce change. Industrial directors will not spontaneously, out of some humane insight, abandon their discriminatory practices; if the change is not profitable, why make it? Neither wil the action (or more precisely, the pronouncements) of the government be anything but pacifiers for dissatisfied women. HEW is, for example, apparently easing up on even the most villainous institutions. Social change requires social mobilization, or organization of people, ideas, and tactics. Discrimination against women must be fought wherever it occurs.

However, the complete liberation of women will require a basic change in every aspect of the society, from the economic structure to the nuclear family. Sexist attitudes and practices will begin to collapse only when the victims, the women themselves, get together to define the problems and exert strong and visible pressure for radical social change. As our sisters demanded in the 19th century newspaper, *The Revolution*:

> *Principle, not policy; Justice not favors—*
> *Men, their rights and nothing more;*
> *Women their rights and nothing less.*

FOOTNOTES

1. Alice Rossi, "Women in Science—Why So Few?" *Science*, 148, May 28, 1965, pp. 1196-1201

2. C. F. Epstein, *Women's Place: Options and Limits in Professional Careers*, Berkeley, University of California Press, 1970, p. 56.

3. N. Weisstein, "Kinder, Kuche and Kirche," in Robin Morgan, ed., *Sisterhood is Powerful*, New York, Vintage Books, 1970.

4. *Ibid.*, and Epstein, *op. cit.*

5. Weisstein, *op. cit.*

6. Epstein, *op. cit.*, p. 169.

7. "Few Women in Academe," *Chemical Engineering News*, May 10, 1971, p. 21.

8. Susan M. Ervin-Tripp, letter entitled "Women with Ph.D.'s," *Science*, 174 December 24, 1971, p. 1281.

9. Arie Lewin and Linda Duchin, "Women in Academia," *Science*, 173, September 3, 1971, pp. 892-895.

10. The figures and the information in the remainder of this section were taken from "Discrimination Against Women: Hearings by the Committee on Education and Labor," House of Representatives, 1970.

11. Extramural forum of the National Institutes of Health, May 26, 1971.

12. Richard H. Bolt, "The Present Situation of Women Scientists in Industry and Government," in Matt Seld and Van Aken, ed., *Women and the Scientific Profession*, Cambridge, Massachusetts, MIT Press, 1965, p. 145.

13. Ervin-Tripp, *op. cit.*

14. *Ibid.* and Extramural Forum of NIH.

15. Ervin-Tripp, *op. cit.*

16. Patricia Albjerg Graham, "Women in Academe," *Science*, 169, September 25, 1970, pp. 1284-1290.

17. Rita Simon, Shirley Clark and Kathlene Galway, "The Woman Ph.D.: A Recent Profile," *Social Problems*, 14, 1967, p. 221.

18. H. S. Astin, *The Woman Doctorate in America*, New York, Russell Sage, 1969.

The Scientist as Worker
—André Gorz

A question seldom asked is, Are scientists workers or are they allied with the capitalist class? In this article the author discusses this question in light of the "proletarianization" of the scientist in industrial societies.

In the back of our minds, we still find it quite hard to believe—or even outright shocking—that a person with a degree in science should be considered a worker just like a person with a "degree" in plumbing, drafting, toolmaking or nursing.

To most of us, whatever the political convictions we possess, there is still an essential difference between a scientific worker and, for instance, a metal worker: the adjective "scientific" does not, in our subconscious, refer to a *skill*, a *craft* or an *expertise* like any other; it refers to a *status*, to a position in society.

Most scientific workers had expected their training in science to earn them an interesting, well-paid, safe and *respected* position. They felt entitled to it. And they felt entitled to it because most of them were brought up in the traditional belief that knowledge is the privilege of the ruling class and the holders of knowledge are entitled to exercise some sort of power. If we are honest with ourselves, we have to admit that most of us had, or still have, an inherently elitist view of science: a view according to which *those in possession of knowledge* are and must remain a small minority. Why must they? Because science *as we know it* is accessible only to an elite:

everyone can't be a scientist or have scientific training. *This is what we have learned at school.* Our whole education has been devoted to teaching us that science cannot be within the reach of everyone, and that those who are able to learn are superior to the others. Our reluctance to consider ourselves just another type of worker rests upon this basic postulate: science is a superior kind of expertise, accessible only to a few.

This is precisely the postulate which we must try to challenge. Indeed we must ask: why has science—or systematized knowledge generally—so far been the preserve of a minority? I suggest the following answer: because science has been shaped and developed by the ruling class in such a way as to be compatible with its domination—i.e., in a way that permits the reproduction and the strengthening of its domination. In other words, our science bears the imprint of bourgeois ideology and we have a bourgeois idea of science. I do not mean by this that science itself is something bourgeois, or that we have to discard all the special knowledge and expertise we may possess because we consider it an undue privilege and a result of bourgeois education. Rather, when I say that our idea of science and our way of practicing it are bourgeois, I have in mind the following three aspects of its class character: (1) the definition of the realm and nature of science; (2) the language and object of science; (3) the implicit ideological content of science.

1. Our society has quite a peculiar view of what is and what is not scientific: it calls "scientific" the knowledge and skill that can be systematized and incorporated into the *academic* culture of the ruling class, and it calls "unscientific" the knowledge and skill that belong to a popular culture which, by the way, is dying out rapidly. Take a few striking examples:

• In medicine, in France (among other bourgeois countries), reliance on synthetic drugs is considered scientific, whereas acupuncture and plant medicine, which spring from ancient popular culture, are considered unscientific and are condemned by the medical profession.
• When the research department of a large automobile firm puts a new engine on the market, this engine, of course, is the product of scientific expertise. But when a group of amateurs or craftsmen who have never been to a university build an even better engine, using hand-made parts, this, of course, is something unscientific.
• When experts in industrial psychology organize the work process

in order to divide the workers and to make them work to the limits of their physical capabilities, this is something scientific. But when workers find a way of uniting, of striking the plant and of reorganizing the work to make it as pleasant as possible, this, of course, is something unscientific.

What are the criteria behind these distinctions? Why is acupuncture considered a "skill," but use of synthetic drugs "scientific?" Why do we call an invention by a mechanic or a toolmaker the product of her or his "craftsmanship," and the same invention, when it is presented by an engineering firm, the product of "science and technology?" Why is the management psychologist a "scientific expert": and the shop steward or the militant nothing of the kind when (s)he expertly turns the tables on the expert? Why does one speak of "the scientist as worker" and never of the "worker as scientist?"

The answer, I suggest, is this: our society denies the labels of "science" and "scientific" to those skills, crafts, and types of knowledge which are not integrated into the capitalist relations of production, are of no value and use to capitalism and therefore are not formally taught within the institutional system of education. Therefore, these skills and this knowledge, though they may rest on extensive study, have no status within the dominant culture. They are not institutionally recognized "professions," and they often have little or no market value: they can be learned from anyone who cares to teach them. Our society, however, calls "scientific" only those notions and skills that are transmitted through a formal process of schooling and carry the sanction of a diploma conferred by an institution. Skills that are self-taught or acquired through apprenticeship are labeled "unscientific" even when, for all practical purposes, they embody as much efficiency and learning as institutionally taught skills. And the only explanation for this situation is a social one. Self-acquired knowledge, however effective, does not fit into the pattern of the dominant culture; and it doesn't fit into it *because it does not fit into the hierarchical division of labor* that is characteristic of capitalism.

Just suppose for a moment that a boilermaker or a toolmaker in a factory were credited with as much expertise as a university-trained engineer: the latter's authority and thereby the hierarchical structure would be placed in jeopardy. Hierarchy in production and in society in general can be preserved only if "expertise" is made the preserve, the privilege, the monopoly, of those who are *socially selected* to hold both knowledge *and authority*. This social selection is performed through the schooling system. The main—though hidden—function of school has been to restrict access to knowledge to those who are socially qualified to exercise authority. If you are unwilling or unable to hold authority, either you will be denied access to knowledge or else your knowledge will not be rewarded by an existing institution.

To sum up: in our society, the relationship between authority and knowledge is the inverse of what it is supposed to be: authority does not depend on expertise; on the contrary, expertise is made to depend upon authority: "the boss can't be wrong."*

2. This social selection of the knowledgeable and the expert is performed principally through the way in which scientific knowledge and expertise are taught. Teaching methods and curricula are designed to make science inaccessible to all but a privileged minority. And this inaccessibility is not due to some intrinsic difficulty of scientific thinking but rather to the fact that in science—as in the rest of the dominant culture—the development of theory has been divorced from practice and from ordinary people's lives, needs and occupations. We may even say that science was *defined* socially as being *only* the kind of systematized knowledge that has no relevance to the daily needs, feelings and activities of people.

Modern science was initially conceived of as being impermeable and indifferent to human concerns, and concerned only with dominating nature. It was not intended to serve the mass of people in their daily struggle; it was meant primarily to serve the ascending bourgeoisie in its effort at domination and accumulation. The ethics and ideology of a Puritan ruling class clearly shaped the ideology of science and generated the notion that the scientist must be as self-denying, insensitive and inhuman as the capitalist entrepreneur.

In this sense, there has never been anything like "free" or "independent" science. Modern science was born within the framework of bourgeois culture; it never had a chance to become popular science or science for the people. It was confiscated and monopolized by the bourgeoisie; and scientists, like artists, could only be a dominated fraction of the ruling class. They could enter into conflict with the rest of their class but they could not break out of bourgeois culture. Nor could they go over to the working class: they were—and still are— separated from the working class by a cultural gulf.

This gulf is reflected in the semantic divorce of the experts

---

*On this point, see Herbert Gintis, "Education, Technology and the Characteristics of Worker Productivity," *American Economic Review*, Vol LXI, 5/71.

from everyday language. The semantic barrier between scientists and ordinary people must be seen as a class barrier. It points to the fact that the modern development of science—like that of modern art—has, since the beginning, been cut off from the overall culture of the people. Capitalism has sharpened the division between practice and theory, between manual and intellectual labor; it has created an unprecedented gulf between professional expertise and popular culture.

In recent decades, it has achieved something even more astounding: as a reult of its need for increasingly huge amounts of scientific and technical expertise, it has cut this expertise into such minute fragments and so many narrow specializations that they are of little use to the "experts" in their daily life. In other words, to the traditional bourgeois scientific culture has now been added are of little if any use to the "experts" in their daily life. In other words, to the traditional bourgeois scientific culture has now been added a new type of technical and scientific subculture that can be used *only* when combined with other subcultures in large industrialized institutions. The possessors of this specialized expertise are professionally as helpless and dependent as unskilled or semi-skilled workers. The kind of scientific expertise which most people are taught nowadays is not only divorced from popular culture, but impossible to integrate into any culture: it is culturally sterile or even destructive.

Here we come to the central aspect of the class nature of modern science; whether theoretical or technical, comprehensive or specialized, so-called scientific knowledge and training are irrelevant to people's lives. There has been a tremendous increase in the quantity of knowledge and information available to us; individually and collectively, we know a great deal more than in previous times. Yet this enormously increased quantity of knowledge does not give us a greater autonomy, independence, freedom or effectiveness in solving the problems we meet. On the contrary, our expanded knowledge is of no use to us if we want to take our collective and indivual lives into our own hands. The type of knowledge we have is of no help to us in controlling and managing the life of our communities, cities, regions, or even households.

Rather, the expansion of knowledge has gone hand-in-hand with a diminution of the power and autonomy of communities and individuals. In this respect, we may speak of the schizophrenic character of our culture: the more we learn, the more we become helpless, estranged from ourselves and the surrounding world. Society controls us by the knowledge it teaches us, since it does not teach us what we'd need to know in order to control and shape society.

3. This brings us to the third aspect of the class character of modern science: the ideology that underlies the solutions it offers. Science is not only functional to capitalist society and domination through the division of labor which is reflected in the language, definition and division of its disciplines, but also in its way of asking certain questions rather than others, of *not* raising issues to which the system has no solutions. This is particularly true in the field of the so-called sciences of man, including medicine: they devote much effort to finding ways of treating the symptoms of illness and dissatisfaction. They devote much less effort to finding ways of preventing illness and dissatisfaction. And they devote no effort at all to finding ways of dispensing with all the "health and welfare" experts, although the only sound solution would be precisely this: to enable all of us—or at least all those who wish to—to cure the common diseases, to shape housing, living, and working conditions according to our needs and desires, to divide labor in a way we find self-fulfilling and to produce things we feel are useful and pretty.

Western science, as it presently exists, is inadequate to all these tasks. It does not offer us the intellectual and material tools to exercise self-determination, self-administration, self-rule, in any field. It is an expert science, monopolized by the professionals and estranged from the people. And this situation after all is not surprising—Western science was never intended for the people. Its main relevance, from the beginning, was to machinery that was meant to dominate workers, not to make them free.

*   *   *   *   *   *

What makes the situation so complicated is the fact that intellectual workers are both the beneficiaries and the victims of the class nature of Western science and of the social division of labor that is built into it.

Whether we like it or not, we are beneficiaries of the system since we still hold significant, though dwindling, privileges over the rest of the working class. Manual, technical and service workers rightly feel that scientific workers belong to the ruling class. As carriers of bourgeois culture, scientific workers are bourgeois at least culturally. Scientific workers in the manufacturing and mining industries may be considered bourgeois socially as well. In France, for example, the engineers of the state-owned coal mines are one of the most reactionary and oppressive groups in the French bourgeoisie. In most factories, production engineers as well as management experts are distrusted and hated by the workers and regarded as their most immediate enemies: not only are these technical and scientific experts relatively privileged as regards income, housing and working conditions, but they are also the ones who engineer the oppressive order of the factory and the hierarchical regimentation of the labor force.

It must be recognized that the class character of the capitalist division of labor and the class conflict between production workers and scientific and technical personnel will not disappear from the factory floor through mere public ownership of industries. Public ownership will not destroy class barriers and antagonisms, even if it were accompanied by extensive wage equalization and change of attitudes. Class distinctions in the factories will disappear only with the disappearance of the hierarchical capitalist division of labor itself, a division which robs the worker of all control over the process of production and concentrates control in the hands of a small number of employees. The fact that these employees—whom Marx called officers and petty officers of production—are themselves part of the "total worker" (*Gesamtarbeiter*) is quite irrelevant as regards their class position. They are in fact paid to perform the capitalist's function, which can no longer be performed by one boss and owner. And the job they perform for a salary is in fact perceived by the workers as being instrumental to their exploitation and oppression.

This oppression will persist, regardless of who owns the factory, as long as the technical, scientific and administrative skills required by the process of production are monopolized by a minority of professionals who leave all the manual tasks and dirty jobs to the workers. Whatever the political views of these professionals, they, in their roles, embody the dichotomy between intellectual and manual work, conception and execution; they are the pillars of a system which robs the mass of workers of their control over the production process, and which embodies the function of control in a small number of technicians who become the instruments of the manual workers' domestication.

Of course, the technical staff in factories are themselves oppressed; they too are victims and not only instruments of the capitalist division of labor. But being oppressed is not an excuse for oppressing others, and oppressed oppressors are in no way less oppressive. Moreover, while engineering and supervisory personnel doubtless are oppressed or exploited, they are not oppressed by the workers whom they are dominating and cannot expect sympathy from that quarter.

I insist on this point because there can be no unity and no common struggle of the various sectors of the working class as long as those who possess the scientific knowledge and skills do not recognize that they in fact have an oppressive role vis-a-vis the manual workers. There is a significant proportion of highly skilled personnel who believe they are anti-capitalist and socialist because they are in favor of self-management, i.e., in favor of running the plants themselves without being controlled by the owners. In truth, there is nothing socialist in this technocratic attitude. Doing away with the owners and their control would not abolish the hierarchical structure of the plant, laboratory or administration; it might serve only to alleviate the oppression suffered by employees in responsible positions, without diminishing the oppression these employees inflict on production workers.

Those who ignore the class nature of the present division of labor and the class division between intellectual and manual workers are in fact incapable of envisioning a classless society and of fighting for it. All they can envision is a technocratic society— that may be branded "state-capitalist" or "state-socialist," as you wish—in which essentially capitalist relations of production will prevail (as indeed they do prevail in Eastern Europe and the USSR).

When I say that intellectual workers are in fact privileged and are objectively in an oppressive role, I do not mean to imply that in order to be socialists they must renounce any specific demands and serve the working class' interests with guilty selflessness. On the contrary, I am convinced that the abolition of the capitalist division of labor is in the intellectual workers' own interest because they are as victimized and oppressed by it as the rest of the working class.

The scientific workers' "proletarianization" began in Germany some 90 years ago when Carl Duisberg, who was research director at Bayer, first organized research work according to the same division of labor as production work. This industrialization of research has since become universal. As industry discovered that science could be a force of production, the production of scientific knowledge was subjected to the same hierarchical division and fragmentation of tasks as the production of any other commodity. The subordination of the laboratory technician or anonymous researcher to his or her boss, and of the latter to the head of the research department, is not very different, in most cases, from the subordination of the assembly-line worker to her or his foreman and of the foreman to the production engineer, etc. The industrial-

ization of research has been responsible for the extreme specialization and fragmentation of scientific work. The process and the
scope of research have thereby become as opaque as the process of
production, and the scientist has in most cases become a mere
technician performing routine and repetitive work. This situation
has opened the way for the increasing use of scientific work for
military purposes, and these military applications, in turn, have led
to a further hierarchizing and specialization of research jobs. Not
only is science militarized in regards to its uses and orientations,
but military discipline has invaded the research centers themselves
as it has the factories and administrations.

In short, since the early nineteenth century, scientific work
has undergone much the same process as production work; in order
to control and discipline production workers, the early capitalist
bosses fragmented the work process in such a way as to make each
worker's work useless and valueless unless combined with the
work of all others. The function of the bosses was to combine the
labor they had first fragmented, and the monopoly of this function
was the base of their power; it was the precondition for separating
the workers from the means of production and from the product. In
the production of science, control and domination of the scientific
labor force are even more vital than in other commodity production:
should the production of knowledge escape the control of the
ruling class, the holders and producers of knowledge might take
power into their own hands and establish a more or less benevolent
or tyrannical type of technocracy. The bourgeoisie has been
persistently haunted by this danger during the second half of the
nineteenth century. To make their power safe, capitalists had to
make sure that knowledge could not wield autonomous power and
would be channelled into uses compatible with, or profitable for,
capital.

There were of course two obvious ways to bring science—and
knowledge generally—under the power of the capital-owning
class:

1. The first way, which is widely practiced in the universities,
is the socio-political selection—and promotion—of scientists. Scientists in responsible positions must belong to the bourgeoisie and
share its ideology. During and after their schooling process,
appropriate steps are taken to persuade the ambitious that their
interest lies in playing the establishment's game. In other words,
scientists tend to be bought off, to be co-opted into the system.
They will be given positions of power and privilege provided they

identify with the established institutions. And their power, which is both administrative and intellectual, has a definitely feudal aspect: the big bosses of medicine or science departments in the universities hold the discretionary powers of a feudal landlord in earlier times. The hierarchy in the production of science is as oppressive as that in factory production. The big bosses of science must be seen as watchdogs of the bourgeoisie, whose particular function is to keep the teaching, the nature and the orientation of science within the bounds of the system.

The domination of these bourgeois scientists over science would be impossible, of course, without the consent of those whom they rule. As usual, two instruments are used to manipulate young scientists into submission to the bosses: (a) ideology and (b) competition.

a) There is not much point in going into great detail on the current ideology of science, with its claim that science is value-free, its pretense that science has no purpose other than to accumulate knowledge—the result being that it accumulates any kind of knowledge, i.e., 90 percent useless knowledge and 10 percent that is useful only to the system. What I want to stress here, however, is that unless s/he accepts this ideology, a young scientist won't get far; s/he won't make a career but will be eliminated by the institution.

b) Such elimination is made possible by the vast abundance of candidates seeking to do research work. The bosses of science, and through them the system, are basing their domination on the tremendous surplus of students that can be found in all industrialized societies. This surplus of students enables the bosses to organize the rat race. In other words, the potential surplus of scientific labor has the same effect as the reserve army of the unemployed in industrial labor; it strengthens the boss vis-a-vis the workers and enables him/her to play them off against each other.

But competition between researchers has an even more important consequence; it leads to the most extreme forms of specialization. The reason for this is obvious; to make a career, a research scientist must produce something original. This can best be done by pushing research into the most hairsplitting details of an otherwise trivial field. The aim of academic research is not to produce some knowledge relevant to a concrete problem, but only to prove the researcher's capability, a "value-free" and "neutral" capability.

2. The extreme specialization of competing scientists is pre-

cisely what capital needed to make its own domination safe. Competing, over-specialized and hairsplitting scientific workers are not likely to unite and translate knowledge into power. Furthermore, the overabundance of scientific talent enables the capitalist class to pick those people who seem best suited to serve the interests of the system. This situation also enables the bourgeoisie to stiffen the division of labor in scientific work in order to keep control over the production of science and to prevent scientific communities from pooling their knowledge and becoming a major force in their own right.

All the modernistic talk about the scientific workers being destined to win power within society because, so the story goes, knowledge and power can't indefinitely be separated—all this talk is pure rubbish. The scientific workers are in no position to claim or to conquer power because so far they have been incapable of uniting on a class basis, of evolving a unity of purpose and of vision encompassing the whole of society. And this inability is not accidental: it merely shows that the type of knowledge held by scientific workers, individually and collectively, is a *subordinate* knowledge, i.e., it is a type of knowledge that cannot be turned against the bourgeoisie because it inherently bears the imprint of the social division of labor, of the capitalist relations of production and of capitalist power politics.

The *immediate* interests of scientific workers therefore are no more revolutionary or antagonistic to the system than the *immediate* interests of any other privileged segment of the working class. Quite the contrary: the present specializations of a majority of scientific and technical workers would be totally useless in a socialist society. The fact that large numbers of scientific and technical workers are unemployed or underemployed as of now, under capitalism, does *not* mean that a socialist society would have to or would be able to employ them in their present specializations. People with scientific or technical training are *not* victims of capitalism because they can't find creative *jobs*—or any kind of job—in their capacity; they are victims of capitalism because they have been trained in the first place in specializations that (1) make them incapable of earning their living, and (2) are useless in this and in any other type of society. And they have been so trained for three reasons:

1. to hide the fact that their labor is not needed by the system, i.e., that they are structurally unemployed and unemployable;
2. because it would be dangerous not to let them hope that through

studying they can win a skilled and rewarding job.

3. because a reserve army of intellectual labor performs a useful function under capitalism.

Therefore, the first step toward the political radicalization of intellectual labor is *not* to ask for more and for better jobs, mainly in research, development and teaching, in order to fully employ everyone in his or her capacity. No, the first step toward political radicalization is to quesion the nature, the significance and the relevance of *science itself as it is practiced now*, and to question thereby the role of scientific workers.

The scientific workers is both the product and the victim of the capitalist division of labor. S/he can cease being the victim only if s/he refuses to be its product, to perform the role s/he is given, and to practice this kind of esoteric and compartmentalized science. How can s/he do this? As a matter of principle, by refusing to hold a professional monopoloy of expertise and by struggling for the reconquest and reappropriation of science by the people. The few Western examples of a successful implementation of this line of action have usually drawn their inspiration from the Vietnamese and Chinese experience. The most important aspect of this experience is the following moral and political option: the goal is not for a few specialists to achieve the highest possible professional standards; the goal is the general progress and diffusion of knowledge within the community and the working class as a whole. Any progress in knowledge, technology and power that produces a lasting divorce between the experts and the non-experts must be considered bad. Knowledge, like all the rest, is of value only if it can be shared. Therefore, the best possible ways of sharing new knowledge must be the permanent concern of all research scientists. This concern will profoundly transform the orientation of research and of science itself, as well as the methods and objects of scientific research. It will call for research to be carried out in constant cooperation and interchange between experts and non-experts.

These basic principles must be seen as radical negations of the basic values of capitalist society. They imply that what is best is what is accessible to all. Our society, on the contrary, is based on the principle that what is best is whatever enables one individual to prevail over all others. Our whole culture—i.e., science as well as patterns of consumption and behavior—is based on the myth that everyone must prevail somehow over everyone else, and therefore that what is good enough for all is no good for anyone. A communist culture, on the contrary, is based on the principle that what is good for all of us is best for each and every one of us.

There can be no classless society unless this principle is applied in all fields, including the field of science and knowledge. Conversely, if science is to cease to be bourgeois culture it must not only be put at the service of the people, but become the people's own science. Which means that science will be transformed in the process of its appropriation by the people. Indeed, science, in its present form, can never become the people's own science or science for the people: you can't make a compartmentalized and professionalized elite culture into the people's own. Science for the people means the subversion of science as it is. As Steven and Hilary Rose put it:

> This transformation carries with it the breaking down of the barrier between expert and non-expert; socialist forms of work within the laboratory, making a genuine community instead of the existing degraded myth, must be matched by the opening of the laboratories to the community. The Chinese attempts to obliterate the distinction of expertise, to make every man his own scientist [sic—eds.], must remain the aim...*

All the talk about the "proletarianization" of scientific workers just demonstrates one fact: most scientific workers still do not feel they are part of the working class. If they did feel part of it they would not discuss their proletarianization. Do we discuss the proletarianization of the chemical workers, or the engineering workers, or the electricians, the printers, the service workers? We do not.

Why then do we discuss the proletarianization of scientific workers? For a quite simple reason: our minds are still not quite reconciled to the fact that the words *scientific* and *proletarian* fit together.

---

*"The Radicalisation of Science," by S. and H. Rose, *The Socialist Register*, 1972, pp. 105-132.

IV
TOWARDS A
LIBERATING SCIENCE

Introduction

Forge simple words that even the children can understand...words which will enter every house like the wind/fall, like red hot embers on our people's souls.

—Brother Jorge Rebelo (Frelimo)

What Otto Rank said of psychology has to be said of every other discipline, including the "neutral sciences"; it is not only man-made...but masculine in its mentality.
—Adrienne Rich in Women and the Power to Change

There is a growing awareness that science for science's sake is not enough. People see medical care becoming increasingly expensive and unattainable. More and more people are questioning nuclear plants and nuclear energy. Awareness and concern over ecological issues have increased dramatically in the last decade. At the same time, we have become increasingly critical of scientists who continue playing their "pure science" games and evince little concern for the relevance and application of their research to social/

political realities. We see educational institutions fostering a dehumanized science, siding with powerful corporate interests and displaying an indifference to humane values.

Is this the only science possible? Is this all we should expect within the present socio-economic framework? What if we were to engage ourselves in the creation of a different kind of science? What if we were to demand that human relevance be built into the paradigm of science itself? What if science became a force for human survival, and ethical and political issues became an integral part of this science? Science as it is today is *not* all that it can be.

We also need to think of new ways of healing the split between scientists and non-scientists—a split that keeps the vast majority of people alienated from and mystified by science. We have to make clear that science should be *for* and *of* people, not *about* people. We need to proclaim that a holistic and humane science is a reasonable expectation to have and that we will not settle for less.

How can work be organized differently in science? Can science be taught differently so that the role of the scientist is dramatically changed? What is it that gets taught and is called science? How will the insights of feminism affect what areas of scientific inquiry are developed? Can science be a part of the struggle for black liberation in this country and abroad? What happens when people get together and start challenging the prevailing mode of work in science? Will that challenge also affect what scientists choose to work on? Is it the very definition and logic of science which must be altered?

The articles in this section represent an attempt to deal with some of these questions. They do not simply ask for change in the present structures, but for new definitions and a science that will be responsive to people's needs and that will become a tool in the struggle for human liberation.

Declaration: Equality for Women in Science
—Women's Group from Science for the People

This is a short declaration made by a group of women scientists at the 1969 meetings of the American Association for the Advancement of Science. The resolutions are as valid today as they were ten years ago.

The stated goals of the American Association for the Advancement of Science (AAAS) are:

to further the work of scientists,

to facilitate cooperation among them,

to improve the effectiveness of science in the promotion of human welfare, and

to increase public understanding and appreciation of the importance and promise of the methods of science in human progress.

None of these objectives can be realized while women in science are relegated to second class status. Female scientists do not escape the oppression faced by all women in our society. They are oppressed economically and culturally—trained for inferior roles and exploited as sex objects and consumers.

Such sexual discrimination is no accident. It serves, in a variety of ways, the interests of those who dominate the economy of this country. It provides them with a source of ideologically justified cheap labor, and as a consequence drives all wages down. It establishes "wives" as unpaid household workers and child raisers, as well as a body of willing consumers. At the same time, the limitations on the creative development of women deprive society of the full contributions of over one half its members.

It is important to note that sexual oppression is both pervasive and institutionalized; within the scientific community it takes many forms. Educational tracking by sex from elementary school on channels women into subordinate roles and stereotypes. While men are trained to develop "logical" patterns of thought, women are encouraged to be "intuitive." Math and science are seen as male prerogatives. Vocational counselling in high schools and colleges pressures women into family roles, clerical work and, if professions are considered, into the service fields: teaching, social work, nursing, etc. Those few women who manage to transcend such socialization and choose scientific careers, encounter a vicious circle of exploitation. Quotas are placed on graduate school admissions and justified by the self-fulfilling prophecies that most women will be unable to finish because they will marry, have children, and lack the emotional stability and drive to meet the arduous initiation rites of the profession. The still fewer who complete their training continue to find themselves between family and profession, while men never have to make that choice.

As scientists, they are limited by being placed in subordinate positions, rarely being given their own labs or first authorship on papers, and, the most glaring inequity, being paid less than their male colleagues for equal work. They are automatically and illegally barred from certain jobs, particularly in industry, and cut off from tenured and supervisory positions.

Moreover, the psychological harassment is constant and debasing. Casual remarks continually define the female scientist simply in relation to her sex, from compliments on her looks to "you think like a man." She is placed in the schizophrenic position of being treated as either a dehumanized worker or a feminine toy.

Universities hold a strategic position with regard to all manifestations of this problem, since they help create and transmit the ideology of male supremacy.

Moreover, the practices of sexual discrimination which permeate all institutions where AAAS members work and study are contradictory to the declared goals of the AAAS. Clearly we cannot "further the work of scientists" while denigrating in so many ways the contributions and potential of women in the profession. Sexual discrimination makes "cooperation among scientists" an ironic platitude. The "effectiveness of science in the promotion of human welfare" is hardly furthered by denying half of humanity the opportunity to pursue scientific careers, or by wasting this tremendous reservoir of talent.

We therefore propose the following resolutions be adopted at the general Council meeting of the AAAS, and be fought for by AAAS members where they work.

1. That universities and other institutions where AAAS members work be immediately required to comply with the law of the land and pay equal wages for equal work to women and men.

2. That graduate school departments and medical schools admit 1/2 women and 1/2 men, regardless of the proportion of applicants, and that they take whatever steps are necessary to recruit sufficient women to comply with this demand.

3. That vocational counselling in high schools and colleges be totally reoriented so as not to channel women into low-status, low-potential occupations.

4. That universities and other institutions give priority to the hiring and promotion of women, increasing the proportion of women to 50% at all levels.

5. That birth control and abortion counselling be provided by university and company health services to all women.

6. That the curriculum of courses in psychology, sociology, anthropology, etc., be thoroughly revamped by women, to end the perpetuation and creation of male supremacist myths.

Further, that sex inequality be added as a topic to all courses and texts which cover social inequalities, and that new courses be created by women about their history and oppression.

7. That universities and government sponsor programs to investigate and change the subordinate status of women in our society.

8. That it be recognized that the actual practices of hiring, promotion and tenure discriminate against women, and that institutions have not accepted their responsibility for such inequalities.

As a first step in the right direction institutions should provide:

a) parenthood leave and family sick leave for all employees, female and male.

b) half-time appointments for mothers and fathers who want them must be considered. (Since child-rearing is a social responsibility, it is preferable for both parents' work to be slowed down than for the mother's to be stopped entirely.)

c) free child care centers should be provided for all children. These day care centers should be open to the communities where the institutions are located, controlled by the parents, staffed equally by male and female teachers, open 24 hours a day, 7 days a week for infants to school age children and after school for older children.

While we realize that the ultimate liberation of both women and men in our society will only come with a total social and economic revolution, we feel that it is important for us to make steps now toward destroying false notions of superiority which do not serve science, scientists, or humanity.

Science and Black People
Editorial from *Black Scholar* Magazine

The uses of scientific knowledge cannot be separated from the society in which those uses occur. The myth of "pure" science, of science as a detached, ivory tower pursuit, has been exposed. Science is enmeshed in the prevailing social ideologies. The choice of what subjects to investigate, which experiments to undertake, what methods to employ, which results to emphasize as important, to whom to report results, how to use results, etc., all these and countless other decisions made by scientific investigators are colored by ideology. Ideology is not simply a nebulous cloud hanging in the social atmosphere. It is the assumptions underlying scientific education and training; it is the prod held by the public and private bureaucracies which fund research; it is the personal ambition of scientists who live in a bourgeois materialist society.

Consequently, it comes as no great surprise that black people have been largely excluded from the world of science and technology—both as practitioners and as factors relevant to decision-making. Of course, there have been outstanding black scientists and inventors, and their achievements are worthy of emulation, but the reality of racism has excluded most of us from the pursuit of science. Regarded as degraded beings, prisoners of undisciplined emotions and suitable only for manual labor, black people have for generations been barred from white institutions of scientifc training.

Moreover, racism has meant that white scientists have regarded us as an undifferentiated part of the environment, a given, rather than a subject active in changing the environment. Our cranial capacity and social institutions may be investigated from time to time, but our brain-power and social needs are seldom considered relevant when important scientific and technological decisions are being made. To the white world of research and development, we are indeed invisible.

Unfortunately, our agrarian heritage and general exclusion from the world of science has generated an anti-science, anti-intellectual attitude among many of us. In colleges today many young black people regard the pursuit of scientific training as "copping out," as individual "tripping" at the expense of the struggle. Certainly it cannot be denied that some black scientists and technicians have dropped out of the struggle, but this must be attributed to their subservience to the individualist and materialist values of this society; it is not a result per se of being a scientist,

It is imperative for us to realize that, despite its abuses in this country, science is key to the material development of society. Industrial advancement and social progress would be impossible without scientific research and development. While it is true that science cannot be separated from ideology, it is also true that ideologies can be changed and hence the uses of science can be redirected. It is no accident that the developing countries of the Third World have given highest priority to building scientific and technological institutes and training young scientists and engineers. Their future survival and independence hinge on developing a culture that encourages science and makes it truly serve the people. (And it is with the intention of undermining their struggle against underdevelopment that the capitalist West has instituted a "brain drain" of scientists and technicians from these Third World nations.)

Scientific training and well-informed inputs into the process of technological decision-making can be a progressive addition to the general struggle for black liberation. This is the message that must be taken to black youth today. We are oppressed by our exploiters, but to the extent that we remain ignorant and apathetic we are accomplices in our own oppression. No struggle can be undertaken without knowledge, and it is not enough simply to blame others for our ignorance. Like our brothers and sisters in Africa and the Third World, we must encourage the pursuit of useful knowledge—physical, biological and social—that will enable us to intelligently take command of our own destiny.

Science Teaching:
Towards an Alternative
—Science Teaching Group
from Science for the People

A group of science teachers questions the values of teaching science that is disconnected from the real lives of their students, and they suggest approaches for doing it differently.

In the last few years many of us have begun to question various aspects of our jobs as teachers. In part this has been due to an awakening consciousness among all teachers about the author-itarian nature of schools and the socializing function they perform. It has been due also to the broad recognition now that continued growth of science and technology may be at best a mixed blessing in our present society. And it has been due also to our difficulty to motivate students and interest them in the science we teach. Of course, all these problems are related to one another in the end, and we have been trying to uncover the basic thread tying them together, and in so doing to find alternatives that will make our teaching more rewarding.

What kind of students leave our classrooms? Are they critical, free-thinking individuals who have learned to respect and under-stand one another and to work together and act for their common good? Or are they mute, compliant youngsters who have learned how to respect authority and to compete with each other for the limited positions they are being schooled to fill? Granted these are stereotypes, but is it not the nature of schools and of the curriculum to produce the latter, and does it not require the exceptional teacher, working against the educational system, to produce the former? Our experience, our frustrations, leave little doubt in our minds.

The reason lies simply in the fact that an educational system reflects the purposes of the society it serves. In the United States we have a highly competitive society built around certain myths of freedom, choice, democracy, and justice. Because of the nature of the social and economic system, most people live, in fact, in a state of financial insecurity and political impotence. The schools are meant to maintain that system. They generate and reinforce the popular belief that poverty and alienation are the result of people's own stupidity—rather than products of this society's social and economic structures. But that's not the worst of it. The schools perpetuate the social class, race and sex role divisions in American society. Through IQ, aptitude and personality testing, through a multitude of teaching mechanisms and many other discriminatory means, students are labelled, and set one against the other, according to the system's definition of ability and achievement. The school's occupational channeling thus fosters competition among youngsters for positions in what many people now recognize as an irrational, hierarchical, and largely oppressive occupational structure. And of course, in providing these functions, it is not the role of the schools to shatter the Great American Dream with the nightmarish realities of genocide, poverty, racism, sexism, political repression, and other forms of institutional violence and injustice.

To make these rather general statements more specific, suppose we look at science education *per se*. How are the materials, methods, and curricula geared toward perpetuating the values and structure of our present society? Consider first the training of the scientist. In the classroom and laboratory the myth of an apolitical, benevolent science prevails. Graduate school, and often undergraduate education, involves a near total submersion of the student in technical material with little if any historical or philosophical perspective. Research productivity is the measure of worth as the student acquires skill in a specialized field. Technical questions are isolated from their social and economic context (e.g., the use of science) except for perhaps consideration of the prestige and financial status of the researcher. Thus the end product of this training is a narrow specialist—one taught to perform scientific miracles without considering their political implications, a reliable tool of the power structure.

Another aspect of this training is an ingrained sense of elitism. Courses are designed to select and separate out potential scientists from their fellow students. Those who succeed are led to view themselves as members of an elite intellectual class. They take patronizing and condescending views of other people's opinions and aspirations, an attitude which reveals an underlying commitment to undemocratic structures. The elitism emerges for example when scientists attribute social problems to incompetence and irrationality—with the implication that a few intelligent people who really understand technology could solve all the problems (technocracy).

The training just described is, of course, only the final stage of a very long educational process begun in grade school. Not surprisingly, the early educational experience is instrumental in developing those values and attitudes which become important later on. We find that elementary and high school science teaching strongly reflect the character and needs of the advanced training programs.

Note for example the large number of curriculum reforms and new programs which have been generated in the U.S. since Sputnik. This large-scale curriculum development was funded as part of a broad program of support of science through agencies such as the National Science Foundation (NSF), the National Aeronautics and Space Administration (NASA), the National Institute of Health (NIH), and the large foundations, and was closely coordinated with increased levels of support by the Pentagon. These programs were designed to produce the technical expertise necessary for the development of an ever more sophisticated war machine.

But the effect of such programs has been to profoundly influence how science is taught in the schools. The curricula are geared to preparing students for professional study. They emphasize basic theory and mathematics to the near exclusion of practical science, that is the understanding of how everyday things work. For example the Physical Science Study Committee (PSSC) physics curriculum was designed for a narrow segment of high school students for the purpose of getting them to think like scientists and thus instill in them the values of working physicists. It was designed largely at M.I.T. by a group of physicists with long experience in designing nuclear weaponry, whose familiarity with science education was almost exclusively in training professional research scientists.

This emphasis of science curricula on "pure science" (removed from everyday experience) as opposed to practical science constitutes in effect a rigid tracking system in schools. The division between "academic" students who take chemistry in preparation for college, and those who take shop instead, is complete. The chemistry student does not learn how to harden a steel tool, and the shop student does not learn about crystalline structure. In ways like this, the framework for extreme division of labor and perpetuation of the social class structure is built into schools. One result is that the future scientist is denied freedom, as is the mechanic. The former is dependent on the existence of a class of workers to perform such tasks and learn quite early in school that this is the way things should be.

The number of women who become scientists is very small, as is the number of blacks, Chicanos, Puerto Ricans, Native Americans, and other oppressed peoples. Aside from the obvious form of discrimination which places many of these minority children in schools that are often below standard, there are also the additional biases on the part of teachers themselves. Blacks, Puerto Ricans, Chicanos, and Native Americans, for example, are often assumed not to have the intelligence or industry to become scientists; women are supposed to be too emotional for the rational process of science. All these myths become self-fulfilling prophesies. As youngsters, members of these groups are not encouraged or helped to become scientists, and so they don't.

Other attitudes also contribute to this process. Teachers may consider the student who performs well to be an anomaly—someone who is exceptional and unlike the others. Or the teachers may find unacceptable the use of "non-standard" English or cultural behavior patterns that are incompatible with the norms that the school system is geared to establish. Thus, although a child may have ability, teachers may discount it because they evaluate the child's overt behavior as not suitable to an up-and-coming professional.

For women the discrimination begins with their orientation as young girls away from "unfeminine" vocations. Girls are channeled into secretarial or domestic skills or into the liberal arts, while the mechanical world is left to men. In this way most women grow up to confront an increasingly mechanized world with no real understanding of how machines work—resulting in their helplessness and dependency on men.

For students who don't pursue scientific studies, the curriculum is structured in such a way as to leave them feeling mystified, frustrated, and helpless against the enormous power of technology and those who control it. In many of the materials available there is a "hidden curriculum" which conveys the social myths that perpetuate the control of people through technology. For example, in "educational" films provided by oil companies the telephone company, or NASA, scientists are portrayed as infallible experts. The implicit message is clear: the corporations and military, through the enormous power of technology, are omnipotent. Scientists are high priests who work in closed and forbidden sanctuaries such as government and industrial laboratories. Their speech is ritualized, devoid of humor and meant to impress people with the awesome nature of their projects, such as sending a man to the moon. This is an old ritual that has been performed since at least the time of the pyramids. In all societies in which power has been concentrated in the hands of a very few, the rulers have found it useful to display their power through a high priesthood of "experts" who maintain their privileged position because they intimidate people with mystifying rites. And we, who are utterly dependent on their benevolence, are right to be frightened, because in this case the holy objects may be capable of destroying life on earth.

Social Aspects of Science

One characteristic of most science curricula is their limitation to purely technical or descriptive material—in contrast to its vast social ramifications. We are living in an age when the atomic bomb is already taken for granted, and science now raises the possibilities of genetic engineering, behavior control technology, complete automation of work, and universal tagging and manipulation of people. Yet these problems and the whole question of the use of science in our society are rarely broached as part of the standard science education. For people to have a real understanding of science and how it affects their lives, they must view it in its social context.

Over the last few years, in light of the Indochina war, the space program and the emergence of environmental pollution problems, those who spoke of "pure" science have come to recognize that all science is tied in one way or another to applications. We now realize that in the U.S. science and technology are developed to serve profit and stability needs of corporate enterprises. Each year, for example, billions of dollars of government and industrial funding are spent on research and development. Research and development for what? To develop technology for health, housing, and transportation? Some is, but the majority of resources is directed toward the development of more sophisticated weapons and counterinsurgency technology (to protect corporate economic interests abroad) and toward automation, information-handling technology, and technologically-induced obsolescence (to maintain the viability of the economic system at home). The use of this technology results in death and destruction, in the waste of natural and human resources, in the fouling of the environment, and in the increased manipulation of society.

Control over science ultimately rests in the hands of a powerful ruling elite which functions in its behalf. Science and technology are used as tools for extending the present social and economic system: they have served to increase the wealth and power of the few at the expense of the many. In this context science can never be considered politically neutral.

The number of examples of how science has been used directly against people is limitless. In Southeast Asia infrared devices, computer monitoring systems, anti-personnel weapons, and bio-chemical warfare are used to crush every popular liberation movement. In the United States, a whole new catalogue of surveillance devices, riot control weapons, instant identification systems, data banks and other counterinsurgency technology are being used or are under development to suppress "dissidents," who most often are people like us. While these examples are perhaps obvious, we are beginning to realize the more subtle ways in which science is used to rationalize the social order. Take, for example, the use of social science to justify in "scientific" terms the oppression engendered by the class structure of our society—the claim that social standing is based on genetic differences (Herrnstein), or that blacks are inherently inferior (Jensen). These are rather grotesque instances of the generally more pervasive use of modern social science to "explain" social injustice.

Faced with the political character of science, the least we can do as science teachers is to help clarify these social relationships. To accomplish this task requires not only that we make available to students information on how science is used, but that we also foster and help students develop the critical attitude that characterizes real science. We must help to expose the scientific and human irrationality of a system that reduces people to objects, that consumes human and natural resources, and that pits people against one another when all could prosper.

Towards an Alternative

What we realized is that as science teachers we are caught in a double bind: not only are we part of an educational system that functions to socialize people into the society, but in addition, we are part of a scientific and technological complex which serves to maintain the existing distribution of economic and political power. This situation makes it difficult to pose alternatives to our present teaching practice. On the broadest level, if educational and scientific institutions did not work to maintain the status quo, they would not be supported. In the same fashion, if we as teachers don't serve such a purpose, we too would not be supported! Or to put it differently, we are highly constrained in our schools by the material we must teach (often packaged curricula) and the way in which we must teach it. How much liberty do we have, for instance, not to give tests, not to award grades, nor to teach the required curriculum? All these structural features are part of what education is all about—especially in large schools or school systems with a sizeable bureaucracy. The fact of the matter is that we teachers have relatively little freedom to do what we and our students might decide is most desirable. After all, we do not control the allocation of educational resources.

We are left with the frustration and exasperation of unwittingly or unwillingly contributing to the maintenance of authoritarian structures, social class divisions, elitism, mystification, powerlessness, and alienation of people from science. But we know that science and technology can provide the tools for people's liberation. We can envision a society in which science teaching would help free people from want and help provide the know-how necessary for people to reach their fullest human potential. But that is a very different society from the one in which we live. It is a society that won't just happen. We must work to make it happen.

Our role as science teachers in that struggle for change has many facets. First we can help people master the technological world around us. To a large extent, people's feelings of powerlessness stem from the fact that they don't know how to do things for themselves. Their ignorance forces them to rely upon others, upon the experts who know. But people can learn how to eat good foods and how to take care of their bodies so that they remain healthy and don't rely upon doctors. Women can learn to do their own pregnancy testing. Students can learn the facts about drugs. People can learn how their automobiles and refrigerators work so that they don't have to rely on repairmen.

Second, as science teachers we can make it our practice always to raise the social aspects of scientific studies with our students. How empty, for example, is the discussion of ecology without discussing the politics and economics of consumption and waste? How can we mention transistors and computers without discussing the national data bank and the automated battlefield? What is the meaning of cell biology and genetics without talking about the health care delivery system, ethnic weapons, genetic manipulation, and sickle cell anemia? Of what significance is the subject of energy without a discussion of electrical power demands, pollution and radioactive waste disposal—or the issue of centralization/decentralization of energy power? How can the study of human anatomy ignore the questions of contraception, abortion, and the discrimination in many forms against women? How can mechanics and machines be considered independently of automation and the alienation of people in advanced industrial society? How can chemistry be taught without reference to the activities of oil companies in the Third World, the marketing practices of the pharmaceutical drug industry, and the variety of chemical additives appearing in our foods?

Third, we can seek to expose the hidden curricula in our methods and to uncover the ideological and political premises of the materials we have used in our teaching. Students can be tremendously helpful in cutting through ideological bullshit. But in addition to such critiques, which increase our understanding of our role as teacher, it is important to gather and prepare new materials to use as substitutes for those in present use. There is now a substantial alternative collection of written materials and films dealing with education and problems of technology which science teachers will find useful.

The alternative, then, in science teaching lies for the present in our practice—in how we teach, in what we teach, in how we relate to our students, our fellow teachers, and to everyone else, of course. Let there be no illusion: fundamentally changing the way science is taught in our schools and used in our society is only possible as part of a fundamental change in its political and economic structures. This won't occur without a struggle. But as teachers, we can be exemplary in our practice, in our critical attitudes, in our democratic spirit, and in encouraging others to their fullest abilities. We can work with students and scientists as part of a national and local organizing effort to share materials and work together, and to make science and the schools serve the people.

People's Science
—Bill Zimmerman et al.

Presented here is an overall perspective on the place of science in U.S. society. Written at the height of the anti-war movement, this article is both a historical document and a current challenge to all people interested in science and social change.

In the 15th century, Leonardo Da Vinci refused to publish plans for a submarine because he anticipated that it would be used as a weapon. In the 17th century, for similar reasons, Boyle kept secret a poison he had developed. In 1946, Leo Szilard, who had been one of the key developers of the atom bomb, quit physics in disillusionment over the ways in which the government had used his work. By and large, this kind of resistance on the part of scientists to the misuse of their research has been very sporadic, from isolated individuals, and generally in opposition only to particular, unusually repugnant projects. As such, it has been ineffective. If scientists want to help prevent socially destructive applications of science, they must forego acting in an *ad hoc* or purely moralistic fashion and begin to respond collectively from the vantage point of a political and economic analysis of their work. This analysis must be firmly anchored in an understanding of the U.S. corporate state.

We will argue below that science is inevitably political, and in the context of contemporary U.S. corporate capitalism, that it contributes greatly to the exploitation and oppression of most of the people both in this country and abroad. We will call for a reorientation of scientific work and will suggest ways in which scientific workers can redirect their research to further meaningful social change.

Science in Capitalist America

Concurrent with the weakening of Cold War ideology over the past 15 years has been the growing realization on the part of increasing numbers of Americans that a tiny minority of the population, through its wealth and power, controls the major decision-making institutions of our society. Research such as that of Mills (*The Power Elite*), Domhoff (*Who Rules America*), and Lundgren (*The Rich and the Superrich*) has exposed the existence of this minority to public scrutiny. Although the term "ruling class" may have an anachronistic ring to some, we still find it useful to describe that dominant minority that owns and controls the productive economic resources of our society. The means by which the U.S. ruling class exerts control in our society and over much of the Third World has been described in such works as Baran and Sweezy's *Monopoly Capital*, Horowitz's *The Free World Colossus*, and Magdoff's *The Age of Imperialism*. These works argue that it is not a conspiracy, but rather the logical outcome of corporate capitalism that a minority with wealth and power, functioning efficiently within the system to maintain its position, inevitably will oversee the oppression and exploitation of the majority of the people in this country, as well as the more extreme impoverishment and degradation of the people of the Third World. It is within the context of this political-economic system—a system that has produced the Military-Industrial complex as its highest expression and that will use all the resources at its disposal to maintain its control, that is, within the context of the U.S. corporate state—that we must consider the role played by scientific work.

We view the long-term strategy of the U.S. capitalist class as resting on two basic pillars. The first is the maintenance and strengthening of the international domination of U.S. capital. The principal economic aspect of this lies in continually increasing the profitable opportunities for the export of capital in order to absorb the surplus constantly being generated both internally and abroad. With the growing revolt of the oppressed peoples of the world, the traditional political and military mechanisms necessary to sustain this imperialist control are disintegrating. More and more the U.S.

ruling class is coming to rely openly on technological and military means of mass terror and repression which approach genocide: anti-personnel bombs, napalm, pacification-assassination programs, herbicides and other attempts to induce famines, etc.

While this use of scientific resources is becoming more clearly evident (witness the crisis of consequence among increasing numbers of young scientists), the importance of scientific and technological resources for the second pillar of capitalist strategy is even more central, although less generally accorded the significance it deserves.

The second fundamental thrust of capitalist political economic strategy is to guarantee a steady and predictable increase in the maintenance of the profitability of domestic industry and its ability to compete on the international market. Without this increase in labor productivity it would be impossible to maintain profits and at the same time sustain the living standard and employment of the working class. This in turn makes it possible to sustain the internal consumer market and to blunt the domestic class struggle in order to preserve social control by the ruling class.

The key to increasing the productivity of labor in the United States is the transformation and reorganization of our major industries through accelerated automation and rationalization of the production process (through economy of scale, the introduction of labor-saving plant and machinery, abolition of the traditional craft prerogatives of the workers, etc., such as is now occurring in the construction industry). This reorganization will depend on *programmed* advances in technology.

There are basically two reasons why these advances and new developments cannot be left to the "natural" progress of scientific-technological knowledge, why they must be foreseen and included in the social-economic planning of the ruling class. First is the mammoth investment in the present-day plant, equipment and organizational apparatus of the major monopolies. The sudden obsolescence of a significant part of their apparatus would be an economic disaster which could very well endanger their market position. (One sees the results of this lack of planning in the airline industry.) Secondly, the transformation of the process of production entails a major reorganization of education, transportation, and communication. This has far-reaching social and political consequences which cause profound strains in traditional class, race, and sex relationships, which have already generated and will continue to generate political and social crises. For the ruling class to deal with these crises it is necessary to be able to plan ahead, to anticipate new developments so that they do not get out of hand.

In our view, because planning and programmed advances in technology are absolutely central to ruling class strategy, an entirely new relationship is required between the ruling and the technical-scientific sectors of society, a relationship which has been emerging since the Second World War, and which, deeply rooted in social-economic developments, cannot be reversed. If one looks at the new sciences which have developed in this period—cybernetics, systems analysis, management science, linear programming, game theory, as well as the direction of development in the social sciences, one sees an enormous development in the techniques of gathering, processing, organizing, and utilizing information, exactly the type of technological advance most needed by the rulers.

It is no accident that two of the most advanced monopolistic formations, advanced both in their utilization and support of science and in the efficiency and sophistication of their internal organization, are Bell Telephone and IBM. They represent to capitalist planners the wave of the future, the integration of scientific knowledge, management technique and capital which guarantees the long-term viability of the capitalist order. They also represent industries which are key to the servicing and rationalizing of the basic industries as well as to the maintenance of the international domination of U.S. capital

* * *

The ruling class, through government, big corporations, and tax-exempt foundations, funds most of our research. In the case of industrial research, the control and direction of research are obvious. With research supported by government or private foundations, controls are somewhat less obvious, but nonetheless effective. Major areas of research may be preferentially funded by direction of Congress or foundation trustees. For example, billions of dollars are spent on space research while pressing domestic needs are given lower priority. We believe that the implications of space research for the military and the profits of the influential aerospace industries are clearly the decisive factors. Within specific areas of research, ruling-class bias is also evident in selection of priorities. For example, in medicine, money has been poured into research on heart disease, cancer and stroke, major killers of the middle and upper classes, rather than into research on sickle cell anemia, the broad range of effects of malnutrition (higher incidences of most diseases), etc., which affect mainly the lower classes. Large sums of money are provided for study of ghetto populations but nothing is available to support studies of how the powerful operate.

Second, on a lower level, decisions by which an individual gets research money are usually made by scientists themselves, chosen to sit on review panels. The fact that these people are near the tops of their respective scientific hierarchies demonstrates a congruence between their professional goals and the scientific priorities of the ruling class. This kind of internal control is most critical in the social sciences, where questions of ideology are more obviously relevant to what is considered "appropriate" in topic or in approach. This same scientific elite exert control over the socialization of science students through funding of training grants to universities, through their influence over curricula and textbook content, and through their personal involvement in the training of the next generation of elite scientists. Thus, through the high level control of the funding now essential for most scientific research, and second, through the professional elites acting in a managerial capacity, ruling class interests and priorities dominate scientific research and training.

***

The same government-corporate axis that funds applied research that is narrowly beneficial to ruling-class interests also supports almost all of our basic, or to use the euphemism, "pure," research; it is called pure because it is ostensibly performed not for specific applications but only to seek the truth. Many scientific workers engaged in some form of basic research do not envision any applications of their work and thus believe themselves absolved of any responsibility for applications. Others perform basic research in hopes that it will lead to the betterment of humankind. In either case, these workers have failed to understand the contemporary situation.

Today, basic research is closely followed by those in position to reap the benefits of its application—the government and the corporations. Only rich institutions have the staff to keep abreast of current research and to mount the technology necessary for its application. As the attention paid by government and corporations to scientific research has increased, the amount of time required to apply it has decreased. In the last century, fifty years elapsed between Faraday's demonstration that an electric current could be generated by moving a magnet near a piece of wire and Edison's construction of the first central power station. Only seven years passed between the realization that the atomic bomb was theoretically possible and its detonation over Hiroshima and Nagasaki. The transistor went from invention to sales in a mere three years. More recently, research on lasers was barely completed when engineers began using it to design new weapons for the government and new long-distance transmission systems for the telephone company.

The result is that in many ways discovery and application, scientific research and engineering, can no longer be distinguished from each other. Our technological society has brought them so close together that today they can only be considered part of the same process. Consequently, while most scientific workers are motivated by humane considerations, or a detached pursuit of truth for truth's sake, their discoveries cannot be separated from applications which all too frequently destroy or debase human life.

Theoretical and experimental physicists, working on problems of esoteric intellectual interest, provided the knowledge that eventually was pulled together to make the H-bomb, while mathematicians, geophysicists, and metallurgists, wittingly or unwittingly, made the discoveries necessary to construct intercontinental ballistic missiles. Physicists doing basic work in optics and infrared spectroscopy may have been shocked to find that their research would help government and corporate engineers build detection and surveillance devices for use in Indochina. The basic research of molecular biologists, biochemists, cellular biologists, neuropsychologists, and physicians was necessary for CBW (chemical-biological warfare) agents, defoliants, herbicides, and gaseous crowd-control devices.

Anthropologists studying social systems of mountain tribes in Indochina were surprised when the CIA collected their information for use in counterinsurgency operations. Psychologists explored the parameters of human intelligence-testing instruments which, once developed, passed out of their hands and now help the draft boards conscript men for Vietnam and the U.S. army allocate manpower more effectively. Further, these same intelligence-testing instruments are now an integral part of the public school tracking systems that, beginning at an early age, reduce opportunities of working-class children for higher education and social mobility.

Unfortunately, the problem of evaluating basic research does not end with such obscene misapplications as these. One must also examine the economic consequences of basic research, consequences which flow from the structure of corporate capitalism under which we live. Scientific knowledge and products, like any other products and services in our society, are marketed for profit—that is, they are not equally distributed to, equally available to, or equally useable by all of the people. While they often contribute to the material standard of living of many people, they are channeled through an organization and distribution of scarcity in such a way as to rationalize the overall system of economic exploitation and social control. Furthermore, they frequently become the prerogative of the middle and upper classes and often result in increasing the disadvantages of these sectors of the population that are already most oppressed.

For example, research in comparative and developmental psychology has shown that enriching the experience of infants and young children by increasing the variety and complexity of shapes, colors, and patterns in their environment might increase their intelligence as it is conventionally defined. As these techniques become more standardized, manufacturers are beginning to market their versions of these aids in the form of toys aimed at and priced for the upper and middle classes, and inaccessible to the poor.

Research in plant genetics and agronomy resulted in the development of super strains of cereal crops which, it was hoped, would alleviate the problems of food production in underdeveloped countries. However, in many areas, only the rich farmers can afford the expensive fertilizers required for growing these crops, and the "green revolution" has ended up by exacerbating class differences.* Studies by sociologists and anthropologists of various Third World societies have been used by the U.S. government to help maintain in power ruling elites favorable to U.S. economic interests in those countries. Real-estate developers in California have used the mapping studies of geologists, carried out in the interests of basic research, to lay out tract-housing developments that mean massive profits for the few and ecological catastrophe for the entire state.

---

*See Commoner's article pp. 76-91.

On a larger scale, nearly all of the people and most organizations of people lack the financial resources to avail themselves of some of the most advanced technology that arises out of basic research. Computers, satellites, and advertising, to name only a few, all rely on the findings of basic research. These techniques are not owned by, utilized by, or operated for, the mass of the people, but instead function in the interests of the government and the large corporations. The people are not only deprived of the potential benefits of scientific research, but corporate capitalism is given new tools with which to extract profit from them. For example, the telephone company's utilization of the basic research on laser beams will enable it to create superior communication devices which, in turn, will contribute toward binding together and extending the American empire commercially, militarily, and culturally.

The thrust of all these examples, which could easily be elaborated and multiplied, is that the potentially beneficial achievements of scientific technology do not escape the political and economic context. Rather, they emerge as products which are systematically distributed in an inequitable way to become another means of further defining and producing the desired political or economic ends of those in power. New knowledge capable of application in ways which would alleviate the many injustices of capitalism and imperialism is either not created in the first place or is made worthless by the limited resources of the victims.

If we are to take seriously the observation that discovery and application are practically inseparable, it follows that basic researchers have more than a casual responsibility for the application of their work. The possible consequences of research in progress or planned for the future must be subjected to careful scrutiny. This is not always easy, as the following examples might indicate.

Basic research in meteorology and geophysics gives rise to the hope that people might one day be capable of exerting a high level of control over the weather. However, such techniques might be used to steer destructive typhoons or droughts into "enemy" countries like North Vietnam or China. As far back as 1960, the U.S. Navy published a paper on just this possibility and the need to develop the requisite techniques before the Russians did. (One has premonitions of future congressmen and presidential candidates warning us about the weather-control gap.) Rain-making techniques were used in Indochina, according to some reports, to induce cloudbursts over the Ho Chi Minh Trail.

Physicists working in the areas of optics and planetary orbits have provided knowledge which the American military was, and might still be, considering for the development of satellites in stationary orbit over Vietnam equipped with gigantic mirrors capable of reflecting the sun and illuminating large parts of the countryside at night. While scientific workers perform experiments on the verbal communication of dolphins, the Navy for years has been investigating the possibility of training them to carry torpedoes and underwater cameras strapped to their backs. Not surprisingly much of the support for basic reasearch on dolphins comes from the Office of Naval Research.

Neurophysiologists are developing a technique called Electric Brain Stimulation, in which microelectrodes capable of receiving radio signals are permanently implanted in areas of the brain known to control certain gross behaviors. Thus radio signals selectively transmitted to electrodes in various parts of the brain are capable of eliciting behaviors like rage or fear, or of stimulating appetites for food or sex. The possibility of implanting these electrodes in the brains of mental patients or prisoners (or even welfare recipients or professional soldiers) should not be underestimated, especially since such uses might be proposed for the most humane and ennobling reasons. Again, the list of examples could be extended greatly.

Science is Political

An analysis of scientific research merely begins with a description of how it is misapplied and maldistributed. The next step must be an unequivocal statement that scientific activity in a technological society is not, and cannot be, politically neutral or value-free. Some people, particularly after Hiroshima and Nuremberg, have accepted this. Others still argue that science should be an unbridled search for truth, not subject to a political or a moral critique. J. Robert Oppenheimer, the man in charge of the Los Alamos project which built and tested the first atomic bombs, said in 1967 that, "our work has changed the conditions in which men live, but the use made of these changes is the problem of governments, not of scientists."

The attitude of Oppenheimer and others, justified by the slogan of truth for truth's sake, is fostered in our society and has prevailed. It is tolerated by those who control power in this country because it furthers their aims and does not challenge their uses of science. This attitude was advanced centuries ago by people who assumed that an increase in available knowledge would automatically lead to a better world. But this was at a time when the results of scientific knowledge could not easily be anticipated. Today, in a modern technological society, this analysis becomes a rationalization for the maintenance of repressive or destructive institutions, put forth by people who at best are motivated by a desire for the intellectual pleasure of research, and often are merely after money, status, and soft jobs. We believe it would be lame indeed to continue to argue that the possible unforeseen benefits which may arise from scientific research in our society will inevitably outweigh the clearly foreseeable harm. The slogan of "truth for truth's sake" is defunct, simply because science is no longer, and can never again be, the private affair of scientists.

Many scientists, even after considering the above analysis, may still feel that no oppressive or exploitative technology will result from their particular research. Two arguments are relevant here. First, even research without foreseeable practical application serves to advance the field generally, and to provide a more sophisticated background from which technology may be derived. The Department of Defense recognizes this and annually invests millions of dollars in such "impractical" research, knowing that in the long run it pays off. The preferential funding of certain areas of basic research makes it more likely that those areas and not others will advance to the point where the emergence of this technology becomes more probable. Second, while formerly scientific activity consumed only an infinitesimal amount of society's resources, the situation has changed drastically in the last 25 years. Scientific activity now commands a significant amount of social resources, resources which are in short supply and are necessary to meet the real needs of the majority of the people. The point here is not that scientific activity should cease, but rather that it should truly be a science for the people.

Some scientists have recognized this situation and are now participating in nationally coordinated attempts to solve pressing social problems within the existing political-economic system. However, because their work is usually funded and ultimately controlled by the same forces that control basic research, it is questionable what they can accomplish. For example, sociologists hoping to alleviate some of the oppression of ghetto life have worked with urban renewal programs only to find the ultimate priorities of such programs are controlled by the city political machines and local real estate and business interests rather than by the needs of the people directly affected by such programs. Psychologists, demographers, economists, etc., worked on a Master Plan for Higher Education for New York City that would guarantee higher education for all. In practice open enrollment was restricted to the lowest level which channeled students into menial jobs set by corporate priorities while the main colleges remained virtually as closed as before.

Behavioral and clinical psychologists have tried to develop procedures for applying conditioning techniques to human psychopathology. Their work is now used in state hospital programs which, under the guise of "therapy," torture homosexual people with negative reinforcement, usually electrical, in order to convert them forcefully to heterosexuality. (There are still 33 states in which homosexuals may be "committed" under archaic sexual psychopath laws for *indefinite* sentences.) No one is impugning the motives of Pavlov or Skinner, but this is what it has come to in the United States. Thus the liberal panacea of pouring funds into social science research to create Oak Ridge-type institutions for the social sciences is no more likely to improve the quality of life than the namesake institution has. The social sciences are not performed in a political vacuum any more than the natural sciences are. They all ultimately serve the same masters.

Even medical research is not without negative social impact. The discovery of a specific disease cure or preventive measure invariably depends upon prior basic research which is frequently linked to nonmedical misapplications, often before it is used to produce disease cures. For example, the work of microbiologists who are decoding the DNA molecule gives hope for the genetic control of a wide variety of birth defects. Already this research has been used by government and military technicians to breed strains of virulent microbes for germ warfare. Further, it is not unreasonable to expect that someday this research will lead to genetic engineering capable of producing various human subpopulations for the use of those who are in technological control. These might include especially aggressive soldiers for a professional army, strong drones to perform unpleasant physical labor, or "philosopher kings" to inherit control from those already possessing it.

Applied medical research, as well as the more basic variety typified by DNA work, is no less free of the possibility of misapplication. More than purely humane consequences could emerge from one of the latest dramatic medical advances, organ transplantation. Christian Barnard has publicly urged that people be educated to "donate" their organs. It is not overly visionary to imagine that society's underclass, whose labor is decreasingly in demand, might be nourished as a collective "organ bank." If this occurred, it would most probably be on a *de facto* rather than *de jure* basis, as is the case with other forms of class and racial oppression. That is, monetary and other incentives would be instituted to encourage "volunteers" so that direct coercion would be unnecessary. Models for the poor selling parts of their bodies already exist in the form of wet nurses, indigent professional blood doners, and convicts and colonial peoples serving as subjects for experiments. An example of the last was the use of Puerto Rican women to test birth control pills before they were considered safe to market in the United States. (And now evidence that had been suppressed by the drug companies, the government, and the medical profession indicates that they are not safe after all—see J. Coburn in *Ramparts*, June, 1970.)

The misapplication of medical or premedical knowledge is, however, only half of the problem. The tragically overcrowded and understaffed city and county hospitals of our large metropolitan areas testify to the inequities and class biases in the distribution of medical knowledge as well. People here and throughout the world needlessly suffer and die because the money to pay for, the education with which to understand, or the physical proximity to, modern medicine has been denied them. By virtue of this, much of medical research has taken place for exclusive or primary use by the affluent.

Some medical discoveries have been equitably and, at least in our society, almost universally distributed. The Salk and Sabin vaccines are one example. Yet one is forced to wonder if this would have occurred had polio been less contagious. If the people who are in charge of our public health services could have protected their own children without totally eradicating polio, would they have moved as fast and as effectively? Witness their *ability* to prevent or reverse effects of malnutrition, while thousands of children within our borders alone suffer from it. In fact, while polio vaccines may have been an exception, the gravest problem we face in terms of disease is not discovering ways of equitably distributing the medical knowledge we already possess, and that, ultimately, is a political problem.

What Is To Be Done?

In this society, at this time, it is not possible to escape the political implications of scientific work. The U.S. ruling class has long had a commitment to science, not merely limited to short-range practical applications, but based on the belief that science is good for the long-term welfare of U.S. capitalism, and that what is good for U.S. capitalism is good for humanity. This outlook is shared by the trustees of universities, the official leaders of U.S. science, the administrators of government and private funding agencies. Further, they see this viewpoint as representing a mature social responsibility, morally superior to the "pure search of truth" attitudes of some of the scientists. But they tolerate the ideology since it furthers their own aims and does not challenge their uses of science.

We find the alternatives of "science for science's sake" and "science for progress of capitalism" equally unacceptable. We can no longer identify the cause of humanity with that of U.S. capitalism. We don't have two governments, one which beneficently funds research and another which represses and kills in the ghettos, in Latin America, and in Indochina. Nor do we have two corporate structures, manipulating for profit on the one hand while desiring social equity and justice on the other. Rather there is a single government-corporate axis which supports research with the intention of acquiring powerful tools, both of the hard- and soft-ware varieties, for the pursuit of exploitative and imperial goals.

Recognizing the political implications of their work, some scientists in recent years have sought to organize, as scientists, to oppose the more noxious or potentially catastrophic schemes of the government, such as atmospheric nuclear testing, chemical and biological warfare development, and the anti-ballistic missile system. Others shifted fields to find less "controversial" disciplines: Leo Szilard, who had been wartime co-director of the University of Chicago experiments which led to the first self-sustaining chain reaction, quit physics in disillusionment over the manner in which the government had used his work, and devoted the rest of his life to research in molecular biology and public affairs. In subsequent years other physicists followed Szilard's lead into biology, including Donald Glaser, the 1960 recipient of the Nobel prize in physics. Yet in 1969, James Shapiro, one of the group of microbiologists who first isolated a pure gene, announced that for political reasons he was going to stop doing *any* research. Shapiro's decision points up the inadequacy of Szilard's response but is no less inadequate itself.

Traditional attempts to reform scientific activity, to disentangle it from its more malevolent and vicious applications, have failed. Actions designed to preserve the moral integrity of individuals without addressing themselves to the political and economic system which is at the root of the problem have been ineffective. The ruling class can always replace a Leo Szilard with an Edward Teller. What is needed now is not liberal reform or withdrawal, but a radical attack, a strategy of opposition. Scientific workers must develop ways to put their skills at the service of the people and against the oppressors.

Political Organizing in the Health Fields

How to do this is perhaps best exemplified in the area of health care. It is not by accident that the groups most seriously dealing with the problem of people's health needs are political organizations. A few years ago the Black Panther Party initiated a series of free health clinics to provide sorely needed medical services that should be, but are not, available to the poor, and the idea was picked up by other community groups, such as the Young Lords, an organization of revolutionary Latins and Puerto Ricans. Health and scientific workers, organized by political groups like the Medical Committee for Human Rights and the Student Health Organization, have helped provide the necessary professional support, and over the past decade literally hundreds of free people's health centers have sprung up across the country.

Health workers, organized into political groups, can provide more than just diagnosis and treatment. They can begin to redefine some medical problems as social problems, and through medical education begin to loosen the dependence of medical people on the medical profession. They can provide basic biological information, demystify medical sciences, and help give people more control over their own bodies. For example, in New York, health workers provided a simple way of detecting lead poisoning to the Young Lords Organization. This enabled the Young Lords to serve their people directly through a door-to-door testing campaign in the Barrio, and also to organize them against the landlords who refused to cover lead-painted walls, often with the tacit complicity of the city housing officials.

It is this kind of scientific practice that most clearly characterizes Science for the People. It serves the oppressed and impoverished classes and strengthens their ability to struggle. The development of People's Science must be marked by these and other characteristics. For example, any discoveries or new techniques should be such that all people have reasonably easy access to them, both physically and financially. This would also militate against their use as a means of generating individual or corporate profit. Scientific developments, whether in the natural or social sciences, that could conceivably be employed as weapons against the people must be carefully evaluated before the work is carried out. Such decisions will always be difficult. They demand a consideration of factors like the relative accessibility of these developments to each side, the relative ease and certainty of use, which will of course depend on the demand, the extent to which the power balance in a specific situation could be shifted and at what risk, and so forth. Finally, scientific or technological programs which claim to meet the needs of the people, but which in fact strengthen the existing political system and defuse their ability to struggle, are the opposite of People's Science.

There is a wide range of activities that might constitute a Science for the People. This work can be described as falling into six broad areas:

1. Technical assistance to movement organizations and oppressed people.

The free people's health centers have already been described as an example of this approach. Another example would be designing environmental poisoning detection kits for groups trying to protect themselves from pollution and trying to organize opposition to the capitalist system which hampers effective solutions to pollution problems. The lead poisoning test was such an effort, and other kinds of pollution are equally amenable to this approach. These kits would have to be simple to operate, easy to construct, and made from readily available and cheap materials.

Research to aid student and community struggles for free, decent higher education is being conducted by the New University Conference and other groups. Of interest are answers to questions involving the economy of higher education, such as what classes pay what share of the tax bill, how are educational resources apportioned among the classes, how is higher education differentially defined in different types of schools, how does discrimination operate against women and Third World people in education, what role do corporations play in setting up program priorities, especially in the working-class junior colleges. Research also needs to be done on the possibilities for open enrollment in various school systems and on the test instruments and the tracking system which channel students and distribute educational privilege on the basis of social class.

Research could be performed which would assist rank-and-file groups now attempting to organize politically in the factories. Useful information might include the correlation between industrial accident rates and the class, race, and sex of the work force, the mechanics of the unemployment compensation and accident compensation programs which more often make profits for insurance companies than help workers, the nature of union-management contracts, how they have served to undermine workers' demands and how they might be made more effective, and so on. All of these projects would be examples of Science for the People as technical assistance.

2. Foreign technical help to revolutionary movements.

American scientific workers can provide material aid to assist struggles in other countries against U.S. or other forms of imperialism, or against domestic facism. For example, the Popular Liberation Movement of Angola, fighting against Portuguese domination, has requested help in setting up medical training facilities. These are sorely needed in those areas of Angola that have been liberated and are undergoing social and economic reconstruction.

Similarly, Americans can aid revolutionary regimes abroad. The effects of the U.S. blockade of Cuba could be reduced by North American scientific workers going there to do research or to teach, as some have already done. Or, they could do research here on problems of importance for development in Cuba, such as on sugar cane and rice production, tropical pest control, and livestock breeding. At a minimum, U.S. scientists should be encouraged to establish regular contact and exchange reprints and other information with their Cuban counterparts.

Another example of this kind of foreign technical assistance was a Science for Vietnam project, which involved collaboration between scientists from the U.S. and the Democratic Republic of Vietnam and the Provisional Revolutionary Government of South Vietnam on such problems as locating plastic pellets in human flesh (several years ago the U.S. Air Force increased the terrorizing effect of anti-personnel bombs by switching from metal fragmentation devices to plastic pellets, which do not show up on x-rays), reforestation techniques, how to decontaminate herbicide-saturated soils, and many other problems now facing the Vietnamese as a result of the U.S. intervention there.

This kind of foreign technical assistance has important political significance in addition to its material consequences, for it is the most direct way one can oppose the imperialist policies of the U.S. government, undermine its legitimacy, and go over to the side of the oppressed people of the world. If an important sector of the population, like scientific workers, begins to act in this way, it may encourage similar action by workers in other areas.

3. People's research

Unlike the technical assistance projects described above, which are directly tied in with on-going struggles, there are areas in which scientists should take the initiative and begin developing projects that will aid struggles that are just beginning to develop. For example, workers in the medical and social sciences and in education could help design a program for client-controlled day care centers which would both free women from the necessity of continual child care and provide a thoroughly socialist educational experience for the children. As such, it would be useless to those who are trying to co-opt the day care struggle into an extension of social control or as a means of making profits.

For use in liberation struggles, self-defense techniques could be developed that would be readily available to the people, and useless to their highly technological opposition. Biologists and chemists, for example, could develop an all-purpose gas mask for which the necessary materials are simple, easy to assemble, readily available, and inexpensive.

Physiologists and others could perform definitive research in nutrition and disseminate their findings so that poor and working-class people would have information on how to get the most nourishing diet for the least cost. Furthermore, such research could aid them in avoiding the possibly dangerous food additives and contaminants that are now found in most packaged foods.

As a minimal effort, medical researchers could begin to concentrate their work on the health needs of the poor. The causes of the higher infant mortality rates and lower life expectancy of a large part of the working class, particularly racial minorities, should get much more research attention. Occasionally funds are available for this kind of research but the class background and biases of many researchers often predispose them toward work on other problems. In addition, new ways of distributing and utilizing medical knowledge, especially with respect to prevention, must be designed.

4. Exposes and power structure research.

Most of the important political, military and economic decisions in this country are made behind closed doors, outside the public arena. Questions about how U.S. corporations dominate foreign economic markets and governments, how corporate conglomerates control domestic markets and policy making, how party machines run city governments, how universities and foundations interlock with military and various social-control strategies, how the class struggle in the U.S. is blunted and obscured, etc., must be researched and the conclusions published to inform all the people.

Exemplary work of this kind has already been performed by research collectives like the North American Congress on Latin America (NACLA), the National Action Research on the Military Industrial Complex (NARMIC), the Africa Research Group and others. These groups have provided valuable information for community and campus groups in campaigns such as those against university collaboration with the Indochina War and exploitation in various Third World countries, against anti-personnel weapons manufacturers (like Minneapolis Honeywell), and against specific corporations involved in particularly noxious forms of oppression (like Polaroid's large investments in South Africa and their current contract to provide the government there with photo-ID cards for all citizens which will help that government to implement more effectively its racist apartheid policy).

There is growing need for research in the biological and physical sciences to expose how the quest for corporate profits is poisoning and destroying irreplaceable and critical aspects of our environment. This information, in a form anyone can understand, should be made available to action-oriented community ecology groups.

5. Ideological struggle.

Ruling-class ideology is effectively disseminated by educational institutions and the mass media, resulting in misinformation that clouds people's understanding of their own oppression and limits their ability to resist it. This ruling-class ideology must be exposed as the self-serving manipulation that it is. There are many areas where this needs to be accomplished. Arguments of biological determinism are used to keep blacks and other Third World people in lower educational tracks, and these racist arguments have recently been bolstered by Jensen's focusing on supposed racial differences in intelligence. Virtually every school of psychopathology and psychotherapy defines homosexuals as sick or "maladjusted" (to a presumably "sane" society). These definitions are used to excuse this society's discriminatory laws and practices with respect to its large homosexual population, and have only recently been actively opposed by the Gay Liberation Movement. Similarly, many psychotherapists and social scientists use some parts of Freudian doctrine to justify sexist treatment of women.

The elitist biases of most American social scientists oppress students from working-class and poor backgrounds, as well as women and minorities, by failing to adequately portray their history and culture. Instead, bourgeois culture and ruling-class history are emphasized as if they were the only reality. This laying on of culture is particularly heavy-handed in community and working-class colleges (for an elaboration of this point, see J. McDermott, *Nation*, March 10, 1969). To combat this, the social scientist should work to make available to the people their true history and cultural achievements.

This kind of Science for the People as ideological struggle can be engaged in at several levels, from the professional societies and journals to the public arena, but for it to be most effective it should reach the people whose lives it is most relevant to, and who will use it. Those in teaching positions especially have an excellent opportunity to do this. For example, courses in any of the biological sciences should deal with the political reasons why our society is committing ecological murder/suicide. Courses in psychopathology should spend at least as much time on our government officials and our insanely competitive economic system as they do on the tortured victims incarcerated in our mental "hospitals," many of whom would not be there in the first place if they lived in a society where normality and sanity were synonymous. Within these and many other disciplines, individual instructors can prepare reading lists and syllabuses to assist themselves and others who are interested in teaching such courses but lack the background or initiative to do the work themselves.

6. Demystification of science and technology.

No one would deny that science and technology have become major influences in the shaping of people's lives. Yet most people lack the information necessary to understand how they are affected by technological manipulation and control. As a result they are physically and intellectually incapable of performing many operations that they are dependent upon, and control over these operations has been relinquished to various experts. Furthermore, these same people undergo an incapacitating emotional change which results in the feeling that everything is too complicated to cope with (whether technological or not), and that only the various experts should participate in decision making which often directly affects their own lives. Clearly, these two factors are mutually enhancing.

In the interests of democracy and people's control, the false mystery surrounding science and technology must be removed and the hold of experts on decision making must be destroyed. Understandable information can be made available to all those for whom it is pertinent. For example, the Women's Liberation Movement has taken the lead in teaching the facts about human reproductive biology to the people who need it the most for control over their own bodies. An example of this is a group of women in the Chicago Women's Liberation Union who have written a series of pamphlets on pregnancy and childbirth, giving complete medical information in language everyone can understand. Free schools and movement publications teach courses and run articles on medical and legal first-aid, self-defense, effective nutrition, building houses, repairing cars and other necessary appliances, and so on. Much more of this kind of work needs to be done. In addition, the relevant scientific information on issues that have important political repercussions, such as radiation poisoning and pesticide tolerance, should be made available to the public.

Part of the job of demystification will have to take place internally, within the scientific community. Scientific workers themselves must expose and counter the elitist, technocratic biases that permeate the scientific and academic establishments. One vechicle for doing this has been the publication *Science for the People* (see pp. 369-382). Attempts to demystify science must take place at many levels. The doctrine that problems of technology can be met with technological rather than political solutions is increasingly being incorporated into the ruling ideology. The counter argument should be made that only political reorganization will be effective in the long run, and this argument will need to be bolstered by more research. On the level of daily practice, elitist tendencies can be undermined in laboratories and classrooms by insisting that *all* workers or students participate in decision making that affects what they do and by creating conditions that insure them the information necessary to make those decisions. The elitism and hierarchical structuring of most scientific meetings and conventions can be opposed by members forcefully insisting that they be given some control over the selection of speakers and that all scheduled speakers address themselves to the political implications of their work. This is already happening with increasing frequency as radical caucuses begin to form in many of the professional associations.

The practice of Science for the People is long overdue. If scientific workers and students want to overcome the often alienating nature of their own work, their impotence in bringing about meaningful social change, their own oppression and that of most of the other people in the world, they will have to relinquish their uncritical devotion to the pursuit of new knowledge. Scientific workers must reorganize scientific work, not in terms of the traditionally defined disciplines, but according to the real problems they consciously set out to solve. The old breakdown into separate disciplines, which produces "experts" who can barely communicate with each other, must give way to new structures which allow more cooperation and flexibility, and which will undoubtedly demand the acquisition of new skills. Such work can be as intellectually stimulating as the work we do now, with the added satisfaction that it is meeting real needs of people.

If projects like those described above are to constitute a réal Science for the People, they must achieve more than their immediate technical goals. They should relate to issues around which people can organize to act in their own self-interest. Research projects should both flow out of the needs and demands of the people, and be relevant to their political struggles. This requires consulting with and relying on the experience of community and movement groups, and taking seriously the criticisms and suggestions that they put forth. Scientists must succeed in redirecting their professional activities away from services to the forces and institutions they oppose and toward a movement they wish to build. Short of this, no matter how much they desire to contribute to the solution, they remain part of the problem.

Towards a Liberated Research Environment
—Thimann Laboratory Group

A group of scientists and science students at U/California, Santa Cruz, describe their efforts to develop a more human, creative and liberating work environment by restructuring their work life along lines of cooperative support.

INTRODUCTION

For the past three years we have struggled to create a working environment which would enrich our scientific and personal lives. We have rejected traditional ways of working together because we feel that the intense competition and static hierarchical structure commonly associated with scientific research hinders our growth and development as scientists and people. In this article we want to tell you about the first steps in this struggle: how we are trying to change the way we deal with each other, and how we are trying to make our workplace physically less alienating. Because this process is still in its infancy for us, we aren't ready to present in any detail our thoughts on the possible "export" of our approaches to other laboratories, nor will we describe how our efforts compare to other attempts to collectivize a laboratory. Our ultimate goal is not simply to develop a better working situation for ourselves, but rather through experimentation within our group, to create a model that others will find exciting and useful. This paper documents our first efforts to share more equally the pleasures, tasks, and frustrations of running a research laboratory.

The Problem

Our laboratory structure has evolved from attempts to answer the basic questions: "What kind of working relationships and environment will nurture our intelligence, creativity, and sense of cooperation?" We present an example of a poor laboratory structure as a point of departure in order to show later on how interpersonal problems and the physical workplace might be improved.

A Typical, Unliberated Laboratory

In their relationships, scientists share many of the problems of other workers in the United States. These problems are complex, but arise directly or indirectly from profit-seeking and competition among individuals as well as between groups of individuals. Working relationships are often damaged by attitudes which place personal gain above overall human welfare, while the cultural biases of our society, e.g., sexism, racism, classism, ageism, etc., further isolate people from one another. Most laboratories are run very much like businesses. There are products to market, labor to divide, and profits and rewards to gain for work of high quality and quantity. Like any small business, the number of people involved is not so great that individuals at different levels of the hierarchy necessarily become locked away from each other. There are variations in functioning of individual laboratories just as there are some variations in business management. We have chosen to present a typical laboratory structure to emphasize the problems. We realize other laboratories have successfully dealt to some degree with various aspects of these problems.

What is a typical laboratory like? Usually there is one principal investigator/professor who is "boss" over the entire operation. Most people in this position are Caucasian males. The professor makes all the decisions except for ones s/he deems trivial. These are delegated to someone else whose time is "less valuable" than the professor's. The professor receives more than her/his fair share of credit for the work done by the lab group, but is also overly blamed for any errors in published experiments. In short, s/he acts like a boss and is treated like a boss both by other workers in the lab and by the rest of the scientific community.

Next in command and in order of prestige and salaries are research associates and postdoctoral fellows. Usually these people have already had several years of research experience and are semiautonomous in day-to-day research functioning. They are often given at least a limited choice of what project to work on in the lab, but are often forced to bow to the professor's final decision. Men often fill this position as a stepping stone to becoming a professor, while women are more commonly in this position as a permanent job.

After the research associates and postdocs come graduate students in order of the stage of completion of their doctoral thesis projects. Graduate students are paid very low salaries either from the research grant awarded to the professor or from the university in exchange for acting as teaching assistants. Their work is often directly supervised by a postdoc or by the professor, and they seldom receive the credit they deserve for their work.

Underneath the youngest graduate student are occasional undergraduates who complete research projects often directly supervised by graduate students. Undergraduates typically receive no money for their work but many receive course credit or may fulfill a senior thesis requirement for graduation by working in a research lab.

The bottom rung of this hierarchy is filled by people performing various supportive services which are essential for the functioning of the lab. Secretaries, maintenance personnel, and glassware washers are often not even considered members of the research group. They are once again primarily female and/or ethnic minorities.

Our Research Group: Some Background

Our story began in 1975 when J. Fred (Paulo) Dice was hired at the University of California at Santa Cruz as an assistant professor of biology. Paulo was interested in exploring an alternative laboratory arrangement which would be consistent with his politics. A "cooperative" model seemed appropriate, and with unanimous support within our lab we began the task of liberating our research environment. Towards this end, we have tried to maintain an awareness of the biases built into the research "business" and to combat them through affirmative action.

The people in our lab range widely in experience and age. Paulo and Gary have both had about ten years of research experience, while graduate and undergraduate students have had 0-3 years of research. The composition of the lab group is continually changing, but has been weighted towards undergraduates due to UCSC's emphasis on undergraduate research experience. The usual composition is one faculty member, one research associate, two graduate students, and six undergraduate students. We have attempted to make sure that our research group is well balanced both racially[1] and sexually, although we feel we need to put even more effort in this direction. The disparity in experience and age probably makes it more difficult than it would otherwise be to develop an egalitarian laboratory structure.

An important question, especially heard from non-scientists interested in our approach, concerns the choice of research. Who decides? In one sense it is the state of scientific knowledge which determines what will be studied next. In another sense it is the people controlling research funds who determine what areas will be researched. (See pp. 171-191) But other factors such as personal preferences and interests also play a role. We study the control and mechanism of protein degradation. Proteins are vital structural and control elements in all living things, and understanding their breakdown represents a major unresolved question in modern biology. The current state of biomedical research "posed the question" of how breakdown is controlled, and Paulo chose to explore the mechanism(s) involved. We stick to this research subject for many reasons. First of all, we have limited funds for our research and enjoy working on problems in the same area so we can "brainstorm" together. Also, although we don't feel that this area of research is *more* relevant to human welfare than any other area, it certainly has relevance. For instance, understanding protein degradation is probably going to lead to better treatment of malnutrition, starvation, diabetes, arthritis, thyroid disorders, nervous disorders, and various genetic diseases.

OUR EFFORTS TO IMPROVE ON
THE TYPICAL, UNLIBERATED LAB

Relations to One Another

We are attempting to share equally the responsibility for running our laboratory. For example, we share the responsibility for directing our laboratory group meetings, teaching new lab members research methods, ordering supplies and equipment, and cleaning up the lab. We also decided to share office space, and, perhaps our most unusual innovation, to share our salaries during the past two summers.

We feel the key to whatever success we may have lies in our attempts to support each other in our scientific pursuits as well as on a personal level. "Personal support" is hard to define, but practically speaking it means trying to look beyond a co-worker's social "veneer," expressing an interest in their life outside the lab, being available to help with a project, or just to talk. Ways to express concern for a person's well-being are varied. The goal, however, is to treat each other with understanding and respect. The most important step in creating a personally supportive environment is to recognize its importance and supply the effort it takes to create and sustain it. It was interesting for us to think about specific instances where personal issues required other lab members to be understanding and helpful. In the past three years, in order of their frequency, the most common problems were: (1) doubts about whether a career in science is really "where it's at"; (2) relationship problems with friends/lovers; (3) problems with health; (4) demands from other classes and/or administrative duties. Within this framework of what we hope is a nurturing environment are some of the ways we try to help each other and to look after our workplace, the lab.

Lab Group Meetings

The most unusual aspect of our lab meetings is our discussion of personal matters. Typically, with a different person leading us each week, we all have the opportunity to share the group thoughts, feelings and even a general overview of how our lives have been during the past week. This information is especially useful and, if listened to carefully, may suggest things that might be done to make a fellow worker's life happier in the coming week. After about an hour of this we turn to general items abut running the lab. For example, a decision may be required on what equipment to buy, which person to hire to work with us, or whether or not to write an article such as this one. Finally, we may discuss a current research paper, a new idea or research scheme, or have a "round-robin" on our own research results from the previous week(s). Whatever we are discussing, we try to keep in mind that almost everyone censures his or her creative thoughts for fear of appearing stupid. The reason for this is that we all have been victims of "put-down" attitudes somewhere in our upbringing, either at home or at school, often in both places. Therefore we try to treat each person's presentation with respect, and to encourage people to ask questions and make comments.

Decision Making

We try to make important decisions by discussion and general consensus. Taking on a new lab member, for example, is discussed from several points of view: pressure on space and supplies, whether the people responsible for training a new person can manage the added time demands, how the person would benefit from the experience. For less interesting decisions we try to divide up the tasks. In purchasing new equipment, for instance, one person may call company representatives, read several brochures, and then report his or her findings to the group. One person doesn't have to do all the work and everybody has the opportunity to be involved in the decision making.

Teaching New People

The general idea is for each lab member to teach what she or he knows best. This procedure broadens the responsibility for teaching to include all lab members and helps the new person to regard everyone in the lab as a resource. A typical sequence of events for someone new joining the lab goes something like this: Whoever knows her/him best describes the general research interests of the group and our general experimental approaches to the question we are asking. S/he is encouraged to hang around the lab and get acquainted with everyone else at this stage. Then the new lab member may be asked to watch several experimental procedures over the next few weeks and to take notes, ask questions, and read about the theory of the techniques. Next, the person will do each type of experiment on his/her own with the person responsible for teaching close by to observe and correct mistakes or to answer questions. By this time, the new lab member can usually carry out the experimental procedures entirely independently and s/he has learned about the projects going on in the lab. S/he is then encouraged to choose a research topic among the several possible ones being discussed in the lab.

Salaries

During the school term students in the lab receive course credit for their research. Everyone has rather extensive obligations and activities outside the lab from September to June, so we decided to accept the University's decision regarding our salaries for this time. This doesn't mean that the University's salaries are appropriate, but we decided that we couldn't handle the potential complications of trying to redistribute wages during this time.

In the summer, we all work approximately full-time and have more or less equal duties. For this reason, we pooled our summer salaries in 1976 and 1977 and shared them equally. This brought up interesting discussions about money. Questions which have arisen and remained unsolved include: Should salary depend on hours spent in the lab? On previous research experience? On need? What are "appropriate" and "inappropriate" needs? Food, shelter, and health care seem like reasonable needs, but what about travel, a television set, etc?

The issue of money has been especially difficult for many of us. It was a real conflict for those who would have earned more by the University's rules to give up part of their fair earnings, and it seemed equally hard for those who received "extra" money to avoid feeling guilty and in debt because of it. We feel the attempt to redistribute salaries has at least raised issues which seem to be avoided in most labs.

The Workplace

The importance of physical surroundings in influencing moods has generally been overlooked by scientists. Research labs are often dreary, dirty and impersonal. How can bright, creative ideas emerge from an environment of gloom? We have taken several steps to improve lab cleanliness and appearance.

One Saturday, as a group project and picnic, we painted some of the walls and plumbing fixtures with bright pastel colors in order to cover the original institutional "eggshell" white. We also tacked up colorful pictures of plants and animals. Several lab members brought in their own art work such as weavings, paintings and drawings which not only beautify the lab but serve to remind us that we can be creative people. To supplement sight with sound, we invested in a stereo system with a tape deck. People have been encouraged to bring in tapes of their favorite music. In general, the lab looks and sounds wonderful, and our spirits can't help but be positively affected by the surroundings.

Cleanup is shared more than it usually is in a research lab. Each person is reponsible for cleaning her/his own equipment, bench space and large glassware items. We also have a weekly cleaning campaign to cover common space such as our office. Washing glassware is vital to a biology lab—dirty containers can ruin weeks of work so this job must be done carefully. Since we found that our glassware was not being cleaned thoroughly in the central washing facility, we decided to hire a biology student to wash glassware for us. We all regard this person as an equal in every sense, and teach him all about our experiments. Recently he has begun his own research project.

Office and Lab Space

Faculty members usually have their own private offices while other members of the lab have desk space allotted according to the traditional hierarchy: graduate students get to share office space with research associates and post-docs; undergraduates get nothing. In keeping with our egalitarian goals, we put three desks in the faculty office space and use it as a common room. There is also desk space in the lab which is available for each lab member. Rather than have assigned bench space for each lab member, we work next to the equipment we need and make certain the space is clean for the next user. So far, this system has worked well, possibly because we have similar standards for cleanliness, also because we feel our system encourages people to accept responsibility for keeping the whole lab neat.

Equipment

Understanding the operation of equipment, explaining its use to new lab members, and taking care of routine inspection and maintenance are also tasks that are shared by all in our group. If a malfunction in something is found, for example, whoever notices it first assumes the responsibility to see that it is fixed. This may mean actually tinkering with the item or simply calling it to everyone's attention, delegating the job and seeing to its completion.

FUTURE DIRECTIONS

We have purposely avoided being very vocal about our attempts to reorganize our laboratory. Our reasoning is that we first had to get our own working relationships in good order. Until we had ideas which we knew to work, there seemed to be little use in preaching about theoretical changes. The most solid way to reach out to other labs with our ideas seemed to be to wait until they asked us what we were doing differently. Explanation could then be tailored to each individual. For example, one person may need to be approached at first with no political rhetoric at all but with a hardnose analysis of how to increase worker productivity by giving lab workers more decision-making power, whereas another person may respond to a theoretical political argument about "how science will be done after the revolution."

We are fairly close to several other research teams here at the University of California at Santa Cruz, and have helped other labs make small reorganizations in their structure. At this point there are no other labs at UCSC who are actively pursuing a goal of egalitarianism, however. This will be our next level of outreach now that our own lab is proceeding in the right general direction.

Since our research group is quite young, we have not yet had much experience relating to the national and international scientific establishment. We have just received our first major grant, and members of the lab have attended only a few scientific meetings. Areas which we are in the process of addressing or have discussed in the past include:

1. Where should we apply for financial support of our research? This was a hot issue especially when we had only small one-year grants in 1976 and 1977. We identified problems and advantages of having federal support, but in our particular case we have not compromised our project goals at all to receive federal money.

2. Who should be co-authors of scientific papers? Should *all* people involved with the project be co-authors? This list would include several lab members in addition to those directly conceiving and carrying out the experiments and would include many people in supportive roles such as grant administrators, secretaries, custodians, etc. How can we properly indicate what was done by whom in such a long list of co-authors?

3. How can we reach other labs working in our particular field to promote noncompetitive, cooperative attitudes? The first step here is to get rid of our own fears of "being scooped" etc., which cause people to compete. By our attitudes at scientific meetings and in correspondence with other workers, we hope to promote cooperation. We suspect that much organizing at this level must be done on an individual basis.

4. How can we demystify our research so it is under-

standable to everyone? Is writing popular articles a possible solution? What are our social responsibilities to see that our research is not misapplied?

EVALUATION AND CRITICISM OF OUR EFFORTS

Overall we should be pleased with the fact that we are a group of scientists who are struggling with important issues concerning how research is conducted in the U.S. We have made definite progress in the past 3 years toward our goal of non-hierarchy, but it has often been a hard battle. There have been hurdles along the way placed by our own upbringing and individual problems as well as by the nonegalitarian institution where our lab resides.

There are several areas we would like to work on. Most important perhaps is our outreach to other labs and our resolving issues about how we are going to relate to the scientific establishment. But along with these efforts, we will need to push for even better relationships within our lab as well. For example, there are still vestiges of hierarchy in our interactions. Although lab group meetings and decision making are more equally shared than in any other lab we know about, people higher up in the traditional hierarchy still tend to take more active roles.

Some of us would like to discuss our political views more extensively and try to come to agreement about being involved in political actions. Others prefer the political action to remain on an individual basis. It's not clear to us whether struggling toward a common political view would bring us closer together or would destroy the working relationships we've worked to create.

We certainly have a long way to go for our goal of establishing an alternative model for how scientists may work together, but we hope it is clear that we are challenging some of the assumptions about how scientific research should be done. We have uncovered more questions than we have answers, but are struggling in an exciting area where there is plenty of room for new, alternative models. We welcome and encourage comments and questions, and would appreciate correspondence from other laboratories which are dealing with these issues.

Science, Technology and Black Liberation
—S. E. Anderson and Maurice Bazin

This article discusses the role of science and technology in oppressing black people and the need for black scientists to organize in order to avoid following the example of the alienated while middle class male scientist.

Forge simple words that even the children can understand...words which will enter every house like the wind/fall, like red hot embers on our people's souls.
                            Brother Jorge Rebelo (Frelimo)

What are science and technology? What are their origins? How do they function within the United States? What are the goals of black liberation? And finally, what are the relationships between our liberation struggle and the political, economic and technical developments within the U.S.'s science and technology?

These are some questions that exist at this point in our revolutionary development. There has been very little effort in the United States to answer these questions, fundamentally because science and technology have been traditionally viewed by the white left as an appendage to the socio-economic system of capitalism with, until recently, an apparently apolitical function. However, today science and technology are put into their proper perspective: a necessary political agent for the development of monopoly capitalism and imperialism, and maintenance and rationalization of white racism (personal and institutional).

Science today is property and therefore, like all property, it is used for the benefit of those who own it. In the USA and in other imperialist nations, the major part of scientific effort is dedicated to the twin purposes of (1) extraction of profits and (2) the maintainance of the control which permits that extraction.[1]

The key innovation is not to be found in chemistry, electronics, automatic machinery, aeronautics, atomic physics or any of the products of these science-technologies but rather in the transformation of Science itself into capital.[2]

Specifically, within Afro-America—in spite of one's political position—science, technology, scientists and technicians are consciously or subconsciously placed on an enviable pedastal: "S/he is heavy because s/he is into the sciences"; "Man, the sciences, engineering and medicine are deep or heavy subjects." That is, more complex and disciplined than the subject matter or work in the bourgeois humanities and social sciences. In short, because of our fear, mystification and envy of the natural sciences, technology and scientists, our ideological development and day-to-day struggles are similar to and have virtually the same results today as black reconstruction did in 1874.

But first let us set a political and historical framework from which we can begin to understand what is and what needs to be done.

What are Science and Technology?

J.D. Bernal—a British marxist physicist and science historian—has written a most comprehensive and readable analysis of the evolution of science and technology in human history called *Science in History*. Rather than give a pat "*Webster Dictionary*-type definition of science (and technology) Bernal states:

Science has so changed its nature over the whole range of human history that no definition could be made of it. Although I have aimed at including everything called science, the centre of interest ... lies in natural science and technology because ... the sciences and society were at first embodied in tradition and ritual and only took shape under the influence and the model of the natural sciences ... Science stands as a middle term between the established and transmitted practice of men who work for their living, and the pattern of ideas and traditions which assure continuity of society and the rights and privileges of the classes that make it up.[3]

Science, then, becomes both an ordered technique and rationalized mythology.[4] Given this reality and given the reality of its complete fusion within a contemporary capitalist or socialist society, we need to perceive science (and technology) as (a) an institution, (b) a method, (c) a cumulative tradition of knowledge, (d) a major factor in the maintainance and development of production, and (e) as one of the most powerful influences shaping beliefs and attitudes toward the universe and humanity.[5] It is within this context that we will use "science and technology" or just the term "science."

The Origins of Science

An essential factor in aiding us to overcome the psychological barriers against understanding and working with the sciences is to place science in its correct historical context. Black folk have to struggle against a double psychological barrier: science as divine and mysterious and science as non-black in the socio-historical sense. Keep it in mind that there has been no societal development without scientific and technological development. One of the earliest revolutionary scientific developments that moved humanity in general and our African ancestral brothers and sisters in particular, into a new social order (the city-state) was the development of iron smelting in Zimbabwe (Rhodesia) at least 40,000 years ago.

This scientific and technological achievement could not have occurred without other scientific achievements in the area of agriculture. It implies that farming (a technological achievement, primarily developed by women) was necessary for the development of a more stable and non-nomadic society. This recent discovery of a 40,000 year-plus old Zimbabwe iron smelting,[6] implies that scientific and technical information moved from Southern Africa to Northern Africa and then on to Europe. This would completely shatter influential racist historians like Arnold Toynbee and his *A*

Study of History where Egypt is depicted as a white nation and the rest of Africa a mere savage pre-historic footnote.[7]

More important than the necessary refutations of racists is a proper analysis of the developments and dynamics of science within those societies before the onslaught of European slave trade, colonialism and neo-colonialism.

> The African continent reveals very fully the workings of the law of uneven development of societies. There are marked contrasts between the Ethiopian empire and the hunting groups of Pygmies in the Congo forests or between the empires of the Western Sudan and the Khosian hunter gatherers of the Kalahari Desert. Indeed, there were striking contrasts within any given geographical area. The Ethiopian empire embraced literate feudal Amharic noblemen as well as simple Kaffa cultivators and Galla pastoralists. The empires of the Western Sudan had sophisticated, educated Mandinga townsmen, small communities of Bozo fishermen, and Fulani herdsmen. Even among clans and lineages that appear roughly similar, there were considerable differences.[8]

Then we have arrested development and underdevelopment in Africa (and Asia and Latin America) for capitalist development in Europe and the United States. Various social factors congeal to determine when a society makes the dialectical leap from small-scale craft technology to a technology which uses labor more efficiently. One primary social factor is the demand for more products than can be produced by hand. Hence, technology and technicians respond to a clearly defined social need, clothes for example. With the penetration of cheap, readily available European cloth, African producers either lost their skills or struggled along, subsisting with the traditional, slow manual weavers, creating pieces for a localized and subsequently small market. Therefore, we witness continental technological arrest, stagnation and regression. The use of European capitalist technology and its associated products forced thousands of Africans (and Asians and Latin Americans) to forget not only the complex techniques of their ancestors but even the simpler ones. When one understands the centrality of smelting to development, one can see that the crucial technological regression for Africa occurred when traditional iron smelting was abandoned.[9]

In short, science and technology during the pre-capitalist periods were advancing on a global scale. But European and U.S. capitalism made scientific and technological development occur only in Europe and the United States.

Hence previously independent societies throughout the Third World today become dependent societies trapped at best in the eighteenth and nineteenth centuries. The more scientific achievements occurred within Europe and the U.S., and the more efficient and exploitative capitalism became, the more the burden of capitalism was placed on black and Third World people.

Looking within the growth of capitalism, we find that during the primitive capitalist stage science was virtually non-dependent upon the capitalist for financial support. On the other hand, the mere existence of a primitive capitalist society allowed a few individuals from merchant families or other non-working class groups to pursue the time-consuming work of scientific inquiry (as opposed to a hit-and-miss or non-experimental inquiry.)

We need only to look at the socio-economic roots of Isaac Newton's *Principia* to understand that this work evolved out of the needs of early capitalist development and its attendant technology. In short, *Principia* was not, and could not be, a work of "pure science" by a value-free or "neutral" scientist.[10] As the capitalists forced their way across Europe, Africa, Asia and the Americas, their system demanded a larger output of goods, which in turn demanded larger, more efficient machines to extract and transport the raw materials. Thus, for example, the invention of the cotton gin initially meant a growth both in African slave labor in the South and European immigrant labor in the North, and in Britain the forcing of peasants into an industrial proletariat.

But, as the cotton gin paved the way for more mechanized agriculture, it also aided in the forcing of black Americans out of agriculture at a time when European immigrant workers were occupying the production jobs in northern industries. Again we see an apparent contradiction: as capitalism provided/demanded more leisure time for the bourgeois Europeans, it encouraged the development of scientific inquiry, and it used the results of this methodology for its own ruthless and grotesque growth:

> In imperialist countries, the scientific venture is devoted, for the most part, to the development of military technology, to mass extermination, and to fascistic control of the behavior of society as well as of the individual. The

objective benefit that humankind might gain from scientific work is of secondary consideration.[11]

Science, then, has become an inseparable part of capitalist development, and capitalist development has become an inseparable part of science. But just as capitalism as a system/civilization is digging its own grave, we have a withering away of capitalism through its prodigal son—modern science:

> The old concept, which goes at least as far back as ancient Egypt, of an educated elite, and a mass of illiterate peasants and workers is bound to disappear, indeed is disappearing already. The archaic methods of control of production and consumption sanctified in the code of free-enterprise capitalism, which has become monopoly capitalism, will have to make way for planned production and to make more and more use of mathematical and computational methods. In words, science implies socialism.[12]

But let us be very clear: that science implies socialism does not mean that there will not be, or should not be, a struggle between the tiny ruling class and the working masses of humanity. It does not mean that all black folk in the Americas and Africa have to do is try and become scientists and then somehow bring about socialism because they are scientists. A people's struggle and the struggle for a people's science are intimately bound up, as victoriously demonstrated by Cuba, North Korea, North Vietnam, Mozambique, Guinea-Bissau and the People's Republic of China.

Black and Third World Dependency and Technology

When we look at science and technology as they function within Third World societies we see two contradictory processes. The first is a liberation process. Science and technology have freed large sections of societies from the killing extremes of labor and diseases. The other is an oppressive and exploitative device. The American and European capitalists have invented mechanisms so subtle and sophisticated that they come disguised as "progress" and "development."[13]

> For this reason, science is like a smoke-screen: while its force appears to be directed at the resolution of the most urgent problems of our peoples, it makes these problems more numerous. It covers up the social roots of "technical"

problems. In the rhetoric of "harmony" it enshrouds the reality of imperialism.[14]

The good that contemporary science and technology can do is overwhelmed by the imperialists' quest for raw materials, new markets and peace at home through mass material consumption. When we turn to black America on this matter there is no improvement. In fact, black America is so technologically backward relative to its position inside of the world's most technologically advanced nation that we are more backward than the newly independent Republic of Guinea-Bissau! More precisely, black America is more technologically dependent upon capitalist America than is Guinea-Bissau. Because of the type of socialist society the liberation struggle is creating in Guinea-Bissau, we will see science and technology evolving *from* and *for* the masses. We will see less reliance upon the Western form of industrialization: products-as-trinkets and pacifiers for bourgeois consumption.

But inside the U.S., inside black America, the dominance of capitalism's technology and racist culture deforms most attempts of our scientists and technicians to create in a human-oriented way. Blood plasma, like Chinese gunpowder, initially non-military, became an essential piece of war equipment so that modern racist-capitalist wars could be fought more efficiently and victoriously (for the capitalists and "manifest destiny"). Technological industrialization in black America has oppressed and exploited black folk in at least three ways outside of the economic super-exploitation:
1. producing goods for the white multinationals and military to be used against ourselves and our Third World brothers and sisters;
2. producing goods for pacification and bourgeois consumption;
3. producing goods that are outdated and/or irrelevant.

Moreover, these types of production only force us to become even more dependent upon the whims of dominant white U.S. Like the countries of Africa, Asia and Latin/Caribbean America, we must be able to judge how desirable a product is by the number of linkages to other production activities within the economy: "Where this is not accomplished and productive technology exists as an enclave, local production is not effectively stimulated and the diffusion of whatever technical advantages may be contained is operationally sterilized."[15] This means to black America (and to Third World nations) growth without development—dependency and a more efficient imperialist machinery.

Let us look at this in more detail. If we are to ever see Third

World people break away from the imperialist stranglehold and advance to a complete liberation, we will all have to confront Western science and its promised technological miracles just as our ancestors should have confronted the Christian missionaries with their talk of miracles and heavenly salvation. This confrontation will have to be face to face, undertaken with the same resolve and firmness of our Vietnamese brothers and sisters when they confronted the naked brutalities of the U.S.'s technological god.

We will have to remember that capitalist technology and its accompanying technical advisers and experts enjoy a lot of publicity and are always up for sale. This technology was designed to serve an individualistic competitive consumer society in the metropolis and imperialist domination elsewhere. Throughout history, the way in which Third World people have come into contact with Western technology has always been through war and capitalist expansion undertaken for profit by a few large interests and guided by the de-emphasis of the human element in science and technology.

Science and technology are not neutral or value free—objects hanging out in space for "natives" to take hold of piece by piece once they are civilized enough to use them. Technology is nothing more than a part of the capitalist package of wares that the Western world uses to exploit and control the Third World. It was manufactured in the heads and institutions of the European bourgeoisie to serve its greed and its requirements for more efficient exploitation of the working class. Technology cannot be separated from the material base where things are produced through labor; what Marx calls a "mode of production" determines which style of science will be favored, which technological innovation will be judged useful. Under capitalism the "valuable" innovations are those which bring profits to the private owners of the means of production, that is to say to the capitalists. When the capitalists looked abroad and built empires, they again decided which technical tools—if any—would be "useful" to have their peanuts raised, their cotton picked or their bananas carried into the big white boats. Imperialists have no special mystical belief in technology, however, when their economic interest is at stake. They may even decide to kill existing indigenous technologies if that is good for profits; thus, just as the looms of Africa were forced into idleness, so were India's as Britain developed mechanically-produced cloth in its ruthless, child-labored sweatshops in Manchester. So, today, when the white man turns around and appears to be generously offering his investments, technical gadgets and know-how as assistance, as the cure-all for what he calls underdevelopment, we had better assume that he has something good (profit) for himself in mind.

What we will accept of it all, what can be salvaged of the technical "successes" of that decadent system, must be carefully chosen and reshaped just as the Cuban revolution salvaged and reshaped the military camps of Batista's buildings.

At this moment the Third World masses, in their struggle for socialist progress, are struggling for themselves, by themselves. We know that the white man's "progress," which drops shiny tin cans on the moon, will not give us food. Less obvious but more important to realize, is the so-called Green Revolution and its transnational agribusiness management will not feed us either.[16]

Screening Any People's Control

Inventions are not made in a vacuum or from the vacuum; "new" ideas do not fall from the sky; they are part of a world of attitudes, ways of thinking and previous ideas about reality; they all merge together into what is called an "ideology." The calculus of probability was invented to help a European prince win at gambling! Today, in the citadel of world imperialism the choices of technologies to develop are still made by the princes of science, who sit on the President's Science Advisory Committee and outline the virtues of fragmentation bombs and lethal gas. The reigning ideology allows bourgeois scientists to feel at ease in such a role and to identify their work on weapons with the pursuit of science itself; thus the distinguished chemist Louis Fieser, on whose textbook many chemists were brought up and who happens to be the inventor of napalm, wrote a book which describes the development of this particular weapon under the distant, depersonalized and universal title, *The Scientific Method*.[17]

In contrast, our world is the one in which, for example, the struggle of Frelimo (the Front for the Liberation of Mozambique) has taken place, in which liberation has been achieved on the Mozambicans' own terms. As Samora Machel stated to the liberation fighters: "Education must give us a *Mozambican* personality which, without subservience of any kind and steeped in our own realities, will be able, in *contact with the outside world*, to *assimilate critically* the ideas and experiences of other peoples, also passing on to them the fruits of our thought and practice."

So there is an absolute need to discriminate, to choose, to exert what we shall call an over-all "cultural screening." By cultural screening we mean the right to choose a style of technological creation that emanates from and is useful to the people, a technology emphasizing the satisfaction of public or mass needs as opposed to individual consumption (that is, private cars, changing styles of clothing, gadgetry, and so on). Screening, however, cannot just be choosing what to let in. More important is how the use of new techniques is kept under the control of the working (peasant and proletarian) population. Furthermore, that control, that taking of matters of choices among technologies and styles of development into the people's hands, is only meaningful if the people themselves are capable of exercising judgements and making decisions, if the adaptation, the introduction of technology holding of trials and instituting of a people's justice. A people's technology is only meaningful if the people at large are involved in discussing it and shaping what it will be. This is being successfully accomplished, as one can readily observe in China, North Vietnam and North Korea.

Third World countries need to be aware that even in the scientific and technological areas there needs to be a close scrutiny of the politico-cultural effects of those technologies which were originally developed in a capitalist framework. Development will not come with a single transfer of U.S. Natural Science Department and Schools of Engineering or Architectural Institutes. Nor can one transfer the practice and concepts of U.S. medical "treatment" on to an underdeveloped country without separating the wheat from the chaff: the humanitarian aspects from the anti-humanitarian capitalist aspects; the part which is directed at helping the sick and the part which uses the sick as research fodder.

For these reasons we maintain that a continuing control over technology's orientation on the part of the peasants and proletariat is absolutely necessary. Otherwise, if technology, under the excuse of being "neutral," is accepted at face value, an elite of technocrats will play a vital exploitative role in determining the socio-political atmosphere. Such an elite will always favor a capitalist-oriented mode of development, which naturally is in its class interest. The struggle to politicize the work of scientists and technicians and the struggle to create a people-controlled science are part of the over-all class struggle, even during the phase of socialist construction.

Struggling Out of Dependency and Underdevelopment

According to the ideology of Western civilization, the way of presenting the current situation in the Third World—as pointed out above—begins by ignoring history. Thus the world appears as a god-given set of nations with some that happen to be more developed than others. There are no causes for this situation. It is never presented as part of historical process, and therefore all solutions offered for development are based on mimickery of the "advanced" countries. "Western thought" takes for granted that capitalism will always exist and that Third World countries will always be behind although they can get a little closer through some fatherly development program. The over-all vision is that since the Industrial Revolution was good for Britain, imitative technical development will be good for the Third World and will "close the gap" between rich and poor. The necessity of technological manufacturing becomes an overwhelming belief, a development status symbol. But one ends up with production of luxury goods for the consumer in the developed world. This orientation is the logical continuation of the historical situation from which today's Third World results. The basic dialectical confrontation remains the same: to have colonized people, one needs to have oppressors. The officially fashionable question, "how can developed nations help underdeveloped ones?" must be recognized as a false issue until it is put into historical perspective by first asking Brother Walter Rodney's question: How [did] Europe underdevelop Africa? In "benevolent" development plans, the Third World, in fact, has only provided cheap labor and in many cases natural resources, but has been led to believe that it did develop industrially because it has some modern (or not so modern) plants to show. Even if some of those plants produce goods of use to the masses of the country itself (like shoes,) the machines remain foreign-built and therefore dependency continues via spare parts and patent rights, and any "modernization" plan would also have to come from abroad. Taking seriously the need to build an on-going people-controlled technology leads us to face the fact that it is the people themselves who must do the development. Just a few foreign-trained specialists will not do. To keep matters in their own hands the people have to rely upon their own forces. This policy has been carried out by Cuba, confronted by the blockade imposed by the United States. It has been adopted by the peasants and workers of China under the name of self-reliance and by the Korean people under the name *Juche* or self-sustenance. Kim Il Sung declared in 1967:

Only when Juche is established firmly in scientific research work, is it possible to bring the initiative and talents of scientists into full play, to accelerate the advancement in science and technology and develop our economy faster by relying upon the resources of our country and our own techniques. Scientists and technicians should concentrate their efforts on the research work designed to promote industrial production with domestic raw materials, tap those raw materials which are in our country and produce substitutes for the raw materials we do not have, and expedite the technical revolution in conformity with our actual conditions so as to free the working people from arduous labour as soon as possible.

Self reliance encourages the development of indigenous techniques and aims at furthering the creativity of indigenous technicians.

On a national scale, self-reliance implies establishing an inventory of a country's own resources. This necessary inventory was never carried out under colonial rule because the colonizer was usually interested in sapping the country of only those natural resources vital to the metropolitan country's industrialization or providing quick cash returns. Today the neo-colonizer—the native bourgeoisie—follows the same dictate: what's good for the metropolitan society is "good" for the neo-colony. For instance, the Ivory Coast and Nigeria were formerly known for their palm oil and their cocoa; today their neo-colonial bourgeoisies are satisfied with cashing in on tourism and crude oil respectively. Meanwhile, Jamaica and Guyana remain the paradise of U.S.-Canadian aluminum companies. In fact, all countries in the Third World, except for the socialist-oriented ones, follow this degenerate pattern.

Having such an inventory allows the peasants and proletariat to use what is nationally available to produce energy, food and building materials. Once the existence of these basic natural assets is publicized there is stimulation for co-ordinated and planned action by the people at the local level. There emerges a national reality within which intiatives for development can be taken. However, means must be found to maintain popular political control over these riches. What is creatively done must evolve through peasant and urban worker leadership within the political party; otherwise there will be no guarantee against local or state level capitalist sabotage or bourgeois sell-outs.

What are some of the advantages resulting from a mass-based

self-reliant way of using technology? First, one should not forget that self-reliance may initially impose great demands upon everyone. Witness those participating in land-reclamation campaigns in Algeria, Cuba or Somalia, being transported—standing shoulder to shoulder—in trucks in the choking clouds of dust from unpaved highways. But the people involved were and are bound together by their common work experience preparing them to successfully confront future struggles demanding their technical innovativeness. Again Samora Machel:

> Releasing the masses' sense of creative initiative is an essential precondition for our victory and one of the chief purposes of our struggle. If the masses are to exercise the power so dearly won, they must display initiative. Colonial oppression, tradition, ignorance and superstitution create a sense of passivity in man which stifles initiative.

> To create a sense of initiative is also to create a sense of responsibility and to make (the masses) feel directly concerned by everything related to the revolution, to our life.

Scientific and technological development and the preparation of scientists and technicians must be governed by the needs of the over-all political economy directed by the socialist ideology of self-reliance. Initially such technical development has to rely upon indigenous techniques. In turn these will be freed from the stagnation imposed by colonialism and from the contempt emanating from the neo-colonial bourgeoisie, drunken on mimicking the West.

Indigenous techniques which have evolved over many generations contain within themselves the answers to many specific local demands and problems. The looseness of the weave of some African cloth, for example, is directly related to clothing needs in tropical areas. The treatment of the fibre has evolved into a process which aids the resistance to mildew. Besides, such unpatented technical realization in indigenous production is also an art involving the participation of various people in cooperation at the village level, from weavers to dyers. This is to a large extent a communal process of production which gives the people a sense of control.

In sum, technological oppression and dependence blocks us from developing (a) revolutionary black scientists, (b) scientific and technological alternatives, and (c) scientific education among the masses of black folk. Given this reality, then, what is a "revolutionary" black scientist, what are scientific and technological alternatives, and what is scientific education among the masses? Ultimately we have to struggle with the function of science and technology in the development of a twenty-first century revolution. Hence what follows is what Brother Amiri Baraka calls "answers in progress."

Black American Scientists for the People

The majority of the few black scientists and technicians that we have are systematically and pathetically caught up in the "special nigger" syndrome. Most of us see ourselves as scientists who happen to be black: the elite and thoroughly enlightened class. Individualism and professionalism are the norm. Not only the number of degrees are important, but where and in what field are vital questions. In short, the typical black scientist or technician is staunchly middle American. We loyally believe in the myth of science as apolitical, pure and "objective," as something above human passion. We try to believe in this myth with such emotional force that we have become the most alienated of blacks from black America. And yet with all of our conscious and unconscious rejection we are still the missing link, the ex-slave, the nigger.

This elite group, which constitutes much less than 10 percent of the 624,400 black professionals and managerials, is for the most part irreversibly fixed in its plastic and crass bourgeois attitudes. The question then is, "how do we struggle to develop a cadre of brothers and sisters who see the need to utilize modern science and technology as weapons and constructive tools?"

Above all else, if we are about to struggle to be serious scientists, then we must recognize the intellectual imperialism of the United States' ideology of an industrialized, bureaucratized and technocratic science: a social process which severs feeling from thought and which, in turn, splinters scientists into tunnelvision specialists who cannot and must not see the whole. This is an alienation from people who are seen as contemptuous objects of experiments. And the people respond with feelings of inferiority and dependency...or respond with revolution. But for us:

> If we do know that there exists a science which is imperialist in its uses, its organization, its method and its ideology, there must exist, and in fact there does exist an anti-imperialist science. It is still in its infancy, and it takes

different forms, according to the conditions it is found in. In colonial countries, dependent countries, or imperialist countries, it begins by exploitation; we denounce the use of science's name in the new pseudo-scientific racism; we denounce the conversion of science into a commodity and of our universities into corporate offices. From denunciation we move to active criticism: we look for means to put our scientific knowledge at the service of the people, and therefore as an instrument of revolutionary national liberation movements.[18]

Over the past four years many brothers and sisters within the National Black Science Students Organization (NBSSO) have been trying to resolve this question by organizing, recruiting and politically educating black students. It is from the hard work of these brothers and sisters that we begin to see the fruition of ideas and actions outlined below.

Blacks in the United States have reached that critical point where they are the leading and most revolutionary force within the U.S. proletariat. The development of history is such that we are within the major industrial metropolitan centers. But, at the same time, industry is becoming more and more dependent on scientific and technical innovations while we are becoming a less skilled, a less scientific people. Our method for developing a revolutionary ideology and our relationship as scientists to the modes of production have been anti-scientific. To resolve this problem , the primary intention for the formulation of a Union of Black Scientists and Technicians (UBST) should be to struggle with our fellow scientists and technicians (from semi-skilled workers to engineers) around developing a scientific methodology, and hence a cadre of revolutionary scientists and technicians:

> We challenge the system of training which tries to continue producing obedient experts. We are beginning to develop a new science on behalf of the whole of technology and society—an integrated science which refuses narrow specialization and idiot realism. We repudiate hierarchical-classist structures in order to search for forms of collective work and more democratic forms in research as well as in training. We repudiate the mystification of a science reinforced by a specialized vocabulary and we launch a campaign to popularize science. As scientists and revolutionaries we unite with anti-imperialist

scientists of the world and with popular movements of our countries.[19]

Let us look at one proposed model of UBST. It proposes the following functions of the Union of Black Scientists and Technicians:

(1) First and foremost is the dire need to support the NBSSO's effort to recruit more blacks into the sciences. Both organizations could help in politically educating our science students.

(2) The Union should present an anti-racist, anti-capitalist stance to both the sciences and world developments.

(3) UBST should formally work with the Scientists and Engineers for Social and Political Action (SESPA)—a predominantly white radical science collective—on major projects of crisis proportion, for example the fuel crisis.

(4) The Union should focus on about three or four major projects which have international implications. For example, the research and development of alternative energy sources for Africa and the Caribbean could be undertaken by brothers and sisters within the Union. We have only scratched the surface on how to cheaply and efficiently tap solar energy, so plentifully available in Third World countries.

(5) UBST must be in the forefront of science curriculum development from kindergarten through to the universities. One of the crucial problems facing black America is that of creating a tradition of dedicated, politically aware and numerous scientists and technicians. We have to break down psychological and skill-deficient barriers by re-analyzing the whole of science and technology and the way it has been taught in the United States.

(6) The Union should give technical assistance to those community organizations that are fighting against the mind control and spying onslaught of cable television. This is a crucial issue that has stunted political and educational development in black America because we are technically unprepared to struggle on the level of the enemy.* The cable television situation is extremely serious. The government today has the technical know-how and the political necessity (creeping fascism to protect imperialism) to make the following scenario a reality:

---

* The enemy is: The Federal Communications Commission, the American Association for the Advancement of Science, Dr. Amitai Etzioni, Director of the Center for Policy Research, New York City Office of Telecommunication Policy, and so on.

The government, in a study of 1,000 known community organizers has developed a profile of their viewing habits over an extended period of time. Computers then scan the output from tens of thousands of cable TV sets over a period of time. Individuals whose viewing profiles are congruent or largely overlap with those of the activists are then selected out for "further study." In this way, "troublemakers" could be identified even before they ever began to make trouble.[20]

(7) Alternative research into the genocidal functions of the birth control system is another vital struggle for the Union. For it has been scientifically shown that the so-called population explosion is a myth and fear of the capitalist. We must relentlessly struggle with our brothers and sisters to understand the myth of the population explosion as clearly as Brother Eduardo Galeano:

The United States is more concerned than any other country with spreading and imposing family planning in the farthest outposts. Not only the government, but the Rockefeller and Ford foundations as well, have nightmares about millions of children advancing like locusts over the horizon from the Third World...Its aim is to justify the very unequal income distribution between countries and social classes, to convince the poor that poverty is the result of the children they don't avoid having, and to dam the rebellious advance of the masses.[21]

(8) Organized black scientists and technicians will be compelled to expose to the masses of black folk the neo-eugenic movement for what it is: a racist pseudo-science. The Union must point out the class (bourgeois) basis of the neo-eugenic thrust as well as the gross misuse of science. Most of the work on exposing the neo-eugenicists has been done by a handful of white radical scientists and activists, but this fundamentally technical work has not been translated into a non-technical language and disseminated throughout the black and Third World communities.

(9) The Union should struggle for a people's implementation of computer-related material. It is clear that black America is one of the sectors of the American society most alienated from computers; yet computers function as necessary mechanisms for oppressing and exploiting us. On the other hand, it is obvious and necessary to see that computers can play an important role in our struggle against racism and imperialism.

The Union of Black Scientists and Technicians, then, is not a union of black folk reacting to the fact that the white scientific groups do not want to integrate at the level of policy and power. Rather, it is an organization of people who assume and know that within the complex machinations of racism and imperialism there are supportive white racist science organizations dedicated to the perpetuation of global imperialism...dedicated to the perpetuation of black oppression, exploitation and genocide. The union is also an organization that must subordinate itself to the necessary Black United Front and ultimately to the revolutionary vanguard multinational party that will guide and develop the proletarian liberation struggle within the United States. We must recognize that the UBST is fundamentally petty bourgeois in its composition and the task—monumental as it may be—is to convert our sisters and brothers to the side of the working-class struggle to overthrow racism and capitalism and struggle to create a socialist U.S. We cannot see ourselves as leaders, for that only leads us down the road of technocratic fascism.

But, meanwhile, what does an individual black scientific worker do? As one of this country's peculiar beings (black and a scientist or technician), we suffer under a barrage of jive propaganda from government, industry and university sources about the fame, money and privileges we can have because of our *peculiarities*. We have to recognize these "niceties" and "privileges" as the carrot of the racist-capitalist. The process by which we transform ourselves into black, conscious and revolutionary beings is an extremely painful and demanding process. Thus, the suggestions that follow are few and broad—but they are things that can be done by every one of us who wants to dedicate his or her life and mind to the liberation of black America through the liberation of America.

Read and discuss among your co-workers, family and friends (1) *How Europe Underdeveloped Africa*, by Walter Rodney, (2) *Science in History*, by J.D. Bernal, (3)*Science for the People* magazine, (4) *Labor and Monopoly Capital*, by Harry Braverman, (5) *China: Science Walks on Two Legs* by a group from Science for the People.

Set up small "Science for Blacks" study groups that would discuss current issues and the development and politics of science. From the small study group construct a realizable community aid project that would necessitate the use of the group's skills.

Encourage more sisterly and brotherly relationships between the worker and the "professional" in a technical area. After all, to the owners and other whites in the industry, we are all first and foremost niggers.

Each of us should encourage at least two sisters or brothers to pursue the sciences for black liberation. Help them find a college and stay in school.

Help construct the Union of Black Scientists and Technicians by supporting every stage of its creation through constructive criticism and work.

In 1968 Black Panther brothers and sisters shouted "the spirit of the people is greater than the man's technology." They were not being anti-science. Those brothers and sisters—in the midst of a black tidal wave—were saying that out of the spirit and will of our people to struggle for liberation there will come a science and technology for and from the people—in spite of the odds against us.

FOOTNOTES

1. Ciencia para el pueblo (Science for the People Group, Mexico), "Towards an Anti-Imperialist Science," in *Science for the People*, 5, September/October 1973, p. 18.

2. H. Braverman, *Labor and Monopoly Capital*, New York, Monthy Review Press, 1974, p. 167.

3. J. D. Bernal, *Science in History*, Vol. I, Cambridge, Massachusetts, MIT Press, 1974, p. 3., also London, C. W. Watts & Co., Ltd., 1954.

4. *Ibid.*, p. 3.

5. *Ibid.*, p. 31.

6. See Y. Agbeyegbe and A. Habtu, *Africa Before the White Man*, New York, Queens College—CCNY, 1971, pp. 4, 9.

7. A. Toynbee, *A Study of History*, Oxford, The Clarendon Press, 1939.

8. See W. Rodney, *How Europe Underdeveloped Africa*, London, Bogle—L'Ouverture, 1972, and his "Problems of Third World Development," *Ufahamu*, 3, UCLA African Studies Center, August, 1972, pp. 27-47.

9. *Ibid.*, pp. 104-105.

10. One needs to read B. Hessen, *The Social and Economic Roots of Newton's Principia*, New York, Howard Furtig, 1971, for a more thorough Marxist analysis/example of the dependence of scientific thought upon the political economy of the state.

11. Ciencia para el pueblo, *op. cit.*, p. 18.

12. Bernal, *op. cit.*, p. 11.

13. R. Girling, "Dependence, Technology, and Development," in F. Bonilla

and R. Girling, eds, *Structures of Dependency*, Palo Alto, California, Institute of Political Studies, Stanford University, 1973, pp. 46-62.

14. Ciencia para el pueblo, *op. cit.*, p. 18.

15. *Ibid.*, p. 50.

16. See *Science for the People* magazine for evidence that the "Green Revolution" feeds U.S. multinational corporations and aids in the genocidal dependency syndrome throughout the Third World. The cumulative index for the magazine will list particular articles.

17. L. Fieser, *The Scientific Method*, New York, Reinhold, 1964.

18. Ciencia para el pueblo, *op. cit.*, p. 19.

19. *Ibid.*

20. R. R., "Cable TV," in *Science for the People*, Vol. VI, January/February, 1974, p. 45.

21. E. Galeano, *Open Veins of Latin America*, New York, Monthy Review Press, 1973, p. 16.

Mr. Babinet is warned by his porter of the comet's visit. —Honore Daumier

Feminism and Science
—Rita Arditti

This article examines what it has been and continues to be like for women to be scientists in the masculinized world of science. It explores the powerful and positive potential of feminism in developing a truly humane science of the future.

"Women drink water while men drink wine"

In *A Room of One's Own*, published in 1929, Virginia Woolf describes a visit to Oxbridge, an imaginary center of learning in England. She is walking through a grass plot when a man stops her: only the Fellows and Scholars are allowed to walk in the turf path; she should walk in the gravel. She steps into the library and instantly a kindly gentleman informs her that "ladies are only admitted to the library if accompanied by a Fellow of the college." Eating in Oxbridge she notices that the male scholars are served delicious foods and their wine glasses are quickly refilled. At the women's dining hall the food is plain and there is only water in the glasses. Why did men drink wine and women water? Why was one sex so prosperous and the other so poor? She reflects that if there had been more support for women, maybe "we could have been sitting at our ease tonight and the subject of our talk might have been archaelogy, botany, anthropology, physics, the nature of the atom, mathematics, astronomy, relativity, geography."

350

All through history while male scholars conversed, debated and contributed to the disciplines of their choice, women provided the services that made that work possible. A non-sexist history of science (still to be written) would show the extent to which our culture has been built on the bodies and labor of women.

***

From the earliest times women have been denied access to formal education. The training of women in ancient Greece was aimed at producing housekeepers, mothers and mistresses—at the very time that Greek philosophers (called sophists, i.e., "wise men") were trying to answer fundamental questions about the nature of the universe, the meaning of life, etc. Greek "democracy" guaranteed equal rights for its male citizens, but women, foreigners and slaves did not have political rights. Praised for their ignorance, relegated to their own quarters, encouraged to be silent when in the presence of men, women were denied opportunities to pursue formal learning.

In accordance with their social reality the Greek philosophers developed a view of the world which was man-centered and dualistic. Pythagoras expressed this view succinctly: "There is a good principle that has created order, light and man, and a bad principle which has created chaos, darkness and women."[1] Despite this prevailing view, women were important members of the Pythagorean communities where early mathematics was developed. In fact, after his death, his wife, Theano, and his daughters continued his teachings at the central school of the order, referred to as a "brotherhood" by most science historians.[2]

Aspasia, the most learned woman in the ancient world (whom Socrates called "my teacher" and who is credited with writing the best speeches of Pericles) fought against the sexual arrangements of her times. She criticized the institution of marriage as it existed in Athens and tried to educate Athenian women and men about the need for equality of the sexes. But the schools of learning, like Plato's Academy, were bastions for selected males from privileged families who engaged in discussions of mathematics, astronomy and philosophy and who were totally convinced of the intellectual superiority of their sex. One of their beliefs was that the penalty for a man who lived badly was to be reborn as a woman in the second generation.[3] Love between males, love "between equals," was considered the most perfect form of love, and male homosexuality was a widespread and accepted practice.

Even midwifery was forbidden to women as a profession, and Agnodice, a famous Greek midwife, was forced to disguise herself as a man in order to practice.[4] Aristotle took the next step by incorporating the social mores of his time into "scientific theories"; he asserted that women had fewer teeth than men and explained procreation as the creative action of the male seed. There was no such thing as female seed; male seed was the active principle, and menstrual blood was the material provided by the female for the growth of the new being. Matter was primitive, undifferentiated (female), and form, an attempt at perfection, was imposed on it by the mind (male). "The male was the carpenter, the female the timber."[5] The ideas of Aristotle had a great influence on later day scientists, and to this very day this hierarchical and dualistic thinking plagues many minds.

Roman women received some rudimentary education. To show an interest in literary matters was acceptable, although there are no records of writings by Roman women. A few women physicians practiced, mostly on women.[6] Hypatia of Alexandria (370-415) escaped the socialization imposed on the women of her time thanks to her father, a mathematician named Theon, who had decided to produce a "perfect being" and had trained her in the arts, literature, science and philosophy. A mathematician and a philosopher, she was also practiced in the physical arts of swimming, rowing, horseback riding and mountain-climbing. She refused offers to marry answering that she was wedded to the truth. She invented an astrolabe and a planesphere, an apparatus for distilling water, another for determining the specific gravity of liquids and one for determining the level of water. She wrote three books on mathematical problems which were respected by her contemporaries. Her teachings, however, were considered heretical to prevailing Christian beliefs, and she was accused of being a promoter of pagan thinking. She died at the hands of a mob of religious fanatics. Because of the unusual circumstances of her death, she is sometimes mentioned in histories of science, but little allusion is made to her work. She appears as somewhat of a legendary figure, and the fact that she was a brilliant mathematician and teacher is relegated to the background.[7]

During the Middle Ages the Church monopolized centers of learning, and almost all intellectual activity took place in convents and monasteries. Formal science was at a low point: geometry, arithmetic and some astronomy were all that was taught. The work of the alchemists produced vast amounts of useful and practical information, but no major breakthroughs in science were made until the 17th century.

The convents were practically the only place where women could get an education. Some of the convents, run by strong-willed abbesses, provided a retreat for women of the upper classes which allowed a few women to explore their talents and creativity. Hildegarde of Bingen, abbess of a convent for more than 30 years, wrote extensively about nature. Much like the Greek philosophers, she asked general questions about the universe and discussed the cosmos, nature, "man," birth and death, the soul and God. Her insights and some of her more elaborate ideas establish her as a remarkable thinker. She recognized that the stars are of different sizes and of different brightness and made a comparison between the movement of the stars and the movement of blood in the veins—an idea that predated the discovery of the circulation of the blood. Other ideas also anticipated later discoveries. She put the sun at the center of the firmament and speculated about the seasons. She argued that if it is winter and cold in one part of the planet, then the other side of the earth should be warm. Her expertise in medical care and herbal medicine attracted the sick from all over Germany.[8] The suppression of the convents in England by Henry VIII signaled the end of organized efforts to educate women. Ironically, when the English convents were closed, their revenues and possessions went to endow Oxford and Cambridge, institutions devoted solely to the education of males.

In Italy, during the Middle Ages, a few women practiced "formal medicine." Women were not allowed to matriculate from medical faculties but could be authorized to practice by passing an examination and obtaining a license. Almost invariably they belonged to the family of a well known male physician. In fact, it was often the desire to keep a secret method of healing inside the family and the lack of a male heir that enabled the wife or daughter of an established physician to get permission to practice. Trotula of Salerno, one such woman physician, gained for herself an impressive reputation as a surgeon in the second half of the XI century; she introduced new methods of suturing and the use of the silk thread.[9] In medieval Europe, large numbers of women were healers. They did abortions, nursed the sick, cultivated healing herbs, delivered babies and travelled from village to village passing on their knowledge and their experience. The suppression of women healers accused of witchcraft and the emergence of a new male medical profession under the protection and patronage of the nobility is one of the bloodiest chapters in the history of women's oppression.[10]

During the Renaissance, the new science that followed from the work of Copernicus, Kepler and Galileo provided an alternative vision of the world and general impetus to question the established social order. Women's education became a lively topic of discussion. A literary controversy ensued called "La Querelle des Femmes" (the dispute about women), with philosophers taking sides on a debate regarding women's capabilities. A few universities opened its doors to women of the aristocracy.[11]

As new ideas in science were developed, strong women inspired and nurtured the male scientists who were creating the basis of science as we know it now. Sister Celeste, Galileo's oldest daughter, followed every detail of his work for eleven years. Her support allowed him to continue his work whenever despair came over him. Her death was a blow from which he never recovered. Kepler relied on his wife, Barbara, to help him keep up his strength and good spirits. Descartes acknowledged the inspiration and encouragement he received from Elizabeth of Bohemia by dedicating his work to her: "...in her alone were united those generally separated talents for metaphysics and for mathematics which are so characteristically operative in the Cartesian system."[12]

In France the Institute of Saint Cyr opened in 1686, the first state school for women. It was a disappointing enterprise. The superficiality of its curriculum quickly became clear: no word was mentioned about philosophy or the natural sciences. The school became known as a center that provided mistresses to the court of the king.

In the same year Newton's monumental work "Principia Mathematica" was published. Emile de Breteuil, marquise de Chatelet, translated Newton's work from Latin into French and added her own scholarly commentary of the work. The translation of Newton's work into French was a significant event because his work contributed to the climate of skepticism that preceded the French Revolution. De Breteuil was a scientist in her own right, and she devoted much of her work to investigating the nature of fire. She had constructed in her castle at Cirey a physical laboratory in which she performed experiments. She was assisted by Voltaire, with whom she shared 16 years of her life.[13]

In the Western scientific milieu a particular ideology and organization evolved following Francis Bacon's ideas. Nature (female) was the enemy, and science was the instrument for its control and domination—a way of recovering the lost dignity of "man."[14] Nature was to be conquered. A group of male scholars, devoted to scientific research and the pursuit of wisdom, would guide society. From this vision, his "House of Solomon," derived the European scientific societies of the 17th century—mainly the British Royal Society and the French Royal Academy. Membership was open to "men of all professions...students, soldiers, shopkeepers, farmers, courtiers and sailors, all mutually assisting each other."[15] Membership in the societies was considered proof of scientific ability.

Outstanding scientists like Maria Gaetana Agnesi and Sophie Germain were denied membership because of their sex. Agnesi's work "Analytical Institutions" had gained her a wide reputation and an appointment as an honorary lecturer of mathematics at the University of Bologna, but the secretary of the French Academy turned down her candidacy with a blunt, "La tradition ne veut pas d'academiciennes" (tradition does not want women academicians). Sophie Germain worked for years and corresponded with Gauss on mathematical topics without letting him know she was a female. She signed her work as "M. leBlanc" in order to "escape the ridicule attached to a woman devoted to science." Ironically, she won a prize from the French Academy because of her work on the vibrations of elastic planes but could never become an official member.[16]

In France in the 17th century a few women from the aristocracy showed an interest in astronomy. They became the target of ridicule in Moliere's play *Les femmes savants* (The learned ladies):

> Get rid of this fierce-looking telescope and all the rest of these gadgets...stop trying to find out what's happening on the moon and mind what's going on in your own house where everything is upside down. It's not decent and there are plenty of reasons why it isn't, for a woman to study and know so much...Women today want to write books and become authors. No learning is too deep for them... and here, in my house, they know everything except what they need to know. In my house, they all know about the moon and the pole star and about Venus, Saturn and Mars, which are no concern of mine...and nobody knows how the pot is cooking..."[17]

In the United States, public education started in Boston in 1642, but females were not accepted into schools until 1789 and then only for half of the year, with teaching restricted to spelling, reading and composition. In private schools, women of the upper classes would also learn French, music and embroidery. Thanks to Emma Willard's persistence and enthusiasm, the Troy Female Seminary opened in 1821, the first endowed institution for the education of women. Among its daring innovations was the teaching of physiology: "Mothers visiting a class at the Seminary in the early thirties were so shocked at the sight of a pupil drawing a heart, arteries and veins on a blackboard to explain the circulation of the blood, that they left the room in shame and dismay. To preserve the modesty of the girls and spare them too frequent agitation, heavy paper was pasted over the pages in their textbooks which depicted the human body."[18] The demand for free education in America was largely based on the rationale that every voter needed to be responsible and intelligent, capable of assessing and discussing information. But since women could not vote, the argument could not be extended to the education of females. Only after the Civil War were free high schools for young women created in Boston and Philadelphia.

The first college to open its doors to women was Oberlin in 1833, but the notion of the inferiority of the female mind prevailed. The Oberlin women students were prepared to be "the mothers of the race" and taught to stay in a well-defined "women's sphere" of activities.

The Declaration of Principles passed at the feminist convention in Seneca Falls in 1848 called attention to the absence of women in the fields of medicine, law, and theology. At the Massachusetts Institute of Technology, Ellen Swallow, the first woman student, received her Bachelor of Science degree in 1873 but was not allowed to enroll in the doctoral program. She was a controversial figure in a conservative milieu. Nevertheless, her work in sanitary chemistry was important to the incipient field of environmental science. Her stressing of the importance of the environment in human development came at a time when heredity was the usual explanation for every major social problem. People were poor or criminal because they were "born that way." She believed that people could be taught to think critically and live in tune with the environment. "We must show that science has a very close relation to everyday life...train (women) to judge for themselves...to think...to reason...from the facts to the unknown results."[19] She arranged for a survey of the health of college educated women in order to refute the myth that education was harmful to women. Her study was published by the Common-wealth.

In 1903 Marie Curie received the Nobel prize for her work on radioactivity together with Henri Becquerel and Pierre Curie. In 1904 she started receiving a salary for the first time in her career. Marie Curie is practically the only woman scientist given the worldwide recognition traditionally accorded to male scientists who are considered competent. Her devotion to scientific research, her inability for small talk and her seriousness made her quite unpopular among many of her colleagues. Her talent elicited mixed feelings. When in 1910 she published her "Treatise on Radio-activity," Rutherford reviewed it favorably in *Nature* magazine, but in a private letter to a friend he expressed his true feelings: "Altogether I feel that the poor woman has laboured tremendously, and her volumes will be very useful for a year or two to save the researcher from hunting up his own literature, a saving which I think is not altogether advantageous."[20] At the height of her career she presented her candidacy for membership in the French Academy of Science. According to the customs of the time she went from house to house and from laboratory to laboratory visiting the members of the Academy and asking for their support. But the Academy was not moved, she was a woman and as such was not eligible. A few days before her name appeared before the Academy as a candidate, its members reaffirmed in plenary session "the immutable tradition against the election of women."[21]

In the United States, women were not admitted to graduate schools until the 1880s. Even so, once admitted, many schools did not allow them to receive advanced degrees.[22] By the beginning of the 20th century the U.S. had become an industrial society. In spite of a generally conservative atmosphere, the needs of the economy for rapid technological development had a positive effect on the position of women. Women started to get out of the home. The number of women in science started to increase and in the 1920s the proportion of women scientists reached an all-time peak. But in the '30s and '40s the depression and WW II made the education of women a low priority. The proportion of women in science decreased and reached an all time low in the 1950s. WW II veterans re-entered school and the workplace, and women retreated to the home.

Current Times

Today, discrimination against women is still overt and socially acceptable within the academic community. Among 207,500 science and engineering PhD's in the U.S. labor force 93.1% are white and 92.1% are male. Only 0.8% are black, 0.6% Latin, 0.04% Native American while East Asians who make up 0.7% of the U.S. population comprise 5% of science and engineering PhD's.[23] Women scientists (8% of the total) have markedly lower salaries than men and are concentrated in certain fields: four-fifths are life scientists, psychologists, or social scientists. The unemployment rates for women in science are two to five times higher than for men in the same field with comparable training and experience. Although women comprise 5 to 8% of physical science researchers, the National Science Foundation gives them only 0.03% of 1% of its awards and grants.

Like the scientific societies of the 17th century, the Cosmos Club in Washington D.C. (an elite club for prestigious scientists) does not admit women scientists to its membership. Among 1,134 living members of the National Academy of Sciences, only 25 are women (of whom 6 were elected in 1975).[24]

Discrimination against females starts early. Young females in the seventh through twelfth grades "tend" to lose interest in mathematics. Later in college they feel they lack the necessary fundamental skills and avoid science courses whenever possible.

In order to become a scientist, both women and men have to learn to behave in a way that will be acceptable and recognizable to other people in the field. Years of informal role learning are necessary before we complete the socialization process that transforms us into "the scientist." During those years—and quite apart from the courses' content—cultural messages concerning the attitudes and expectations for women in science are transmitted. Basically, the message for women is "be a good little girl scientist." While it is accepted that women can efficiently perform technical and data-gathering functions, there is a general feeling that the truly original work, the work that "makes a difference," is produced by male scientists. That truly creative work is beyond women's capabilities. This debilitating thought has in effect prevented many women from ever fully exploring their abilities or believing in themselves. However, such a view of creativity ignores the realities of the organization of work in the scientific world. The image of the distracted and genial scientist, oblivious to practical details, devoting heart and soul to finding a solution to a research problem, is an image that bears little resemblance to what actually takes place. What is essential is a knowledge of how to operate within a certain framework and follow the often implicit rules. Acceptance and recognition from one's peers, a clear understanding and the use of appropriate networks of support, i.e., organizational abilities, the "know-how" of pursuing formal and informal contacts, the mentor system: these are some of the attributes necessary in order to work successfully as a scientist.

Another role expectation concerns competition. Competitiveness, racing to be "the first," is an openly acknowledged characteristic of the scientific world. But a woman who is competitive is received with hostility and mistrust.*

The position of women in research laboratories is suspiciously similar to the position we have in the nuclear family. The laboratories resemble a patriarchal household, with the "head" of the laboratory usually male, women in marginal positions without independent status, job security or benefits, and younger students playing child-like roles. A woman in a research laboratory is often expected to perform "mothering" functions, the supportive and nurturing functions that nobody else will take on. It is still a

---

*Two books in recent years give an excellent portrayal of competition and elite male scientific networks: *The Double Helix* by James Watson and *Rosalind Franklin and DNA* by Anne Sayre. Both books describe well how groups of scientists interact informally, communicating and making "gentlemen's agreements" (literally) about how their work should proceed. However, the books diverge radically in their view regarding Rosalind Franklin. For Watson, Franklin was an argumentative character who made life hard for people around her and who "assisted" Maurice Wilkins. "The best place for a feminist was in another person's lab" is the way Watson summarized his feelings towards Franklin. Sayre, who was a personal friend of Franklin, reports that Franklin was assigned to work on the DNA problem at a meeting in which Wilkins was not even present. She was not an appendage to Wilkins but a co-worker. Franklin was the first to establish the helical structure of DNA, but credit for this discovery has been consistently give to Wilkins. Franklin's basic work on X-ray diffraction patterns on the B chain of DNA was the key to understanding DNA structure. Shortly after Watson and Crick were raised to stardom for their discovery of the structure of DNA, Franklin left King's College and changed the area of her research. She was finally out of the way, and the male network gave her practically no recognition for her findings. She died in 1959, at age 37, leaving the scenario to her co-workers. In 1962, Wilkins, Watson, and Crick received the Nobel prize for their work. In their acceptance speeches Franklin's crucial contribution is lightly acknowledged among a host of other citations.

common attitude that a woman should consider herself fortunate to be assisting an eminent scientist, to be his "second" or his "shadow." Of course, like in a marriage, if the eminent scientist leaves, dies or changes his mind, the woman is left in a poor situation.

Even in the most sophisticated laboratories women are expected to accept the inferior and less prestigious tasks and to be proud to assist the "big" scientist in keeping his energy for superior and directive functions. The sexual dynamics are such that very few women manage to develop the skills and self-confidence necessary to survive in an extremely competitive environment. And very few are encouraged to do so. The scenario is set for the "failure" of the majority and acceptance of a "few" exceptions.

One mechanism that ensures the perpetuation of this exploitative situation is the alienation that many women who do research feel. This alienation originates from the contradiction between the work experience and what one has been told or taught to believe. Having become a scientist, having "made it" in a man's profession, one is supposed to have overcome the disadvantages of belonging to the female sex. On the other hand, the day-to-day experience of being in a subordinate position (even if one is doing scientific work) negates the previous optimistic viewpoint. As a result many women feel ambivalent about their own capabilities. Having been taught that when there is something wrong in a situation it is likely to be our fault, we keep reticent about our confusion. Consequently, we do not join forces to oppose the oppressive situation.

The Personal is Political

One of the central ideas of feminism, the *the personal is political*, allows us to see how the exploitation of women is perpetuated. By reflecting on our personal experiences, we begin to gain an understanding of how oppressive patterns are allowed to run our lives. Nothing clamorous need happen. Things go as usual. Our lives are filled with what is familiar. And the thread of oppression binds us without our even noticing it. We know very little about the experiences of women in the sciences. Only recently have women begun to speak up.*

In my own case it took many long years before I could see what the situation clearly was. I got attracted to biology in the fifties because it seemed a field that offered the promise of scientific rigor and at the same time relevant knowledge useful to my own life. As

*See "The anomaly of a woman in physics" and "How can a little girl like you teach a great big class of men" in *Working It Out*, Sara Ruddick and Pamela Daniels, ed., Pantheon, 1977.

a student at Rome University in Italy, I was struck with the news of the discovery of the DNA structure. The field looked exciting, and I decided to become a geneticist. Genetics related to people; its internal logic, its consistence and its elegance delighted me. In my early years I dismissed manifestations of hostility from some of my teachers and colleagues toward me as "personality" problems, and I immersed myself completely in the genetics of microorganisms. It never occurred to me that my sex had anything to do with some of the difficulties I encountered—this in spite of the fact that I was one of the few women doing that type of research and that I had picked up sexual innuendoes from some of my teachers. I cannot offer any reasonable explanation for my blindness.

As the years went by, however, I began to notice that women scientists had special problems, such as lack of advancement. They seemed to be stuck in the same position while men were moving ahead rapidly. The few women who were around did not show up at meetings or participate actively in them. Family obligations seemed to tear them apart.

Before long, I was one of them, struggling with my own marriage, child care arrangements, and the need to keep focused on my work. At the same time, I was receiving all kinds of mixed messages from husband and colleagues. I also began to notice the particular situation of women scientists married to men in the same field of research. All had secondary positions to their husbands regardless of ability, and their loyalties as wives had led them to accept a precarious work situation in which their research had become dependent on their marriages. They did the day-to-day work necessary to produce a solid piece of research, and the credit would invariably go to their husbands. Naively, I thought I was safe from exploitation because my husband (soon my ex-husband) worked in a different field. I did not know then that all men are potential "husbands." By this I mean that the pattern of accommodations and expectations that I had observed between women scientists married to men in the same field, could and would also arise with male colleagues. This led to confusion and disappointment.

In the late sixties in the U.S. I got involved in the movement against the war in Vietnam and began meeting with a group of scientists who were questioning the role of science in the war. Our analysis soon led us into discussions about the contradictions of trying to do humane science in an inhumane society. Scientific laboratories were supposedly dedicated to the discovery of new knowledge. Which kind of knowledge, we asked, and for what purposes? Who benefits from it? And what are the allegiances of the scientific community? These were exciting meetings, and I felt I was getting in touch with deep questions that would in some way or other alter the course of my life.

At about the same time I started to read and reflect about women. I had once attended a women's group meeting where questions had been raised regarding our position in society and statements had been made about our lives being "political." It took me almost a year before I was ready for another meeting like that, but in the interim I did a lot of thinking and, more importantly, I began to observe everyday situations in a new way. I realized how my professional training had often led me to disassociate myself from other women, especially if they were not professional women. I had grown to accept feelings of inadequacy and isolation as a normal part of life. Next, I sadly discovered that my questioning colleagues so ready to fight against the war did not have any understanding of how their own behavior was a declaration of war against women. They branded me as "difficult" and "oversensitive." "What do you mean women's issues when there is a war going on?" Comparisons were made about whose oppression was "worse," and women were always at the losing end. It finally dawned on me that you do not discuss how to overthrow oppression with your oppressors. I started to reach out to other women, and after an initial period, we found that indeed we had many common experiences and could validate each other's perceptions. We were not crazy. We all had received destructive messages, and we all had deep insecurities regarding our role as scientists. On the surface we had all been led to believe that we were "one of the boys" but we feared being found out: we "knew" we were inadequate. Would they know?

Exploring my feelings in a safe context allowed me to remember experiences that I had buried deep down and I saw them with a fresh eye. I understood why blood had flushed to my cheeks when the head of my department, years ago, running into me one Saturday morning in the cold room of the laboratory had exclaimed affectionately: "What are you doing here? You should be home with your child!" I remembered how tense I would get before making a public presentation about my work, and I realized why. Most of the time I would be judged as much for my appearance as for the quality of my work. I was told many times, jokingly, that with my new awareness I had to give up my female "privileges."

What this meant was that I was not supposed to ask for help if I needed it and that if I did I would be most likely ridiculed for being "weak." Carrying a 25-gallon flask of distilled water was not easy for me (I am a short person) but once my colleagues decided I was "liberated," I was on my own.

Talking with other women made the difference between sanity and insanity. I knew that my own life had to conform to my changing awareness and that the next task was to start looking for alternatives. The elitism of the research laboratories suffocated me. Concomitant with my personal growth was a new interest and delight in teaching. In my search for alternatives I read some of Rachel Carson's work. Her words, as she accepted the National Book Award for her book, *The Sea Around Us*, spoke directly to me:

> Many people have commented with surprise on the fact that a work of science should have a large popular sale. But this notion, that "science" is something that belongs in a separate compartment of its own, apart from every day life, is one that I should like to challenge. We live in a scientific age, yet we assume that knowledge of science is the prerogative of only a small number of human beings, isolated and priestlike in their laboratories. This is not true. That materials of science are the materials of life itself. Science is part of the reality of living; it is the what, the how and the why for everything in our experience.[25]

When I found out that Rachel Carson was looked upon with suspicion by the scientific community because she did not have a PhD and because she was so deeply concerned with educating the public, I laughed and I cried. Here was somebody who had done more than anybody else I was aware of to integrate science with public concerns, and she was mistrusted and put down! My ideas about my own future began to change. It became clear to me that my interest in doing research in genetics of microorganisms was minimal and that I enjoyed working with people. Teaching became a source of deep satisfaction and continuous learning. I decided to put my energies into trying to change some of the oppressive conditions that I and many others had encountered. Learning about the social and political implications of science, with a focus on women, became a priority. Learning about power relations is also learning about how to change them.

Learning about the past and present position of women in science and understanding my own experiences made me realize that feminism and science (my two main interests) did not mix easily, and that some of the fundamental ideas and insights from the women's movement run contrary to the way science is now. The potential of feminism in the development of a truly humane science became an issue to explore.

Feminism and Science

While lack of encouragement and blatant sexism have prevented women from fully participating in the sciences, the dehumanization of science has also played an important role in keeping women away. Women, generally more in touch with their feelings, often raise uncomfortable questions about "detached scientific objectivity." The prevalent mode in science today presents serious problems for people who have human concerns, as many women have. "Objectivity" applied to people often leads to objectifying them, or perceiving only their object aspects. As mentioned before, the scientific community is composed mainly of white males who have been socialized into the professional value system. Professionalism, an elitist concept, provides the means to control others and to maintain privilege over them. It divides economic and occupational groups into the thinkers at the top and the unthinking masses at the bottom.[26] It protects scientists from external evaluation or even egalitarian discussions with the people affected by their work and dependent upon their performance.

As scientists we are taught to approach problems with a purely cerebral attitude and *not* to bother with the consequences or ramifications of our work. We are taught to "keep things separate": scientific inquiry on the one hand and human concerns on the other. This way of working leaves little room for our development as human beings and opens the door to the creation of exploitative technologies. We stand powerless, producing knowledge that can be used against people. Nuclear weapons, chemical weapons, recent advances in the life sciences have all been developed by scientists who gave their energies to the narrow task before them without concern for the larger issues that would affect the community at large. It is clear that scientific inquiry without concern for the pressing social problems of our time will only create new ones. We do not live in a vacuum. Scientists will have to deal with new concepts (like being accountable) and seek closer contact with other parts of the community. More and more people are beginning to question the right of scientists to divest themselves of responsibility for the direction of their work and the use of scientific results.

Because of our experience as women in a patriarchal culture, we know, first hand, that a purely mechanistic approach does not add much to knowledge. Scientists have studied us as the reproductive system of the species, and we have been reduced to our reproductive organs, our secondary sexual characteristics and/or sexual behavior. "Scientific" rationalizations are offered for the secondary status of women, racial minorities, and poor people. Sociobiology, the study of sex differences, and anthropology contain our own cultural myths about women. Sexism is rampant in the hard and soft sciences.[27] For women to take the place of male scientists and not to advocate a humanistic and committed science would be a tragic mistake. A man-centered science serves a man-centered society—we have to question the process by which scientific work is accomplished as well as its product. We have to question the professionalism that keeps people separate.

Out of the experience of support groups in the women's movement some of us have learned that the conditions under which people are able to work creatively and joyfully are practically non-existent in the scientific milieu. We know now that in order to communicate clearly it is essential to feel that one is being listened to with attention and interest, that qualities that one may seem to lack can be developed, and that leadership skills can be learned, if there is an interest in sharing them. All this runs against the competitive patterns prevalent in the research environment.

A feminist perspective in science would involve the creation of an environment that maximizes the development of minds and bodies and encourages positive attitudes towards one's own biological identity. It would involve the conversion from an exploitative "value free" technology to a commitment to a humane technology: to preventive medicine, fair distribution of material goods and educational opportunities. The concept of self-help would be fully accepted and fostered by the scientific community. The whole area of reproductive research, contraception and sex differences would be revamped to eliminate sexist stereotypes. Females would no longer be considered the sole reproductive units of the species.

A feminist perspective would not necessarily hail new technological developments as "liberating" because it would realize that oppression is not the result of biological or natural conditions but of social constructs. Technology would be assessed for the impact it has in bringing meaningful change in social relations.

Feminism is also a special form of "knowing." By eliminating the division between intellect and emotion, scientists can perform intellectual tasks without becoming intellectual robots. We do not know how science would have evolved if women had been full participants in its development. There seems to be a connection between the specialization and reduction, the indifference to values and the masculinization of science.

Since science does not progress only by inductive analytical knowledge, the importance of imagination and emotion in the creative process should be obvious. The role of intuition in science is consistently undervalued in a science which is exploited for corporate, military and political reasons. A feminist perspective would re-introduce and re-legitimize the intuitive approach. The benefits of this in terms of new knowledge might well be incalculable.

Adrienne Rich notes that Virginia Woolf suggested that "women entering the professions must bring with them the education—unofficial, unpaid for, unvalued by society—of their female experience, if they are not to become subject to the dehumanizing forces of competition, money lust, the lure of personal fame and individual aggrandizement, and 'unreal loyalties'."[28] Rich adds: "In other words, we must choose what we will accept and what we will reject of institutions already structured and defined by patriarchal values...We need to consciously and critically select what is genuinely viable and what we can use from the masculine, intellectual tradition, as we possess ourselves of the knowledge, skills and perspectives that can refine our goal of self-determination with discipline and wisdom...In fact, it is in the realm of the apparently unimpeachable sciences that the greatest modifications and revaluations will undoubtedly occur. It may well be in this domain that has proved least hospitable or attractive to women—theoretical science—that the impact of feminism and of women-centered culture will have the most revolutionary impact."

Today, in science we know "more and more" about "less and less." Science as an instrument of wealth and power has become obsessèd with the discovery of facts and the development of technologies. The emphasis on the analytical method as the only way of knowing has led to a mechanistic view of Nature and human beings. We should remember that the concept of evolution, for instance, did not emerge from developments in the field of genetics or biochemistry but from "inspired guesses based on a sort of

Gestalt awareness of complex relationships in natural situations. The modern scientific techniques have served merely to verify the theory and to elaborate its details."[29]

The task that seems of primary importance—for women and men—is to convert science from what it is today, a social institution with a conservative function and a defensive stand, into a liberating and healthy activity. Science needs a soul which would show respect and love for its subjects of study and would stress harmony and communication with the rest of the universe. When science fulfills its potential and becomes a tool for human liberation, we will not have to worry about women "fitting" into it because we will probably be at the forefront of that "new" science.

FOOTNOTES

1. Quoted in Simone De Beauvoir, *The Second Sex*, translated and edited by H. M. Parshley, New York, Bantam Books, 1952, p. 74.

2. H. J. Mozans, *Woman in Science*, Cambridge, Massachusetts, MIT Press, 1974, p. 199.

3. Mary R. Beard, *Women as a Force in History*, New York, Collier Books, 1962, p. 321; Stephen F. Mason, *A History of the Sciences*, New York, Collier Books, 1962, p. 39.

4. Mozans, *op. cit.*, p. 268.

5. Mason, *op. cit.*, p. 45.

6. Mozans, *op. cit.*, p. 271.

7. Lynn M. Osen, *Women in Mathematics*, Cambridge, Massachusetts, MIT Press, 1974, p. 31.

8. Mozans, *op. cit.*, pp. 41-42, 46; also Charles Singer, "The Scientific Views and Visions of Saint Hildegard (1098-1180)," in *Studies in the History and Method of Science*, Oxford, Clarendon Press, 1917.

9. Mozans, *op. cit.*, p. 54; also L. Munster, "Women Doctors in Medieval Italy," CIBA Symposium, 10, Boston, Little Brown, 1962.

10. Barbara Ehrenreich and Deirdre English, *Witches, Midwives, and Nurses: A History of Women Healers*, Old Westbury, New York, The Feminist Press, 1976.

11. Julia O'Faolain and Lauro Martines, Eds., *Not in God's Image: Women in History from the Greeks to the Victorians*, Scranton, Pennsylvania, Harper Torchbooks, 1973, p. 181.

12. Mozans, *op. cit.*, p. 370.

13. Osen, *op. cit.*, pp. 49-69.

14. Joseph Haberer, *Politics and the Community of Science*, New York, Van Nostrand-Reinhold, 1969.

15. Rene Dubos, *The Dreams of Reason: Science and Utopias*, New York, Columbia University Press, 1961, p. 34.

16. Osen, *op. cit.*, p. 83.

17. Quoted in O'Faolain and Martines, *op. cit.*, p. 245.

18. Eleanor Flexner, *Century of Struggle*, Paterson, NJ, Atheneum, 1970, p. 26.

19. Robert Clarke, *Ellen Swallow: The Woman Who Founded Ecology*, Chicago, Follett Publishing Co., 1973.

20. Robert William Reid, *Marie Curie*, New York, Saturday Review Press, 1974.

21. Mozans, *op. cit.*, p. 230.

22. M. Elizabeth Tidball and Vera Kistiakowsky, "Baccalaureate Origins of American Scientists and Scholars," *Science*, 193, August 20, 1976, pp. 646-652.

23. Betty Vetter, "Women and Minority Scientists," editorial in *Science*, 189, September 5, 1975, p. 75.

24. Betty Vetter, "Women in the Natural Sciences," *SIGNS: Journal of Women in Culture and Society*, 1, 1976, pp. 713-720.

25. Paul Brooks, *The House of Life: Rachel Carson at Work*, New York, Fawcett-Crest, 1974, pp. 116-117.

26. Mary McKenney, "Class Attitudes and Professionalism," in *Quest*, 3, Spring 1977, pp. 48-58.

27. Evelyn Reed, *Sexism and Science*, New York, Pathfinder Press, 1978; R. Hubbard, M.S. Henifen, and B. Fried, *Women Look at Biology Looking at Women*, Cambridge, Schenkman, 1979.

28. Adrienne Rich, "Toward a Woman-centered University," in Florence Howe, *Women and the Power to Change*, New York, McGraw-Hill, 1975, p. 15.

29. Dubos, *op. cit.*, p. 116.

History of Science for the People: A Ten Year Perspective
—Kathy Greeley and Sue Tafler

This article tells the story of Science for the People, a ten-year-old organization which continues to raise critical questions and inspire actions around significant and controversial science-related issues in U.S. society. The editors of this book proudly point to this notable group which persists in calling for a science done to serve the interests of all people.

Several months ago, when we were first asked to write a brief history of Science for the People, we accepted without realizing just how difficult that task might be. SftP had done so many different things, it had involved so many different people with differing political opinions at different times and places, that we could have written a book. It was only after we began to organize the material we collected that we realized the history of SftP could not be understood as a simple linear chronology. Rather it is an intricate web of people, issues, politics and activities. What gave this history coherence was the conception of the role of science and technology in our society that has evolved within SftP over the past ten years.

The economic affluence of post World War II U.S. sparked a boom in science and technology, and it became clear that these elements would play an increasing role in our lives. The question was what kind of role would that be, who would have control over it, and who would it benefit. SftP was the only organization that began to develop a radical critique of science and technology. SftP came to understand that science, although grounded in objective, material realities, is nonetheless neither "neutral" nor "value-free." The way science is used—the kinds of questions asked, the kinds of research funded, the application of scientific theories—is determined by those controlling the pursestrings and values of each society. In our society, science serves the class interests, both materially and ideologically, of a small elite group that values profit over people and private property over human well-being. The elite mystique of wisdom and infallibility built up around science and scientists—and the widely held and deliberately fostered belief that only such experts can know what is best for society—effectively serves to obscure the class nature of the practice of science.

Many factors contributed to a general political awakening in the 1960s. It was the Vietnam War though, that really catalyzed the birth of SftP. In January 1969, a caucus of dissident physicists introduced an antiwar resolution at the American Physical Society convention. At the same time, a group of industry-based engineers began meeting in Boston. In March, 1969, scientists and students joined together to force M.I.T. to stop all war-related research on campus. That same year, a California group, Scientists for Social and Political Action (SSPA, later changed to SESPA with the E for Engineers), began publishing a national newsletter. This created a communication network among individuals, caucuses and study groups across the country. From this, a movement was born whose slogan became "Science for the People."

The early activities of this new organization naturally focused on the Vietnam War and weapons research. SESPA members participated in the November, 1969 antiwar march. SESPA members also marched and distributed leaflets against the Anti-Ballistic Missile (ABM). Berkeley SESPA circulated a pledge at various scientific meetings stating:

> I will not participate in war research on weapons production. I further pledge to counsel my students and urge my colleagues to do the same.

In 1972, the Berkeley chapter published the leaflet "Science Against the People," an expose of the Jason project, a symbiosis between the university and the military complex. Jason helped develop the automated battlefield and, in general, cultivated academics as advisors to the Department of Defense. SESPA also supported Karl Armstrong during his trial for the unintended fatality in the bombing of the Army Math Research Center (AMRC) in Madison, Wisconsin, and the Madison chapter published the booklet "The AMRC Papers" demanding the closing of the Center.

While the truths about the war sparked moral outrage among

many, some people began to see it as part of a growing pattern of U.S. imperialism. These people realized that the end of the war did not mean the end of U.S. involvement in Latin America, Africa, the Middle East, or even other Asian nations. Nor would it signal the end of the exploitation of blacks and other minorities, women or working people in this country. The people in SftP wanted to build a movement that would fight sexism, racism and elitism around the world, and specifically within the scientific community.

AAAS

In December, 1969, the American Association for the Advancement of Science (AAAS) invited some young scientists and graduate students to talk about their research at their annual conference in Boston. Instead, this Boston group decided to hold a symposium entitled "The Sorry State of Science." The title itself started a series of confrontations with the AAAS. Attendance at the AAAS brought together these symposium planners, SESPA, and others, vitalizing and enlarging the membership of SESPA in Boston.

The AAAS is the largest and most diverse association of people involved in science and science-related work. While the AAAS represents a bastion of establishment ideology that seeks technological solutions for social and political problems, it was a useful political arena and a place to meet other sympathic scientists. SftP has attended nearly every AAAS convention since 1969. Reorganization of the AAAS itself has been one immediate goal of SftP. Efforts to democratize the structure in order to eliminate the intimidating, elitist nature of sessions have ranged from meetings with AAAS officials to open the conference to the public, to rearranging chairs into circles to encourage more discussion. At the 1970 Chicago convention, resolutions were presented which opposed the use of scientific work for political repression, called for support for leftist scientists and academics fired for being outspoken, and demanded an end to discrimination against women scientists. Eight resolutions toward "Equality for Women in Science" were prepared by a caucus of SESPA women. SftP continued to participate in caucuses of women scientists and at the 1972 American Chemical Society meeting, two SESPA women dressed in lab coats and draped in chains got up and read a statement about the problems of women scientists.

SftP also went to the AAAS meetings to expose particularly

reactionary research. Critiques were distributed at these targeted sessions and members insistently pointed out the political nature of the work. Needless to say, SftP was not welcomed with open arms by the AAAS officials nor by certain participants. In 1971, one SESPA member was attacked with a knitting needle by the wife of Garret Hardin (the population theorist) when Hardin's speech was interrupted. At that same meeting, Daniel Moynihan cancelled his talk to avoid a confrontation. At the Washington meeting in 1972, when some SESPA people tried to set up a literature table, the AAAS called in the police to arrest them. They were released as the police felt SftP did have the right to distribute literature and were not creating a nuisance.

SftP's strategy at AAAS conferences included organizing alternative sessions, although it was not until 1976 that they were officially recognized. These workshops were organized to encourage questions and discussion, to raise political awareness, and to develop the concept of a People's Science. These sessions were enormously successful, often drawing larger participation than the traditional ones. A paper was distributed at the 1970 AAAS that discussed the connection between applied and "pure" research, the interest of government and corporations in research results, and the consequences of so-called neutral research. It tried to begin the development of the idea of what would be a true "science for the people."

People's Science

As the organization grew, so did its scope of activities. In Boston, SftP tried to implement people's science through the Technical Assistance Project (TAP) in cooperation with the Black Panthers and other local groups. The idea behind TAP was to demystify technology by teaching people basic technical skills, like working on automobiles, sound systems, chemical analyses and self-defense mechanisms. Unfortunately, the project was never really successful for a number of reasons, including the fact that TAP members ended up doing all the work themselves instead of teaching others.

Another people's science project was motivated by the desire to contribute scientific services to national liberation struggles. SftP meetings in the summer of 1971 led to the creation of the Science for Vietnam project. Science for Vietnam had chapters in several cities and cooperated with similar movements in Europe by sending textbooks and technical information to North Vietnam.

Another aspect of developing a science to serve the people was demystifying science and explaining it so people could understand how science affected them. SftP's early interest in science teaching brought it to vocal attendance at the National Science Teachers Association (NSTA) conferences in 1971, 1972 and subsequent years. Issues raised at the NSTA included the role of science education and tracking in society, and the "hidden curriculum" of social myths being conveyed in science textbooks. A small group in Boston began to meet in conjunction with SESPA to develop alternative curricula and resources for high schools. Veterans of several NSTA conferences, they began staging their own one-day conferences for local teachers. The March 1973 and December 1974 conferences included workshops on teaching political issues in science classes and creating science and society courses. The April 1977 conference entitled "Inequality and Schools" attempted to respond to the Boston school busing crisis. Science Teaching Groups doing similar work have since sprung up in several other chapters as well.

IQ, Genetics, and Biological Determinism

SftP has applied its analysis to many issues in science and technology: energy and the environment, occupational health, imperialism in the Third World, the plight of scientific workers as well as professionals and science teachers. But the organization has played a particularly important role in the genetics and IQ controversy. Genetics research is the Establishment's latest hope for a technical panacea for society's ills. On the other hand, biological determinism has a long history of trying to find "scientific proof" of the inferiority of women, blacks and working class people. Since 1971, with the outpourings of Herrnstein, Jensen, and others claiming racial bases for intelligence, SftP has led the fight against this latest attempt to justify racism and the status quo. SftP made sure the IQ issue was not ignored at the International Genetics Congress in 1973.

Another use of biological determinism was appearing in the form of assertions that XYY males were genetically predisposed to criminally aggressive behavior. The Genetic Engineering Group of the Boston chapter took action to discredit these assertions and succeeded in stopping Harvard University research with newborn XYY males in the winter of 1974. In 1975, the Genetic Engineering

Group moved from the XYY research issue to a public campaign against the dangers of Recombinant DNA research. They spoke out within the scientific community which was wrestling with the creation of guidelines for research funded by the National Institutes of Health. This SftP group (now the Recombinant DNA group) also testified at the Cambridge, Ma. city council hearings in 1976 on the need for public control of potentially dangerous biological research. The Science for the People chapters in Amherst, Ma. and St. Louis, Mo. were also active in bringing the Recombinant DNA issue to the attention of their city councils.

Biological determinism again reared its ugly head with the publication and popularization of a massive book by Harvard's E.O. Wilson, *Sociobiology: The New Biological Synthesis*. Groups opposed to sociobiology started meeting within the Boston, San Francisco and Ann Arbor chapters of Science for the People. In the fall of 1975 the Ann Arbor chapter sponsored a conference at the University of Michigan on biological determinism. The various speakers attacked this new ideology, disguised as an objective scientific theory, for being merely a rationalization of the bourgeois status quo. The conference speeches were gathered into a book, *Biology as a Social Weapon* (Burgess Publishing Co., 1977). Science for the People groups continued to work at discrediting sociobiology, both as bad science and as reactionary ideology, and sponsored public forums at Stony Brook and Boston in 1977 as well as a symposium at the February 1978 AAAS meeting in Washington, D.C.

China Trip

One of the notable events in SftP history was the visit in 1973 of ten SftP members to the People's Republic of China. China had been the focus of considerable study in SftP because of the Chinese commitment to developing a nonelitist science that would serve the interests of the people. The SftP delegation was one of the first to travel to China from the U.S. after the Cultural Revolution. They visited research institutes, universities, factories, agricultural communes, and even a mental hospital. They discussed social and political issues—who decides what kind of research is pursued, how these decisions are made and how they are implemented. Upon returning from China, the ten authored a book called *China: Science Walks on Two Legs*, that described their experiences and gave many examples of a people's science actually being practiced. A second delegation has recently returned from the PRC. The focus of their trip was agriculture, food production and distribution.

Current Directions

SftP has been fairly successful working within professional academic circles. However, 1977 marked a move, spurred on by the women's caucus, towards activities rooted more in the surrounding communities and less university oriented. More effort has gone into working in coalitions with groups like INFACT*, Mobilization for Survival, and United Farmworkers, as well as local groups. Various chapters have offered workshops and forums in places like food co-ops and public libraries, and some are developing resources for people to use who need information but do not have scientific expertise.

The Magazine

Science for the People magazine is nearly as old as the organization itself. In 1970, with the eighth issue of the SESPA newsletter, people in Boston decided to change the format of the newsletter to that of a "newsmagazine" that would include articles of increased depth and analysis. The new format, it was hoped, would appeal to nonmembers and attract new people to the group while continuing to act as a forum for discussion of organizational activities. Early issues included articles about scientific and technical workers organizing, critiques of establishment science and exposes of the abuses of science and technology.

This more extensive format required more time and organization than the old newsletter. Rotating collectives of 4-6 people took on the responsibility for production of one whole issue from start to finish. This structure ensured that people would learn a wide range of skills (typesetting, layout, etc.) while demystifying the production process and avoiding dependency on or control by any one person or group. Of course, there were varying degrees of commitment to putting out the magazine. Some people worked tremendously hard on one issue, burned out and were never seen again. And although volunteers worked on production on an ad hoc basis, it was because of dedicated members that the magazine always came out.

As the magazine grew in size and circulation, it became increasingly evident that a group of people holding full time jobs could not produce a quality magazine in their spare time without significant help. In 1973, a Magazine Coordination Committee (MC²) was set up, and a paid position of magazine coordinator was created to aid the committee in establishing stability and continuity

---

*INFACT stands for Infant Formula Action Coalition

in the magazine and in organizing production.

Although the editorial collectives seemed to be a good idea, they proved to have many drawbacks in actual practice. Too much time and energy went into recruiting each new collective, while wholesale replacement of magazine personnel was thought to prevent political and stylistic continuity. Moreover, new labor and energy often had to be recruited from outside the organization, and many members objected to the national voice of the organization being determined by new people with little familiarity with its history and goals. For these and sundry other reasons, late in 1974 three separate committees with slowly rotating memberships— editorial, production, and distribution—were set up. This structure has proven more successful than the collectives: full participation in the work of the magazine is far less demanding, there is more continuity issue to issue, people on a committee are able to develop stronger working relationships, and there is more of a chance to plan ahead for future issues.

The magazine has been a crucial activity of the organization as it has tied together the various chapters and isolated individuals scattered throughout the country and has been the primary tool of outreach for the movement as well. It has been the general feeling, however, that it is important for chapters outside Boston to become more involved in magazine work. But the logistics of this have never been worked out satisfactorily. The Stony Brook chapter did produce two issues of the magazine (in 1974) and a number of individuals outside Boston contribute editorial work, but the magazine has predominantly been an activity of the Boston chapter.

Over the years, the content and style of the magazine have changed considerably along with the production process. During the early '70s, many articles naturally targeted war-related research. The language was confrontational, often rhetorical, and the appeal was mainly to people who already shared our political perspective. In the last few years, there has been a real effort to speak to a broader audience. While we still maintain a radical analysis, we now try to avoid articles that are overly technical or steeped in political jargon. We are also concerned to show how the scientific issues to which we address ourselves relate to the conditions and struggles of the nonscientific working population. Feminism has also come to play an important role in the magazine, both in the number of articles that specifically concern women and in the expression of a feminist point of view around a variety of issues.

Organizational Structure

Science for the People is organized as a loose federation of chapters. Each chapter is autonomous in that its activities are determined by the needs and interests of the members. Many chapters are divided into groups that focus on specific issues like energy, nutrition, genetic engineering, etc. But, because there is no mechanism for nation-wide discussion and decision-making, the organization as a whole has not actually taken a position on any of these issues.

Regional conferences have played an important role in binding together and overcoming isolation of chapters. They were originally called in order to define our politics more clearly and to develop a wider organizational base with an eye towards organizing on a national level. The conferences (Eastern—1973, 1974, 1975, 1977, 1978; Western—1973, 1978; Midwestern—1974, 1977, 1978) have built a stronger network of communications between the chapters and have hosted some important discussions around activities and problems of SftP and the magazine. While the changes have been slow, over the years there has been some progress made towards more coordinated actions, and a national conference has been planned for March, 1979.

The establishment of a Science for the People office in Boston has been an important factor in stabilizing the organization. It has functioned more as a central clearinghouse for information than as a national headquarters, however, and has provided a fixed location for production of the magazine. Initially, the "center" was located in donated space in a house owned by one of the members. Office work was done on a volunteer basis. In 1972, after considerable debate about having any "paid workers" and about job description, the membership decided to have a "compensated" office coordinator to organize and teach people about office tasks, define problems and weaknesses, and work with the steering committee (representatives from each activity group plus at-large members). In 1974, the office was moved to rented commercial space and then again in 1976 to its current, more centrally accessible and somewhat larger space.

The office, the magazine and the two part-time staff people (the second being the magazine coordinator, first hired in 1973) have provided the organization a center and a focus. The fact that this activity—and thus much control over the organization as well—is located in Boston has created a considerable imbalance of power among chapters. While people have tried to be sensitive to this, and numerous schemes and resolutions have been proposed to involve other chapters more directly in magazine work and organizational decision-making, it is only recently, with the number, size and strength of other chapters growing, that there is serious (and welcome) challenge to the "Bostocentricity" of the organization.

Politics of Science for the People

At certain points in our history the issue of establishing a definite political identity dominated our time and energy, but for the most part our politics have remained unformalized and implicit in our actions. To some people viewing us we seemed to avoid politics altogether, and yet to others we have seemed overly political.

Since we were founded as a "non-organization" with no constraints on membership, our members have spanned the left spectrum and have included Marxist-Leninists, progressive-liberals, anarchists, democratic socialists and many others. With a few significant exceptions, Science for the People has tried to be an organization in which most left-progressive people would feel comfortable.

However, some of us have felt dissatisfied with our amorphous political image. Shouldn't we have a political program both to present to prospective members and to specify our own priorities? One attempt to establish principles of unity was made in 1974-1976 by a group of members who came to be called the Unity Caucus. The Unity Caucus proposed a draft of principles of unity based on an anti-imperialist and leadership-by-the-working-class analysis. Other members of Science for the People reacted very negatively to the struggle for adoption of these principles, feeling pushed and fearing that any set of strict principles would be used to limit membership and to exclude long-time members from the organization. Also, the concept of working class leadership raised questions about the role professionals should play. Some resistance to the principles of unity came from the academic background of many of our members who were personally uncomfortable with the self-criticism implicit in the principles and who wanted to avoid any group definitions. Many people opposed the *way* the Unity Caucus put forth their politics, their insensitivity and inability to relate them to the particulars of SftP. Some members felt that the Unity Caucus' belief in their own position being the best and only way to define Science for the People would lead to cleavage rather than unity. It is important to note that the Unity Caucus included some strongly committed and influential founding members of the organization. When the Unity Caucus failed in their struggle, they left the organization.

The experience of Science for the People with the Unity Caucus led to an unfortunate backlash against virtually any political discussion from which the organization is only now starting to recover. Reacting to the feeling of being pushed to define a definite political line, members in post-Unity Caucus days have been wary of any sense of the "right way." In this political vacuum, many of us have backed off from sharing our experiences and political views and from defining our organization explicitly even as anticapitalist or prosocialist. As the organization grows larger and other chapters besides Boston grow stronger, however, there is more and more interest in developing a national decision-making structure or at least a stronger network. Increasingly over the years some members have felt the need for the organization at least to develop a more detailed and coherent analysis of science and to define our goals and strategies for reaching them more explicitly. At the same time, we want an organization which allows for the expression and discussion of different points of view. That there is room in SftP for people to differ has been one of the strengths of the organization.

If the growth of a national organization is to continue, some important questions must be raised about our goals and strategies. We do not expect many of these questions to be resolved; maintaining a consciousness of the issues has been an important process for the organization. Through evaluation and self-criticism, we hope to learn from our past experiences and increase the effectiveness of our work.

One question is who should be in Science for the People? In the popular media we are usually portrayed as a small group of professional academic scientists. While some of our members are indeed academics, SftP has a broader base which includes students, high school teachers, health workers, industrial scientists, and many people who work outside science. But, SftP, although committed to supporting women, Third World and working-class struggles, has remained predominantly a white, college-educated, professional organization. This to some degree reflects the class composition of the science world, but it also reflects just how little impact our work in SftP has had on the nonscientific public.

We have supported many progressive struggles such as those of the United Farmworkers, the J.P. Stevens textile workers and the unionizing efforts of technical and medical workers, but with varying amounts of energy and attention and without any systematic approach. We have also looked within the communities we live to discover what issues we want to support, but we have not been integrated into our neighborhoods as a resource for community people looking for help.

To whom are we addressing ourselves? Who is our audience and constituency? We have consistently and successfully targeted the scientific worker, but we are also concerned with talking to other workers about their workplace and community. This task has, to date, received comparatively little of our energy, however, and has been met with a corresponding lack of success.

How should our political viewpoint be reflected in our organizational structure? SftP has long been committed to working as a collective or group of collectives with no one member vested with more power than any other. Decisions are made democratically and almost always by consensus. There are no officers, no directors, no bosses. Working this way entails controversy and struggle. It can be time consuming and frustrating. But it is also a crucial aspect of developing mutual respect for one another, commitment and responsibility to the organization, and generally raising the level of political understanding.

Problems of sexism, racism and elitism have emerged repeatedly in our work as well as in the internal process of the organization. Self-proclaimed radicals can still be elitist, sexist and racist and act in ways that discourage participation of non-whites, women or people without college degrees. We realize that changing this takes time and continual struggle—especially for many of us who come from backgrounds (particularly academic) that foster such attitudes and encourage an isolated, individualistic workstyle. But in order to change society, we must transform ourselves in the process.

Progress has been made, especially in the area of sexism. In the early years, women had to struggle to be heard on this issue. Some women left the organization to devote their energies to fighting sexism elsewhere. More recently, other women have succeeded, through the support of women's caucuses, in making the organization take seriously the whole issue of personal process. It has become clear that workstyle, tone of meetings, and such have been barriers to effective political action in many radical organizations.

It is therefore essential to change the ways of interacting that we bring into the organization from the outside society by concerted and consistent evaluation and criticism of our own process.

Impact of SftP

It is difficult to assess the importance or impact of SftP. It will probably never be clear, for example, what our effect was in the anti-Vietnam war movement. Even our role in the separation of Department of Defense research labs from universities such as M.I.T. is of unknown significance, in light of the continuation of the same research in now independent military research labs.

In the scientific community, our critique of "bad science" has produced controversy and made other people more critical. The AAAS itself has become more open to political issues—a change for the good and one for which SftP deserves much credit. The AAAS will pass resolutions now which ten years ago it would not even discuss. Our effect is also clearly seen in the form letter now received by all organizers of AAAS symposia which spells out how to handle disrupters!

We have tried to show that no one issue in science exists in isolation. Unlike other groups who have fought the IQ and genetics issue, for example, SftP did so not on an issue-by-issue basis but from a general critique of science under capitalism. The same may be said of our work against sociobiology, which has been an important factor in the discrediting of this "discipline" among many scientists. At the same time, it must be noted that school curricula and college texts have started to include sociobiology as the "accepted wisdom" even as academics are backing off.

Science for the People has also had an important impact on its own members. One significant reason for our survival is clearly that the organization has been rewarding enough to some members for them to persist in putting great effort into its continuity. Over and over long-time members have told us such things as "involvement in Science for the People has balanced my professional work" or that "my activity group kept my mind working so I am not just following the typical career path." We have been told that SftP has offered a chance to "develop a new life style and to work according to my politics."

Of course, SftP has been far more than a comfortable support group for its members. In its activities and publications it has served as a forum for the development of a radical critique of science and its applications. This is an important task in a world in which the role of science is enormous and becomes larger every year. But SftP has, at the same time, come to appreciate the vital importance of internal process and the application of its political theory to its own practice.

Most organizations that originated during the Anti-War years have long since come and gone. SftP has not only survived, but has grown, in part because it was based on an understanding of the crucial role of science in supporting that war and the imperialist system generally. Obviously, neither imperialism nor the science and technology used to maintain it ended with the victory of the Vietnamese. But the key strength of SftP may be that it has not only offered a credible analysis, but has also provided alienated scientists, technical workers and others a framework in which to take action on scientific issues as they affect our jobs, our schools, and our communities.

BOOKS
I. The Myth of the Neutrality of Science

Ann Arbor Science for the People Editorial Collective, *Biology as a Social Weapon* (Minneapolis: Burgess Publishing Company, 1977). An excellent collection of articles on race and I.Q., sex roles, aggression, the environmental crisis and Sociobiology.

Bernal, J.D., *The Social Function of Science* (Cambridge MA: M.I.T. Press, 1967); *Science in History*, 4 volumes (M.I.T. Press, 1974); *Science and Industry in the Nineteenth Century* (Bloomington, IN: Midland Books, 1970). These three books provide a thorough history which examines the inextricable ties between the rise of capitalism and the development of modern science.

Ehrlich, Paul, *The Race Bomb, Skin Color, Prejudice and Intelligence* (New York: Ballantine Books, 1978). This book goes to the core of the "nature" versus "nurture" debate on the connections between I.Q. and race by pointing out that race is a political and cultural concept, and not a biological variable.

Herschberger, Ruth, *Adam's Rib* (New York: Harper and Row, 1970). First published in 1948, an early feminist classic exposing the bias of "male" biology.

Hubbard, R., Henifin, S. and Fried, B. editors, *Women Look at Biology Looking at Women* (Cambridge, MA: Schenkman, 1979). A collection of feminist critiques. Papers on evolution, hormones and sex differences, right and left brain, menopause, menstruation and other topics. Includes a lengthy bibliography on women and science and women and health.

Hubbard, R. and Lowe, M., editors, *Genes and Gender II: Pitfalls in Research on Sex and Gender* (Staten Island, New York: Gordian Press, 1979). Critiques the inadequate data base and research methods in much of the present day research into the origins of sex differences in behavior.

Jaubert, A. and Levy-Lebond, J. M., *Autocritique de la Science* (in French) (Paris: Editions du Seuil, 1973). A collection of articles about the scientific establishment and the radical critique of science.

Kamin, L., *The Science and Politics of I.Q.* (New York: Halsted Press, 1974). Exposes the poor science done in connection with the I.Q. issue.

Kuhn, Thomas, *The Structure of Scientific Revolutions* (Chicago: University of Chicago Press, 1970). A very interesting discussion of how new scientific theories emerge, in sexist language. In the postscript written in 1969 Kuhn emphasizes the need of studying the community structure of science.

Ladner, J.A., *The Death of White Sociology* (New York: Vintage Books, 1973). Though the focus is on sociology, many articles discuss also the contribution of the natural sciences towards racism, especially the article "Proving Blacks Inferior" by Rhett S. Jones.

Leibowitz, L., *Females, Males, Families: A Biosocial Approach* (North Scituate, MA: Duxbury Press, 1978). From an evolutionary perspective reviews fascinating cross-cultural data indicating that the family is an instrument of larger economic forces and not vice versa.

Noble, D., *America By Design: Science, Technology and the Rise of Corporate Capitalism* (New York: Alfred A. Knopf, 1977). Describes the history of modern technology in America as part of the rise of corporate capitalism.

Reed, E., *Sexism and Science* (New York: Pathfinder Press, 1978). Deals with sexist biases in anthropology and the pseudoscience of Sociobiology.

Criticizes Levi-Strauss' famous piece "The Elementary Structures of Kinship."

Reiter, R., editor, *Toward an Anthropology of Women* (New York: Monthly Review Press, 1975). Challenges traditional anthropology and covers a wide variety of topics: evolution of sex differences, matriarchal societies, sex bias in anthropology, women and economic development and the status of women in several societies.

Rose, H. and Rose, S., editors, *Ideology of/in the Natural Sciences* (Cambridge, MA: Schenkman, 1979). This is a one volume reprint of most of the articles in the 2 volumes published separately in England by MacMillan, 1976, under the titles "The Political Economy of Science" and "The Radicalization of Science." Introductory essay by R. Hubbard.

Teitelbaum, S.M., editor, *Sex Differences, Social and Biological Perspectives* (New York: Doubleday, 1976). Discusses sex differences from a number of perspectives: biological, anthropological, and social. Elizabeth Fee's article on "Science and the Woman Problem: Historical Perspectives" shows how the social and ideological forces at work in Great Britain and in the U.S. from 1860 to 1920 affected the findings of scientific research.

Tobach, and Rosoff, B., editors, *Genes and Gender I* (Staten Island, New York: Gordian Press, 1978). First in a series on hereditarianism and women. A collection of articles on the Genes and Gender Conference, Spring 1977.

II. Science and Social Control

Boffey, Phillip M., *The Brain Bank of America: an Inquiry into the Politics of Science* (New York: McGraw Hill, 1975). Introduction by Ralph Nader. Shows how the National Academy of Science does not act on the best interests of the common citizen. Covers radioactive waste disposal, defoliation, food, pesticides (The Academy versus Rachel Carson) and other topics.

Braverman, Harry, *Labor and Monopoly Capital, The Degradation of Work in the Twentieth Century* (New York: Monthly Review Press, 1975). Traces the development and consequences of the principles of "scientific management" first proposed by Frederick Taylor in the late 19th century. It is a story of science in the service of industry, creating an ideology of efficiency, "productivity" and "profitability" while reducing work itself to purely mechanical actions.

Dickson, Paul, *The Electronic Battlefield* (Bloomington: Indiana University Press, 1976). Dickson uses the development of the automated battlefield to show how prestigious scientists provided the ideas and the enthusiasm leading to the surgically clean and efficient weapons used in Vietnam.

Greenberg, Daniel S., *The Politics of Pure Science: an Inquiry into the Relationship between Science and Government in the United States* (New York: New American Library, 1967). Clearly shows the relationship between "pure" science and the U.S. government.

Howard, Ted and Rifkin, Jeremy, *Who Should Play God?* (New York: Dell, 1977). Important reading to understand the recombinant DNA controversy.

NACLA East, *Dying for Work: Occupational Health and Asbestos* (151 West 19th St, NY 10011). Special issue of the NACLA report examines the political economy of occupational safety and health.

Vitale, B., editor, *The War Physicists* (Istituto di Fisica Teorica. Mostra d'Oltremare 80125, Napoli, Italy). Documents the European protests against the physicists working for the U.S. Military through the JASON division of the Institute for Defense Analysis.

Women Studies Committee Against Sterilization Abuse, *Workbook on Sterilization and Sterilization Abuse, 1978* (available from Women's Studies/Sarah Lawrence College, Bronxville, NY 10708).

III. Working in Science

Brooks, P., *The House of Life: Rachel Carson at Work* (New York: Fawcett, 1972). How Rachel Carson worked and how little support she got from the science establishment.

Mozans, H.J., *Women in Science* (Cambridge, MA: First M.I.T. Press Edition, 1974). First published in 1913, it is a very interesting account of women's work in the various branches of science. Introduction by Mildred Dresselhaus.

Sayre, A., *Rosalind Franklin and DNA* (New York: Norton, 1975). Written by a personal friend of Rosalind Franklin, it is a must for anyone who has read James Watson's version of the discovery of the double-helix structure of DNA.

IV. Science and Liberation

Anderson, S. and Bazin, M., editors, *Ciencia e Independencia: O Terceiro Mundo Face a Ciencia e Tecnologia* (Science and Independence: The Third World with regard to Science and Technology) (Rua das Chagas 17-1-D, Lisboa—2 Portugal: Livros Horizonte, Ltda, 1977). Two volume set, not yet in English. Edited by two Science for the People members for distribution in the independent African countries that were formerly under Portuguese colonial rule.

Brown, M., editor, *The Social Responsibility of the Scientist* (New York: Free Press, 1971). Sixteen scientists (all male!) discuss topics of interest for the community at large: food additives, biological warfare, nuclear radiation, population, etc.

Commoner, B., *The Closing Circle* (New York: Bantam Books, 1972); *The Poverty of Power* (Bantam Books, 1977). These two books argue powerfully that issues of environmental degradation and energy have their origin in the capitalist order of business. To solve these problems will require changing the private profiteering system with a system of socialist planning.

Dixon, B., *What is Science for?* (New York: Harper and Row, 1973). Dixon wants scientists to accept more social and political responsibility for their work and non-scientists to participate more widely in decisions that affect them.

Dubos, R., *The Dreams of Reason, Science and Utopias* (New York: Columbia University Press, 1961). Using Goya's words, "The dreams of reason can produce monsters," Dubos argues that each scientific problem is also a problem in ethics and that the long-term consequences of every scientific undertaking must be considered. *Reason, Awake: Science for Man* (Columbia University Press, 1970). A very interesting book about science becoming a matter of public concern and the need for scientists to allow non-scientists to be the final judges of the kind of science they want to support. Sexist language, but very good ideas.

Freire, P., *Pedagogy of the Oppressed* (New York: Herder and Herder, 1970). Freire, a Brazilian educator in exile, describes a radical approach to education. He sees education as a powerful political tool. Although his approach is not limited to science teaching, his philosophy is of special value in the natural sciences because of the authoritarianism and mystification that permeate science.

Gyorgy, A., and Friends, *No Nukes: Everyone's Guide to Nuclear Power* (Boston, MA: South End Press, 1979). This book is a beautiful example of "science for the people." It explains the technical concepts underlying nuclear energy in clear English, develops the technical, economic and political critiques of nuclear power and provides a guide to alternatives.

Horn, Joshua, *Away with All Pests: An English Surgeon in People's China. 1954-1969* (New York: Monthly Review Press, 1969). A fascinating account of how medical care began to change in the new China. Horn lived in China for 15 years and worked there as a surgeon.

Howe, F. editor, *Women and the Power to Change* (New York: McGraw-Hill, 1975). A volume of essays sponsored by the Carnegie Commission on Higher Education. Four essays by feminist writers; gives insight on what a feminist vision can bring to higher education and to knowledge.

Maslow, A.H., *The Psychology of Science: a Reconnaissance* (Chicago: First Gateway Edition, Henry Regnery Company, 1969). Though it focuses on psychology, many of its ideas apply also to the natural sciences. Maslow is a strong advocate for a humanistic science and believes that "science as a social institution has goals, ends, ethics, morals, purposes—in a word, values."

Miller, L.B., *Towards a New Psychology of Women* (Boston: Beacon Press, 1977). Miller argues that the subordination of women has distorted human perceptions and experiences. "As other perceptions arise—precisely those perceptions that men, because of their dominant position could not perceive—the total vision of human possibilities enlarges and is transformed." It follows that science has been developed according to a male model and that women can play a crucial role in shaping a different science.

Ruddick, S., and Daniels, P., eds., *Working it Out* (New York: Pantheon Books, 1977). Twenty-three women writers, artists, scientists and scholars talk about their lives and work. Two women scientists "tell it like it is."

Science for the People, *China: Science Walks on Two Legs* (New York: Avon, 1974). In 1973 a ten-person delegation from Science for the People visited China and wrote about what they learned regarding research institutes, the health care system, agriculture and mental health, etc.

Sidel, V.W. and Sidel, R., *A Healthy State: An International Perspective on the Crisis in United States Medical Care* (New York: Pantheon Books, 1977). A fascinating comparison of medical care in Sweden, Great Britain, the Soviet Union and China. The Sidels examine health care in the United States from an international perspective and suggest new initiatives in health care to maximize health and provide the services so badly needed.

Teich, A. editor, *Technology and Man's Future* (New York: St. Martin's Press, 1972). A collection of articles about technology and society, with some good criticism of American society's infatuation with technology.

ARTICLES

I. The Myth of the Neutrality of Science

Barber, Bernard, "Ethics of Experimentation with Human Subjects" (*Scientific American*, February 1976, vol.234, no. 2.) Discusses scientists caught up in the socially structured competitive system of science pursuing peer recognition and overvaluing their work in place of humane treatment of people.

Beckwith, Jonathan, and King, Jonathan, "The XYY syndrome, a dangerous myth" (*New Scientist*, November 14, 1974, pp. 474-476). Points out the social and political values and assumptions upon which XYY research was based.

Bennett, Arlene P., M.D., "Eugenics as a Vital Part of Institutionalized Racism" (*Freedomways*, 1974, vol. 14, no. 2, pp. 111-126). Discusses eugenic sterilization laws in the U.S. and other ways to encourage racist beliefs.

Colman, Andrew, "Psychology and the Legitimization of Apartheid" (*Science for the People*, May 1972, vol. 4, no. 3). Discusses the role of the Psychological Institute of South Africa in supporting racist research.

Elliott, Ruth, "Women and Alternative Technology" (*Undercurrents*, August/September 1976, no. 17). Describes the exclusion of women from technology and the alienation that results.

II. Science and Social Control

Berkeley SESPA, "Science Against the People: the Story of Jason" (a *SftP* pamphlet, Berkeley, CA 94704, 1972). Describes the Jason Group, an elite group of physics professors who consulted with the Institute of Defense Analysis during the Vietnam War and who were responsible for the creation of the automated battlefield.

Bhattacharya, K. R., "Soybeans in India" (*SftP*, January 1975, vol. VII, no. 1). Suggests that the promotion of soybeans as a protein supplement in India primarily serves U.S. business interests.

King, Jonathan, "Biomedical Research, Politics, and Health" (*SftP*, March 1975, vol. VIII, no. 2). Discusses misplaced priorities in biomedical research.

Looney, Mark, "Selling the Rain: Weather Modification as a Weapon of Imperialism" (*SftP*, March 1975, vol. VII, no. 2). Describes the connection between weather modification research and the military and the possibilities for weather modification in secret warfare.

Madison SESPA, "The AMRC Papers" (*SftP*, January 1974, vol. VI, no. 1, pp. 30-35). Excerpts from a book on war research involvement of the Army Mathematics Research Center at the University of Wisconsin.

Singh, Narenda, "Is our Food Problem Due to Overpopulation?" (*BASWI*, April 1975). Suggests that the food problem in India is due to the economic structure of the country.

Sinnette, Calvin H., "Genocide and Black Ecology" (*Freedomways*, vol. XII, no. 1, 1972, pp. 34-46). Focuses on issues of health and population control of blacks in the U.S. and in the Third World.

Spier, Sandra and Skoog, Sam, "First our Land, Now our Health" (*SftP*, September 1974, pp. 26-29). Describes experiments funded by the Defense Department and carried out by the University of Minnesota Medical

School where Native Americans with impetigo were left untreated in order to gather information on the disease.

III. Working in Science

Narek, Diane, "A Woman Scientist Speaks" in *Voices from Women's Liberation*, compiled and edited by Leslie R. Tanner (New York: New American Library, 1970). A personal account of how one woman becomes a mathematician despite lack of support and alienation.

IV. Science and Liberation

Bazin, Maurice, "At the Side of the Workers" (*SftP*, November 1973, vol. V, no. 6). Describes the experiences of the author working in a worker-run metallurgical factory in Chile during Allende's regime.

Brennan, Patricia, "Science Teaching" (*SftP*, March 1976, vol. VIII, no. 2). Describes problems of teachers interested in social change and the support offered by regularly meeting and planning activities with other teachers.

Cavrak, Steve, "Workers Demand Production of People" (*SftP*, November/December 1976, pp. 16-18). Describes how engineers, draftspeople and technicians at a British Aerospace Corporation demanded the right to work, that work be meaningfully structured and that the products of their work be socially useful.

Fluck, Michelle, "Cancer Prevention: Good News for People's Science" (*SftP*, Nov/Dec 1976, pp. 16-18). Describes how engineers, draftspeople describes a simple test that can be used to screen for chemicals with mutagenic properties which are likely to be carcinogens.

Gilbert, Len, with Weinrub, Allen, "Organizing in Silicon Valley" (*SftP*, January 1976, vol. VIII, no. 1). Discusses the difficulties of trying to organize physicists, chemists, engineers, and technicians, despite low pay, job insecurity and arbitrary decisions by management.

S.N.A.P. (Students and Neighbor's Action Program) "People's Science in Philadelphia" (*SftP*, November 1972, vol. IV, no. 6). Black students in the Chemistry Department of a local community college bring their scientific and technical skills to a rat-infested neighborhood.

PERIODICALS

Edcentic, a journal of educational change. P.O. Box 10085, Eugene, OR 97401.

Politics and Education. Wesleyan Station, Fisk Hall, Middletown, CT 06457.

Radical Teacher. P.O. Box 102, Kendall Square, Cambridge, MA 02142.

Radical Science Journal. Published by the Radical Science Journal Collective, 9 Poland St., London W1. England.

Science for the People. Published by Science for the People, 897 Main St., Cambridge, MA 02139.

Science for People. 9 Poland St., London W1. England.

Undercurrents—a British magazine on radical technology. For subscriptions: 12 South Street, Uley, Dursly, Goucestershire, England. U.S. Distributor: Carrier Pigeon, 88 Fisher Avenue, Boston, MA 02120.

abortion, 119-121, 123, 125, 155
accelerator, 219-221
aggression, 35, 36, 39
Academie des Sciences, 18,20,25
Arms Control and Disarmament Agency, 101
Advisory Committee on Sterilization
  Guidelines, 119
Africa, 289. See also Third World, and
  individual countries.
Africa News, 199
Africa Research Group, 316
African
  Asians in, 212
  Coloured, 212
  slave labor, 334
Algeria, 342
Alvarez, Louis W, 183, 230
American Academy of Occupational
  Medicine (AAOM), 142
  asbestos and cancer meeting, 142
American Association for the Advancement
  of Science (AAAS), 283-285, 371, 372, 374,
  381
American Chemical Society, 371
American Civil Liberties Union, 118
American Committee on Africa, 201, 206
American Council on Education, 74, 75
American Management Association, 71
American Men and Women of Science, 173
American Metals Climax Company, 178
American Physical Society, 236, 237, 370
American Telephone and Telegraph (AT&T),
  64-66, 74
Amnesty International, 193
Animal Disease Center, Long Island, 151, 152
Ann Arbor SftP Conference, 374
Anti-Ballistic Missile System (ABM), 370
anti-eugenicist, 56
apartheid, 191-194, 201, 202, 206-212
  policy, 316
Apollo space program, 130
Argonne Zero Gradient Synchrotron, 219, 220
Armed Forces, 198
ARMSCOR (South African Arms Develop-
  ment and Production Corporation), 198-199
Armstrong County, Pennsylvania, 117
Army Mathematical Research Center (AMRC)
  4, 5, 370
  "The AMRC Papers", 370
Asbestos, 130, 131, 133-143, 180
Association for Voluntary Sterilization, 116
atomic bomb, 23, 24, 27. See also weapons,
  nuclear.
Atomic Energy Board of South Africa, 201
Atomic Energy Commission, 221, 223, 225, 226

Baran, Paul, and Paul Sweezy, Monopoly
  Capital, 300
Barnaby, Frank, 97
Bell Labs, 66
Bell Telephone, 302
Bernal, J.D., 21, 32

Bernalians, 21, 22
  Science in History, 331, 347
Bevatron, 222
Big Science, 22, 23, 26
biological determinism, 7, 9, 35, 36, 38, 42
Biology as a Social Weapon, 374
birth control, 80, 82, 83, 119, 121
  abortion, 285
  pills, 310
Black(s)
  America, 336
  dependence on technology, 335
  exclusion from science, 287, 288
  liberation, 288
  Panthers, 312, 348, 372
  people, 13, 92, 201, 245
  revolutionary, 342, 343
  in science, 293
  scientists and technicians, 288, 342, 343, 347
  in South Africa, 193, 194, 211
  trade unions, 202
  and Third World, 335
  workers, 347
bosses
  function of, 275
  in science, 276
Botswana, 197
Braverman, Harry, Labor and Monopoly
  Capital, 347
Britain, 18, 19, 21, 22
British Association for the Advancement of
  Science, 19
British Ministry of Defense, 26
British Royal Society, 18, 19, 355
Brookhaven, 221
  Alternating Gradient Synchrotron, 219, 220
Bromley, Allen, 236
  report, 229, 235, 236
Buck vs. Bell, 115
Bulletin of Atomic Scientists, 96
Burroughs Corporation, 181, 191, 192, 193,
  195, 206

California Institute of Technology, 67, 260
  women on faculty, 261
Cambridge, Massachusetts
  City Council, 148, 163
  Electron Accelerator, 220
  Experimentation Review Board, 162-165
Cape Town, South Africa, 134
capital, 227
capitalism, 19, 41, 269, 271
  277, 331
  laissez-faire, 70
capitalist, 63, 273, 275, 278
  relations, 269
  society, 257
capitalization of science, 64, 65
Carnegie Commission on Higher Education,
  172
Carnegie Foundation, 73
Case Institute, 74

Cavendish, 219
    Laboratory, 229
Center for Disease Control, 118, 153
Central Intelligence Agency (CIA), 200, 216, 304
Cetus Corporation, 149
Chemical Biological Warfare (CBW), 304
Chemical and Engineering News, 262
Chicago Women's Liberation Union, 318
child labor, 337
Children's Hospital Medical Center (Boston), 9
Chile, 208
China, 24, 39, 215, 216, 306
    Cultural Revolution, 24
    Science Walks on Two Legs, 347, 374
    trip of SftP, 374
Church Committee, 42
CIBA-Geigy Corporation, 149
class
    analysis, 273
    barriers, 273
    capital owning, 275
    classless society, 279
    conflict, 273
    interests, 274
    ruling, 58, 59, 268, 275
    working, 272, 275, 278
Coalition for Responsible Genetics Research 116
Cold Spring Harbor, 56
    phage meeting, 162
Cold War ideology, 300
Columbia University, School of Chemistry, 249
Committee to End Sterilization Abuse (CESA), 118. 119
competition in science, 276, 320
computer(s), 92, 94, 191-198, 200, 203-208, 211, 212, 346
    chip, 195
    data, 195
    industry, 192, 193, 203
    "How to Beat the Computer Siege", 204
    manufacturing in South Africa, 205
    military and police, 195, 198
    software, 195
    technology, 191
    "Why Police States Love the Computer", 196
Computer and Business Equipment Manufacturers Association, 209, 210
computer companies
    Anglo-American Corporation, 104
    Computer Automation, 192, 205-206
    Burroughs, 202
    Control Data, 192, 195, 196, 201, 202, 206-209, 211, 212
    Computer Science Corporation, 195, 201
    Data General, 202
    defend their role in South Africa, 206
    ICL, 202
    Mercedes Group, 205

Messina, 205, 206
Congress of Industrial Organizations (CIO), 57
Congressional Record, 60
contraception, 77, 78, 91
Control Data Corp. See computer companies.
consulting, 171-174, 176, 177, 179, 181-184
    academic, 175
Cornell Electron Accelerator, 220
Cornell Medical College, 250
Council for Scientific and Industrial Research, 198
corporate, 65
    capitalism, 75
corporation(s), 64, 69, 70, 73, 75, 172, 175, 182, 186-190, 206, 210
    in DNA research, 149-151
Cosmos Club, Washington DC, 358
CS, CN tear gas, 26
Cuba, 45
    Cuban Revolution, 335, 338, 340, 342
    US Blockade, 314

Data General Corporation, 192, 201
Data Point Corporation, 192
day care centers, 285
decision making in the lab, 325
Defense Documentation Center, 5
Del Monte Corporation, 180
Democratic Republic of Viet Nam, 314. See also Viet Nam.
demystification, 317
Digital Equipment Corporation, 192
discrimination in hiring, 261
Division of Industrial Cooperation, MIT, 68
division of labor, 272, 273
DNA, 145-147
    corporate applications, 149
    molecule, 309
    patents, 157
    recombination, 9, 10, 92
    structure, 361
The Double Helix, 228, 359
Dow Chemical Corporation, 70, 149
Draper Lab, MIT, 109
Dun and Bradstreet's Million Dollar Directory, 174
Dupont Corporation, 64, 65, 70, 149

East Rand Bantu Administration Board, 202
E. coli, 10, 153, 159, 161
    K-12, 161
education
    higher, 65, 69, 71, 73, 74
    during World War II, 73
Einstein, Albert, 20, 27, 231
electronic battlefield, 196
electronics, 240, 241, 243, 244,
    technician, 239, 240-242, 244, 245
Eli Lilly and Company, 11
elite, 339
    ruling, 56
    science, 158
    scientists, 303

elitism, 267, 292, 296, 318, 380
equipment maintenance, 327
Emergency Act of 1921, 52
Energy Research and Development Agency
    (ERDA), 94, 96, 99-101, 105
Enhanced Radiation Warhead (ERW), 6
Environmental Defense Fund, 166
The Environmental Fund, 88
Environmental Protection Agency (EPA), 106
    and DNA, 153
eugenics, 48-60
    funding of, 56
    movement, 55
    Record Office, 53, 54, 114, 115
Europe, 18, 19, 97, 102
    Eastern, 274
    immigrant labor, 334
European Organization for Nuclear Research
    (CERN), 218, 221
"experts", 271, 319
Export-Import Bank, 210

faculty, 172, 176, 177, 179-184, 186, 187, 240
faddism in science, 233
Federal Communications Commission (FCC),
    244
Federal Systems Division (IBM), 198
females
    in public education, 355
    scientists, 284. See also women in science.
feminism, 240, 241
    and science, 364
    and SftP, 376
feminist perspective, 365, 366
Food and Drug Administration, and DNA,
    153, 167
Ford Motor Company, 181
Foxboro Corporation, 192, 199, 200
Freedom of Information Act, 42
Front for the Liberation of Mozambique
    (FRELIMO), 338
Freudian doctrine and women, 317
funding
    of eugenics movement, 56
    policies, 222
    of science, 234, 235
Future Farmers of America, 241

Gallup poll, 115
Gay Liberation Movement, 317
Geiger counter, 230
General Chemical Corporation, 64
General Electric, 64-66, 69, 70, 72, 73
    DNA patents, 157
    DNA research, 146
General Motors, 211
genetic, 48, 50
    code, 145
    engineering, 31, 156, 157, 165, 169
    theory, 35
Genetics and Society, 50
Genetic Engineering Group, 373
genocide, 114

Gerbending committee, 93, 94, 107
Germany, 19-21, 44
gesamtarbeiter, 273
Gordon conference, 167
government and corporations, 304. See also
    US Government and US Corporations
government agencies, 235. See also US
    Government.
W.R. Grace Corporation, 149
Great American Dream, 291
Green Revolution, 11, 12, 159
Gross National Product (GNP), 84-86
group meetings (in the lab), 324
Guidelines on Sterilization (NYC), 124
Guinea-Bissau, 335-336
Guyana, 341

Harvard University, 82, 162-165, 173, 178, 225
    Committee on Research Policy, 163
    consulting, 172
    consulting at Medical School, 178
    faculty, 172
    medical school, 178, 187
    research on XYY males, 373
    women on faculty, 260
Hatch Act, 67
Haymarket Riots, 56
H-Bomb, 304. See also weapons, nuclear
healers, women, 353
Health and Hospitals Corporation (NYC), 119,
    120
Health Policy Advisory Center, 119
Health Research Group, 118
Health right, 119
hereditarian thinking, 60
HEW, See Department of Health, Education
    and Welfare under U.S. Government
Hewlett-Packard Co., 183, 192, 195, 198, 199,
    201
hierarchical division of labor, 269, 274. See
    also sociobiology
hierarchy of science, 276
High Energy Physics Advisory Panel, 226
Hiroshima, 22, 94, 304, 307
Hitachi, 204
Ho Chi Minh Trail, 4, 306
Hoffman-La Roche, 151
Homestead Strike, 56
homosexuality
    male, 351
    as "sick" or maladjusted, 316, 317
    torture of homosexual people, 309
Honeywell (corporation), 192
House Committee on Science and Techno-
    logy, 168
How Computers Work, 195
"How Europe Underdeveloped Africa", 347
Huxley's warning, 156
The Hyde Amendment, 125

IBM (International Business Machines), 192,
    193, 195, 197, 198, 200-202, 204, 206-212, 302
ICL (International Computer Limited), 192,

193, 202
ideology, 275
   of science, 268, 270, 272, 276,
ideological struggle, 316
Ik (Uganda), 40
imperialism, 182, 332, 346
Immigration Restriction Act of 1924, 52, 58
India, 82, 83, 86, 113
   Madras, 86
Indian Health Services (Native American) 113,
   114
Indochina 304, 311
industrial capitalism, 81
Industrial Health Foundation (IHF), 133
industrial revolution, 19
Industrial Workers of the World, 57, 58
industrialists of the l930's, 68
industry, 2, 69-71, 153
   asbestos, 130-134, 136-138, 141
   consulting, 174, 175, 178
infant mortality, 80
Institute of Defense Analysis (IDA), 3
Institute of Occupational and Environmental
   Health (Quebec), 135
Institute of Saint Cyr, 7
Inter-Agency Committee, 45, 153, 154
Interfaith Center on Corporate Responsibility
   196, 198, 201
International Genetics Conference, 1973, 373
I.Q., 7, 8, 43, 44, 48, 51, 55, 60, 115, 159, 291
   biological determinism, 373
   and genetics, 381
Isabelle, colliding beam proposal, 221
ITT (International Telephone and Telegraph),
   206
Ivory Coast, 341
Ivy League, 9

Jason project, 3, 179, 219, 370
Jensen, Arthur, 8, 44, 316
   blacks, 293
Jews, 20, 21, 54
   Polish, 57
job in science, 247
Johns-Manville, 130-132, 137-140, 142, 143
Journal of Heredity, 50, 56

Kalahari Desert, 43
Kalamazoo, Michigan, 149
Kansas State College, 73
Khanna Study, 83
Kim Il Sung, 340
Kings County Hospital (NYC), 121
King William Town, 202
knowledge
   control of, 271
   diffusion of, 278
   as privilege, 267
   restriction, 270
Kodak Corp, 70
Korea, 97
Kuhn, Thomas
   scientific paradigms, 27
!Kung Bushmen, 40

Labeling, 114, 115
Labor
   intellectual, 252
   intensive work,243
   organizing, 57
   unions
      Local 1199, Drug and Hospital Workers
      Union, 252
      Oil, Chemical, and Atomic Worker's
      Union, 166
      United Farmworkers, 375, 380
laboratory, 23, 65, 147, 148, 152, 247, 248, 279
   of Clinical Investigation, 253
   paraphernalia, 168
   technician, 274
   worker's salary, 250
Lawrence Livermore Laboratory, 93, 95, 100,
   101, 103, 105-109
E.O. Lawrence Radiation Laboratory at Berkeley, 229
Lawrence and Livingston, 229
Arthur D. Little Corp., 66, 67
Litton Industries, 168
Los Alamos Meson Physics Facility (LAMPF),
   234
Los Alamos Scientific Laboratory (LASL), 93,
   94, 95, 98, 99, 100, 108, 109
Los Angeles County Hospital, 117
low birth rate nations, 85
Lowveld, 197
Lunar Society, 19
Lundgren, The Rich and the Superrich, 300
lung cancer, 131-135, 142

male hormones, 37
Machel, Samora, 338, 342
Magdoff, Harry, The Age of Imperialism, 300
Malthus, 16, 116
Mamdami, Mahmood, 82, 83
management, 196-198, 201
   applied to science, 231
   experts, 213
Manchester, 337
Manhattan Project, 22, 23
Mansfield Amendment, 219
Manupur, India, 84
Manville, New Jersey, 132
Mariepkop, 197
Mao Tse-tung, 245
Maoist, 24, 25
Marcuse, Herbert, 25
Marx, Karl, 20, 73, 75, 273
   mode of production, 337
Marxism, 19
Massachusetts, 243
Massachusetts Institute of Technology, 22, 66,
   67, 68, 72, 74, 109, 148, 163, 164, 292, 390
   Division of Sponsored Research, 68
   women faculty, 261
masters degree, 249
Master Plan for Higher Education for New
   York City, 309
Maxwell, 231
MC2, See Science for the People

McGill University, 135
Medicaid and Medicare
    and sterilization abuse in
        Aiken, South Carolina, 117
        New York City, 116
Medical Committee for Human Rights, 312
Medical Research Councils, England, 132
Medical College of South Carolina, 131
Medical World News, 149
Mellon Institute, 67
Memorial Hospital for Cancer and Allied
    Diseases, 247
Metropolitan Life Insurance Company, 130,
    131, 141
Mexican Americans, 117
Mexico, 189
Michelison-Morley experiment, 231
"Michurinist" biology, 23
Middle South Utilities System, 177
Miles Laboratory, 147
Military, 3, 4, 16, 95, 191, 193, 195, 196, 199
    See also Department of Defense
    establishment, 109
    electronics training, 244
    technology, 196
    use of computers, 195, 198
Military-Industrial Complex, 300
Mills, C. Wright, The Power Elite, 300
Minneapolis-Honeywell, 316
Mirage jet fighter, 198
Mississippi appendectomy, 116. See also ster-
    ilization abuse
MIT, See Massachusetts Institute of Techno-
    logy
Mobile Oil Company, South Africa, 205
mobile trailer, 152
Mobilization for Survival, 375
Monsanto Corporation, 149
Montgomery Community Action Agency, 117
Mozambique, 335
    border, 197
Mt. Sinai Medical Center, 134, 136
myth of the the neutrality of science, 15, 17,
    18, 23, 91

NACLA, See North American Congress on
    Latin America
NACS, See National Association of Corpora-
    tion Schools
Nagasaki, 2, 3, 304
NAL, See National Accelerator Laboratory
Nambia, 203
NARMIC, See National Action Research on
    the Military Industrial Complex
NAS, See National Academy of Sciences
NASA, See National Aeronautics and Space
    Administration
National Academy of Sciences, 50, 166, 167,
    178, 217, 222, 229, 358
    DNA debate, 167
    DNA forum, 166
National Accelerator Laboratory, 219, 220,
    221, 223
    black employees, 223, 236, 237

National Association of Corporation Schools,
    71, 72, 73
National Black Science Students Organiza-
    tion, 344, 345
National Bureau of Standards, 66
National Cash Register Corp., 192, 199, 202
National Council of Churches, 209
National Institute for Environmental Health
    (of NIH), 142
National Institute for Occupational Safety and
    Health, 131, 135, 138
National Institutes of Health, 10, 145, 148, 149,
    151-154, 158, 161-168, 262, 292, 374
    DNA meeting, 145
    postdoctoral and special fellowships, 263
    research career development awards, 263
    study panels, 263
National Institute of Telecommunications Re-
    search, 198
National Research Council, 68
National Research Council of Canada, 225
National Science Board, 226
National Science Foundation, 221, 226, 292,
    358
    grants awarded by, 263
National Science Teachers Association, 373
National Supplies Procurement Act, 205
Nationalist Party, 203
Native Americans, 113, 117
    in science, 293
NBSSO, See National Black Student's Organ.
NCR, See National Cash Register Corp.
NELA Labs, 66
neocolony, 341
neo-colonizer, 341
nepotism, 261
    university, 262
New Economic Policy, 20
New Scientist, 226
New University Conference, 313
New York City Hospital System, 8
    Health and Hospitals Corps., 119, 120
    Medicaid and Medicare sterilization abuse,
        116, 117
New York Times, 35, 87, 177, 196, 223
    Magazine, 34
Nigeria, 341
NIH, See National Institutes of Health
NIOSH, See National Institute for Occupa-
    tional Health and Safety
Nobel Laureats, 227
Nobel prize, 8, 17, 177
Northern Air Defence Sector, 197
North Korea, 335, 339
    Korean people, 340
North Vietnam, 306, 335, 339
North American Congress on Latin America,
    244, 316
NRC, See National Research Council of Can-
    ada
NSF, See National Science Foundation
Nuclear Non-proliferation Treaty, 199
Nuclear Regulatory Commission, 97, 106
Nuremberg, 245

Oak Ridge, 309
Oberlin College, 356
"objectivity," 364
Occupational Safety and Health
    Act of 1970, 138
    Administration, 130, 138
Office of Naval Research, 307
Oil, Chemical and Atomic Worker's Union,
    166
Oklahoma City, 113
One Hundred Bushel Corn Club, 241
Oppenheimer, J. Robert, 24, 307
oppression, 247, 273, 274
organizing efforts, 253, 255
    technicians, 252
Orwellian, 10

P1, P2, P3, P4 laboratories, 10, 149, 162, 163, 168
patent control, 64
    See also DNA patents
Paterson, NJ, 137
Pelindaba Atomic Research Facility, 200
People's Business Commission, 166
    (formerly People's Bicentennial Comm.)
People's Liberation Army, 39
People's Republic of China, 335, 339, 340
people's research, 315
people's science, 313, 372
Pentagon, 3, 100, 292
    See also U.S. Government
Pentagon Papers, 179
"personal is political," 360
personal support for scientists, 324
Pharmaceutical Manufacturers Association,
    167
Physical Review, 218
Physical Review Letters, 218
Physical Science Study Committee, 292
physicists, 227, 231, 233, 234, 236
    unemployed, 225
physics, 20, 23, 226, 234, 238
    community, 217, 222, 229
    conferences, 226
    establishment, 235, 237
    high energy, 223, 234, 235
    theorists, 237
Physics in Perspective, 217, 221, 222, 226
Physics Letters, 218
Physics Today, 225
Pietersburg, 202
PL 92-463, 173
Planned Parenthood, 124
Plutonium, 106
Polaroid Corporation, 316
Polish Jews, 57
political imprisonment in South Africa, 193
political activity
    activist, 240
    in health field, 312
    radicalization of intellectual labor, 278
    in science, 240
poor people
    "scientific" rationalizations, 365

Popular Liberation Movement of Angola, 314
population, 76-82, 84-89
Populist revolt, 56
Porton Down, 26
Portuguese domination, 314
The Power Elite, 300
    research on, 316
Pretoria, 200
Price-Anderson Act, 169
Princeton-Penn Accelerator, 220
Pritchard, Henry, 67
production of labor, 301
professionalism, 7, 248, 269, 273, 366
    myth of, 251, 252
programmed advances in technology, 301,
    302
Project Houston, 199
Project Igloo White, 3
proletarianization, 274
    of scientific workers, 279
Provisional Revolutionary Government of
    South Vietnam, 314
PSSC, See Physical Science Study Committee
The Psychology of Sex Differences, 35
Puerto-Ricans
    in science, 293
    women, 310
    Young Lords, 312
Puerto-Rico, 123
Pullman Strike, 56
Purdue University, 73
pygmies of Ituri, 40

Quebec Asbestos Mining Association
    (QAMA), 133, 135, 136
quality of life, 78

Randburg, 202
race, 55, 56, 60
Race Betterment Association, 56
racism, 50, 54, 61, 182, 206, 244, 287, 288, 380
    eugenics, 331, 346
radical, 240
    radicals, 182
    students, 252
Sperry-Rand Corporation, 192, 195
Raybestos-Manhattan, 137
recombinant DNA, 146-152, 154, 155, 157-160,
    162, 164, 166, 167
    Research Act of 1977, 168
Red Scare, 57
Regents Nuclear Fund, 109
research, 65, 67, 68, 179, 180, 223-225
    DNA, 145, 146, 150-154, 166
    industrialization of, 274, 275
    laboratory, 320, 321
    medical, 133-136
    "pure," 303
    quasi-independent, 249
    recombinant DNA, 149, 153, 155, 157, 158,
        160, 161, 163, 165, 166, 168, 169, 374
    science, 276
    space (office and lab), 327

research and development
  in the U.S., 2, 3
Rhodesia, 205
Rich, Adrienne, 366
Rift Valley fever, 10
Robben Island, 201
Roche Institute for Molecular Biology, 161
Rochester, New York, 149
Rockefeller Foundation, 56
Rockefeller University, 249, 250, 254
Rocky Flats, Colorado, 105
Roetgen, 230
A Room of One's Own, Virginia Woolf, 350
The Roots of Hunger, 85
Route 128, 210
ruling class, 210
  ideology, 316
  research on, 316
ruling ideology, 318
Rydberg Law, 232

safe shutdown, 106
salaries, 253
  in the lab, 325, 326
  professional, 261
Salk and Sabin vaccines, 310
SALT, see Strategic Arms Limitation Talks
Saranac Research Laboratory, 131
Science Magazine, 22, 220
  article by D. Wolfe, 226
Science Against People, 295, 370
science
  anti-, 24
  basic, 26, 28
  big, 26
  buying, 130
  class character, 268
  class nature, 371
  competition in, 359
  education, 291, 292, 294
  for people, 279
  funding, 3, 4
  hiring, 235
  job hunting, 234
  job security, 238
  jobs, 235
  in the Middle Ages, 352
  mystification of, 158, 165
  is "neutral", 369
  "nominal", 27
  origins, 332
  policy, 26
  is political, 307
  as private affair of scientist, 308
  as property, 331
  "pure", 293, 294
    myth of, 287
  puzzle solving, 29
  reaction against, 24
  revolutionary, 27
  and social ideologies, 287
  teachers, 295, 296, 297
  teaching, 180, 292, 296, 297

  teaching groups, 373
  and technology, 3, 22, 32
    290, 294, 295
    in a capitalist society, 330
    in pre-capitalist society, 334
  tenured positions, 235, 238
  theorists, 235
  values, 1, 7, 15, 17
    funding, 3, 6
    neutrality of, 5, 6, 12, 15, 16, 23, 24, 28, 29
    teacher's, 13
  value-free, 369
  Western, 272
"Science for Blacks" Study Groups, 347
Science for the People, 13, 162, 240, 312, 313,
  314, 315, 317-319, 369, 374
  Amherst Chapter, 374
  Boston Office, 377
  Boston centricity, 378
  collectives, 375, 376
  conferences, 377
  impact of, 381
  magazine, 347, 348, 375
  magazine coordinator, 377
  as a national organization, 379
  organizational structure, 377
  political program, 378
  St. Louis, Missouri, 374
  Stony Brook Chapter, 376
  survival, 382
  Unity Caucus, 378, 379
Science for Vietnam project, 314, 372
"scientific" rationalization, 365
scientific
  research and technology, 304
  unemployment, 277
scientific work
  fragmentation of, 275
  military applications, 275
  redirection of, 300
scientific workers, 274
  proletarianization of, 279
  redirection of, 300
scientists, 2, 3, 5-8, 11, 12, 17, 18, 28, 30, 31,
  92, 100, 218, 276, 279
  and citizens, 11
  bourgeois, 275
  German Jewish, 21, 23
  Marxist, 23
  and political action, 329
  non-industry, 136
  pro-industry, 136, 140
  pro-worker, 140, 142
  resistance, 299
  social aspects, 294
  training, 6
  in U.S. labor force, 358
  women, 358, 364
  work of, 283
  as workers, 267
Scientists and Engineers for Social and
  Political Action, 345, 371-373, 375
school of engineering and architecture, 339

Searle Laboratories, 149
Second General Law Amendment Act of 1974, 205
Second World War, 302
Self-help concept, 365
self-sustenance (juche), 340, 341
Senate Foreign Relations Committee, see U.S. Government
Seneca Falls Declaration of Principles, 356
separation of theory and practice, 210
SESPA, see Scientists and Engineers for Social and Political Action
sex
    differences, 36, 38
    discrimination, 250
    hormones, 36
    psychology of sex differences, 35
    roles, 35, 40
sexism, 216, 241, 242, 244, 380,
    in science, 364, 365
Sharpsville massacre, 206
Singer Business Machines, 192
SIRRI, 97
SLAC, see Stanford Linear Accelerator Center
Alfred P. Sloan Research Foundation, 263
social change, 12, 265
Social Darwinism, 53
social stratification in science, 250
socialism, 12, 19, 20
socialist, 87, 274
Society for Freedom in Science, 21
Societies for Social Responsibility in Science, 25
sociobiology, 9, 33-35, 40, 41, 43
Somalia, 342
"The Sorry State of Science", 371
South Africa, 92, 191-196, 198-200, 203-212
    Air Force, 197
    Airways, 196
    Armaments Board, 199
    Asians, 201
    Atomic Energy Board, 196, 200
    Coloureds, 201
    Defense Department, 197, 201, 204
    Defense Research Institute, 198
    Department of Interior, 201
    Department of Justice, 201
    Department of Labor, 202
    Department of Prisons, 201
    Official Secrets Act, 205
    Soweto, 195
    Whites, 201
South Bend, Indiana, 149
South East Asia, 295
Soviet Revolution of 1917, 20
Soviet Union, 20, 21, 23, 24, 25
Soweto, 195
Society for Promotion of Engineering Education, 71, 72, 73, 74
Sperry-Rand Corporation, 192, 195
Society of Professional Engineers and Scientists (SPSE), 107
Sputnik, 292

SSPA (later SESPA), 370
Squibb-Beech Nut Corporation, 178
E.R. Squibb and Sons, 178
Standard Telephone and Cables of South Africa, 206
Stanford
    Linear Accelerator Center (SLAC), 218, 219
    Research Institute (SRI), 109
    University, 5
    DNA patents, 157
    women on faculty, 261
    Workshop on Political and Social Issues (SWOPSI), 5
"state-socialist", 274
State Department, see U.S. Government
sterilization, 8, 113-125
    abuse, 113, 123, 125
    compulsory, 115
    hysterectomy, 116
J.P. Stevens, 380
Stellenbosch, 202
Stockholm International Peace Research Institute (SIRRI), 97
Stoddard, Lothrop, The Rising Tide of Color Against White Supremacy, 51, 52, 57
Strategic Arms Limitation Talks, 24
strike command (mobile radar unit), 197
St. Simonism, 19
Student Health Organization, 312
superwoman stereotype, 264
SV-40 virus, 161

Taiwan, 139
Tassaday, 38
teaching
    new people in the lab, 325
    science, 270
tear gas, 26
technical assistance, 313
Technical Assistance Project (TAP), 372
technicians
    biological, 248
Technische Hochschulen, 19
"Tech Plan", 68
technology, 3, 9, 11, 24, 63, 64, 65, 75
    DNA, 158-161, 163, 164
    Third World, 337
technological arrest, 333
    regression, 333
test ban treaties
    Comprehensive, 100
    threshold, 100
Theory of Relativity, 231
Third World, 88, 288, 289, 297, 300, 305, 334-338
    countries, 316, 338, 340, 341
    peoples, 313, 316
    solar energy, 345
Threshold Test Ban Treaty, 100
Time Magazine, 195
Toynbee, Arnold, A Study of History, 333
Transvaal, 197
    Provincial Administration, 202
Transkei, 202

triage, 88
Tycho, 232

underdevelopment, 333
undergraduates, 322, 323
Union of Black Scientists and Technicians
  (UBST), 344, 345, 347, 348
Union of Concerned Scientists, 108
USSR, 199, 274
unionizing efforts, 380
unions, see labor unions
United Nations (UN) Security Council, 199,
  203
United States, 21, 24, 30
  capitalist class, 300, 301, 302
  corporate capitalism and science, 299
  corporate state, 300
  corporations, 191, 192, 203, 316
  ruling class, 301-303
    interests of, 302
  commitment to science, 311
United States Government and Agencies
  Air Force, 314
    Office of Scientific Research, 219
  Armed Services Committee, 102
  Arms Control and Disarmament Agency,
    101
  Army, 51, 304
  Atomic Energy Commission, 25
  Department of Commerce, 200, 203, 206,
    209
  Department of Defense, 27, 31, 95, 173, 179,
    195, 219, 308, 370
  Department of Energy, 94
  Department of Health, Education and
    Welfare (HEW), 117, 118
    sterilization guidelines, 119, 122
  Department of State, 100, 200, 203
  Congress, 220, 222, 261
  Joint Committee on Atomic Energy, 101
  General Accounting Office, 110
  House of Representatives, Committee on
    Education and Labor, 261
  Natural Science Department, 339
  Navy, 306
  National Cancer Institute, 139
  Public Health Service, 131
  Senate, 220, 222
    Foreign Relations Committee, 100, 205
  Treasury Department, 205
Universal Foods Corporation, 180
university, 67, 75, 171, 179-183, 186-190, 224,
  245, 248
  consultants, 65
  elite, 70
  enterprise, 66
  hiring policies, 260
  graduate, 249
  and the military, 4, 5
  research, 68
  state, 244
  and women, 285
University of

Bologna, 171, 355
California, 93-110, 174, 176, 178, 180, 186, 187
  at Berkeley, 261
  DNA patents, 157
  Nuclear Weapons Lab Conversion
    Project, 93
  Regents Nuclear Science Fund, 94, 98
  at Santa Cruz, 322, 328
  weapons research at, 93, 94, 96-98, 102-
    104, 106, 108, 109, 111
Chicago, 311
Cincinnati, 72
Michigan, at Ann Arbor, 162
Rhode Island, 9
Stanford University, 5
Wisconsin, 4
Uranium Enrichment Corporation, 192, 200

Valindalba enrichment plant, 192, 200
value-free science, 276
Van de Graafs, 220
Viet Nam, 2, 3, 31, 179, 181, 196, 304, 314
  anti-war movement, 361
  satellites in stationary orbits, 307
  war, 219, 370
Very Large Array (VLA), 221

war, 21, 211
  World War I, 130
  World War II, 219
    women and veterans, 357
War Department
  Committee on Education and Special
    Training, 73
  Committee members, 74
Washington Post, 103, 148
weapons, 24, 295
  electronic battlefield, 196
  modern, 196, 198
  Mirage jet fighter, 198
  nuclear, 93, 94, 96-98, 102, 104, 109, 195
Westinghouse, 64, 69-71, 73
What is to be Done?, 311
white racism, 331, 346
white regime, 194
Who Rules America, 300
Who's Who, 173
Witbank, 202
witchcraft, 353
women, 13, 36, 41, 42, 115-121, 123-125, 216,
  242, 243
  Black, 117, 121, 243
  channeled, 258
  contraception, 77, 78, 91
  discrimination, 264
  education, 354, 356
  free high schools, 356
  in graduate schools, 357
  in industry, 262
  Latin, 243
  in law, 262
  in Medieval Europe, 353
  Roman, 352

in Science for the People, 375
sterilized, 116
Third World, 243
WW II veterans, 357
working, 250
"Women drink water while men drink wine",
    350
women in science, 256, 283
  barriers to, 259
  in chemistry, 257
  children and child care, 255
  employment, 263
  engineers, 257
  and grants, 262
  graduate study, 260
  in labs, 247, 248, 251
  management, 262
  physics, 257, 261
  as research associates, 322, 359
  role of, in science, 12
  science degrees, 259, 260
  socialization away from science, 258
  in universities, 260, 261
  who become scientists, 293
women's caucus in Science for the People, 375
Women's Liberation Movement, 318
women's movement, 240
Woolf, Virginia, 366
Worcester, 202
workers, 168, 273. See also labor unions.
  asbestos, 130, 132, 133, 135-143
  electronics, 243
  DNA and health, 166
  gesamtarbeiter, 273
  Oil Chemical and Atomic Workers Union,
      166
  production, 275
  radicalization, 59
  scientific, 252, 267, 272, 277, 288
  textile worker, 380
  women, 250
  women in science, 255
working class, 239
workplace (in labs), 326
Wyeth laboratories, 149

XYY males, 373, 374

Yale University, 71, 216, 225, 254
  Medical School, 253
  Non-Faculty Action Committee, 255
Y chromosome, extra, 8
Yearbook of Armaments and Disarmaments,
    97
Young Lords, 312

Zimbabwe, 332
Zinner Committee, 99, 103